PHOTODIODES – COMMUNICATIONS, BIO-SENSINGS, MEASUREMENTS AND HIGH-ENERGY PHYSICS

Edited by **Jin-Wei Shi**

Photodiodes – Communications, Bio-Sensings, Measurements and High-Energy Physics
Edited by Jin-Wei Shi

Published by InTech
Janeza Trdine 9, 51000 Rijeka, Croatia

Publishing Process Manager Petra Zobic

Technical Editor Teodora Smiljanic

Cover Designer InTech Design Team

Additional hard copies can be obtained from orders@intechopen.com

Photodiodes – Communications, Bio-Sensings, Measurements and High-Energy Physics, Edited by Jin-Wei Shi
p. cm.
ISBN 978-953-307-277-7

For EU product safety concerns:
IN TECH d.o.o., Prolaz Marije Krucifikse Kozulić 3, 51000 Rijeka, Croatia,
info@intechopen.com or visit our website at intechopen.com.

Meet the editor

Dr. Jin-Wei Shi was born in Kaohsiung, Taiwan on January 22, 1976. He received the B.S. degree in Electrical Engineering from National Taiwan University, Taipei, Taiwan in 1998 and the Ph.D. at the Graduate Institute of Electro-Optical Engineering from National Taiwan University, Taipei, Taiwan in 2002. He was a Visiting Scholar at the University of California, Santa Barbara (UCSB), CA, during 2000 and 2001. In 2002-2003, he served as a post-doc researcher at Electronic Research & Service Organization (ERSO) of Industrial Technology Research Institute (ITRI). In 2003, he joined the Department of Electrical Engineering, National Central University, Taoyuan, Taiwan, where he is now an professor. In 2011, he joins the ECE Dept. of UCSB again as a Visiting Scholar. His current research interests include ultra-high speed/power optoelectronic devices, such as photodetectors, electro-absorption modulator, sub-millimeter wave photonic transmitter, and semiconductor laser. He has authored or co-authored more than 83 journal papers, 130 conference papers and holds 20 patents. He was the invited speaker of 2002 IEEE LEOS, 2005 SPIE Optics East, 2007 Asia-Pacific Microwave Photonic conference (AP-MWP), and 2008 Asia Optical Fiber Communication & Optoelectronic Exposition & Conference (AOE). He served as the technical program committee of OFC 2009-2011. He was the recipient of the 2007 Excellence Young Researcher Award from Association of Chinese IEEE and the 2010 Da-You Wu Memorial Award.

Contents

Preface

The photodiode device structure, which has developed almost simultaneously with Si based p-n junctions, has had a dramatic impact on everyday life, especially in the field of communication and sensing. The last few decades have seen optical techniques come to dominate long-haul communication and photodiode technologies, serving as an energy transducer in the receiver end, which can convert optical data into electrical signals for further processing. In addition to communication, photodiodes have also found some killer applications in advanced high-speed image systems and will eventually replace traditional slow charge-coupled devices (CCD).

This book describes different kinds of photodiodes and several interesting applications, such as for high-speed data communication, biomedical sensing, high-speed measurement, UV-light detection, and high energy physics. The discussed photodiodes cover an extremely wide optical wavelength regime, ranging from infrared light to X-ray, making the suitable for these different applications. Compared with most other published studies about photodiodes, the topics discussed in this book are more diversified and very special. Take the category of high-speed data communication for example; which covers the applications of photodiodes under very-low temperature operations such as for the optical interconnects used in ultra-high speed superconducting electronic circuits. This topic has rarely been discussed in relation to the use of photodiodes for fiber communication. Furthermore, in this category, we also discuss a unique high-speed single photon detection technique based on the use of the low-noise Si-based APD and the photon up-conversion technique, which could be important for the development of the next-generation of quantum communication. Overall, the aim of such book is to provide the reader with information about novel, unique, and practical examples of different photodiodes for several diversified applications without going into detail on complex device physics and math. This should be a very useful "tool-book" for engineers, students, and researchers in different academic fields who want to understand the most advanced photodiode applications.

Jin-Wei Shi
Ph.D. Associate Professor, National Central University,
Taiwan

Part 1

Photodiodes for High-Speed Data Communications

1

Photodiodes with High Speed and Enhanced Wide Spectral Range

Meng-Chyi Wu and Chung-Hung Wu
Department of Electrical Engineering, National Tsing Hua University, Hsinchu 300, Taiwan

1. Introduction

Typically, 0.85, 1.3, and 1.55 μm are wavelengths of interest for fiberoptic communications. Conventionally, photodetectors or photodiodes (PDs) based on different absorption layers are used for corresponding wavelengths. For instance, at 0.85-μm wavelength, GaAs or Si-based PD is preferred, while for 1.3-μm and 1.55-μm wavelengths, InP-based PD is most suitable. A PD served all these wavelengths therefore is desirable for network projection.

Conventional PDs with a broad spectral range can be classified into two configurations as illustrated in Figs. 1(a) and 1(b). One is with a shallow p-n junction directly formed in the absorption layer, either by epitaxial growth, or diffusion, or ion implantation, with the metal contact directly deposited on the p-type absorption layer. The principal drawbacks of this configuration involve excessive surface leakage. The other, which intends for reducing surface leakage, is with a shallow p-n junction formed in the absorption layer by selective acceptor diffusion through a dielectric window and a thin wide-bandgap cap layer and with the metal contact deposited on the diffused cap layer. Although by such a configuration and selective-area-diffusion (SAD) process all the p-n junction periphery in the narrow-bandgap absorption layer is sealed inside and surface leakage is minimized, the device still suffers from problems of low surface concentration, efficiency diminution at shorter wavelengths, and alloy spike. Since the temperature of the SAD process is usually limited by the tolerance of the dielectric mask and the diffusion time is limited by a required shallow diffusion depth, for high (saturated) surface concentration the wide-bandgap cap layer should be thick enough for elongating diffusion time under an optimum temperature. Thick wide-bandgap cap layer results in severe efficiency diminution at wavelengths shorter than the cutoff wavelength of the cap layer. As a consequence, for broad spectral range operation, a thin cap layer and thus a low surface concentration are necessary in this configuration. High contact resistance emerges and, if alloyed, metallurgical spike can be risky.

For the case of conventional InGaAs p-i-n PDs, to be reliable, the InGaAs PDs utilize SAD process and place the outermost junction periphery in the wide-bandgap cap layer, which is usually InP. Either front-illumination or back-illumination sets the lower limit of the device spectral range to be ~0.92 μm, the absorption cutoff of InP. Consequently, the operation spectral range is usually limited to 0.9-1.65 μm. However, devices based on these structures, although has achieved a broad responsivity spectrum, require shallow SAD process, which is rather difficult to achieve satisfactory junction properties. As a consequence, such devices

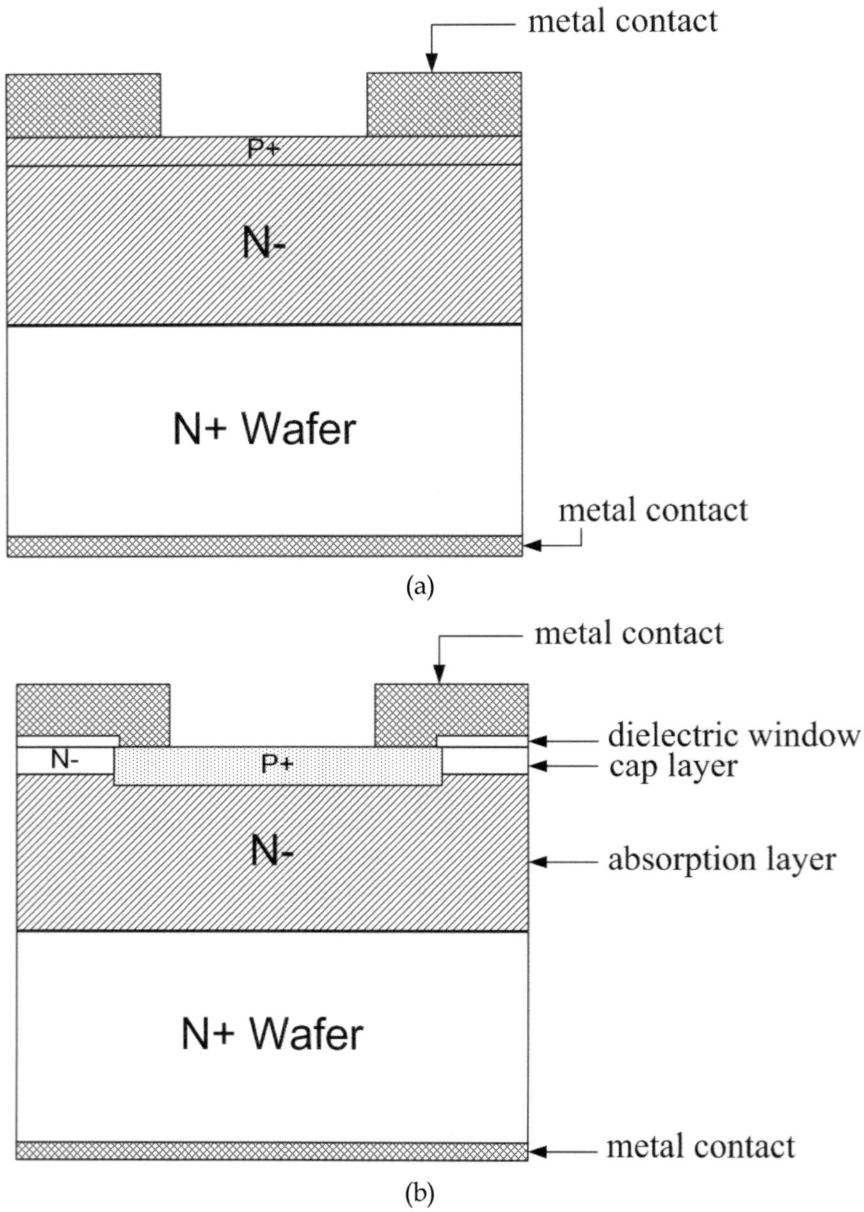

Fig. 1. Two conventional PD structures.

usually suffer from an excessive leakage and a low breakdown voltage. Besides, alloy spikes after contact annealing might electrically short the pn junction and cause device failure. Edge-coupling configuration can also achieve a wide responsivity spectrum if most of light is direct-coupled into the InGaAs region. However, rather accurate alignment is required and a more stringent limit is placed on the pseudowindow thickness for shorter wavelength operation [1]. To achieve a wider spectral range, structures with a thin InP cap (< 0.2 μm) were utilized [2], [3]. As the bandwidth demand goes higher, the coupling loss affects the performance of the fiber-optic link more, due to a smaller period per bit. For the receivers, to reduce the capacitance charging delay, the PD aperture shrinks as the bandwidth rises. Therefore, for high-speed operation, it would be more meaningful if the PD has a large coupling aperture. Conventional top-illuminated 10-Gb/s InGaAs p-i-n PDs typically have an optical coupling aperture of 20-40 μm in diameter, only one-third of the 2.5 Gb/s version. A larger aperture for 10 Gb/s operation, although achievable, must be based on the bondpad reduction or bondpad isolation design [4], which usually invokes wire-bonding difficulties and process complexities, respectively, and thus the yield and reliability concerns. Nevertheless, the small-aperture PD requires an active alignment during device package, which necessitates the use of more complex and more expensive facilities. If the alignment fails, not only the coupling efficiency is lost, but also the bandwidth is deteriorated due to the slow diffusing carriers. To cope with the problem, wet or dry etched backside microlens was proposed to increase the alignment tolerance at the price of more backside processing steps [5], [6]. However, the enhancement of the optical coupling tolerance using such a backside-etched microlens is limited, and not to mention the degraded chip yield due to the backside process. Other fabrication methods for a microlens, such as surface micromachining [7], mass transport after preshaping [8], and photoresist reflow method [9], need complicated processes and are difficult to control the microlens radius. Therefore, using a commercial microball lens integrated with a planar high-speed PD to enlarge the alignment tolerance is a simple and attractive method.

This chapter reports the PD whose configuration is suitable for broad spectral range operation. The device is configured so that light illuminates directly upon the narrow-bandgap absorption layer, while with p-contact metal depositing on a thick wide-bandgap cap layer. Because the p-n junction periphery in the absorption layer is still sealed inside, the device has minimum surface leakage. Besides, a thick wide-bandgap layer facilitates the reach of maximum surface concentration and prevents the effect of alloy spike. With a shallow p-n$^-$ junction inside the absorption layer, the PD ideally can exhibit a wide responsivity spectrum with only the long-wavelength side limited by the absorption-layer cutoff. Beside, to achieve a 10-Gbps PD with wide spectral and spatial detection range, structures with a thin cap can be utilized. We shall first illustrate the spin on diffusion technique and for the applications to InP and GaSb materials. We fabricated the InGaAs/InP and InGaP/GaAs p-i-n PDs by removing the window layer of the conventional InGaAs/InP and InGaP/GaAs PDs [10], [11] on the photosensitive surface. These PDs exhibit a low capacitance, a low dark current, a high speed, and a high responsivity in the enhanced spectral range, which permits applications as PDs for the high-speed communication, optical storage systems CD-ROM, as well as red and blue laser DVDs. Finally, we also accomplish the improvement of the coupling loss of small coupling aperture of 10-GHz InGaAs p-i-n PD by using a simple method of enlarging alignment tolerance of a high-speed PD with integrating to a self-positioned micro-ball lens.

2. Zinc diffusion in InP from spin-on glass

Zn is one of the most typical p-type impurities in the InGaAs/InP system. Diffusion of Zn into semiconductor, such as n-type InP, InGaAs or InGaAsP, and Zn-doped InP epitaxial layers, is an important technique for forming p-n junctions in optoelectronic devices.

The most common low-cost technique for planar junction formation in InP/InGaAs PD manufacturing is the sealed-ampoule Zn-diffusion technique. This process seals the patterned InP wafers under vacuum inside a quartz ampoule, along with some dopant source and a decomposition suppressant. The junction pattern is typically etched through a silicon-dioxide or silicon-nitride mask. The dopant source for p-n junction formation is usually solid-phase zinc, although cadmium also has been used. Zinc and cadmium are p-type dopants, so the InP-based semiconductor material must be n-type for a diode junction to be formed. This type of diffusion is not realistic for InP-based material systems because the temperature requirement far exceeds the decomposition temperature even when a Group-V decomposition suppressant is used.

A relatively new technique for junction formation, at least within the InP/InGaAs material system, is spin-on diffusion (SOD) [12]-[14], a process that has been available for many years for silicon and is still used in some low-cost silicon processes. These days, though, ion implantation and carrier-gas diffusion dominate most silicon production, and spin-on-glass (SOG) doping for silicon has mostly been relegated to undergraduate electronics laboratories because of its relative simplicity and low cost. These same features make the process attractive for fabrication of InP-based diodes. SOG is a type of glass that can be applied as a liquid and cured to form a layer of glass having characteristics similar to those of SiO_2. In general, SOG is mainly used for planarization and as a dielectric material. The process sequence of spin-on diffusion is outlined as below:

1. Silicon nitride diffusion-mask deposition and shallow delineation etching.
2. Application of source on top of silicon nitride layer with open diffusion windows by spin coating of an InP substrate and soft bake on hotplate.
3. Deposition of 1500 Å thick cap layer of silicon nitride.
4. Drive-in process with application of rapid thermal annealing (RTA) at 550°C
5. Removal of excess glass and silicon nitride films in HF: H_2O.
6. Deposition of silicon nitride, followed by lithography and etching steps

It is found that the samples prepared by SOD method economize 100 min than those prepared by furnace diffusion (FD) which does not include the heat clean of furnace system of 2 days. The economical process time of SOD is an advantage for mass production.

To inspect the relationship between diffusion depth and diffusion time, the electrochemical C-V (ECV) measurement was applied. The diffusion-depth test was applied to a 3 μm thick undoped InP epitaxial layer which was grown on an n^+-InP substrate. The diffusion process was performed at 550°C in a RTA with N_2-purged ambient, and the rising ramp rate of temperature was set to 5°C/sec. After driving in of Zn diffusion source and removing residual glass and dielectric, the diffused wafer routinely underwent RTA process for impurity activation while virtually eliminating the potentially damaged interstitial zinc. Fig. 2 shows the concentration profiles for the various thermal treatment condition (ramping rate/temperature/time). The thermal treatment condition used for PD fabrication is 600°C RTA for 25 sec in N_2 ambient. This shows that most Zn atoms are activated and act as acceptors. The net-acceptor concentrations are around 2-4 x 10^{18} cm^{-3} in all of the samples.

Similar SOD technology is also applied to the case of GaSb wafer. The fabricated p-n junction structure is shown in the inset of Fig. 3. The diffusion depth measured by electrochemical C-V profiler is shown in Fig. 3. The junction depth is 0.6 μm and the concentration of surface can be achieved as high as around 5×10^{20} cm^{-3}.

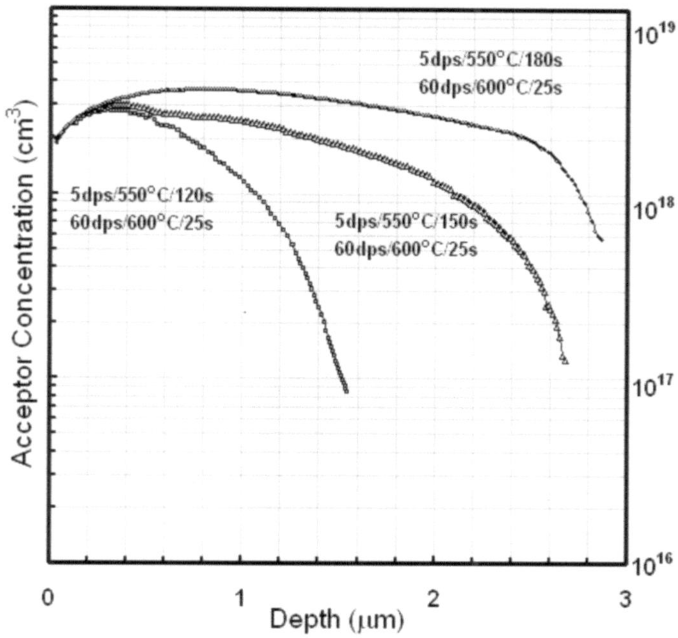

Fig. 2. Concentration profiles of spin-on Zn diffusion in undoped InP after 600°C RTA for 25 sec in N_2 ambient

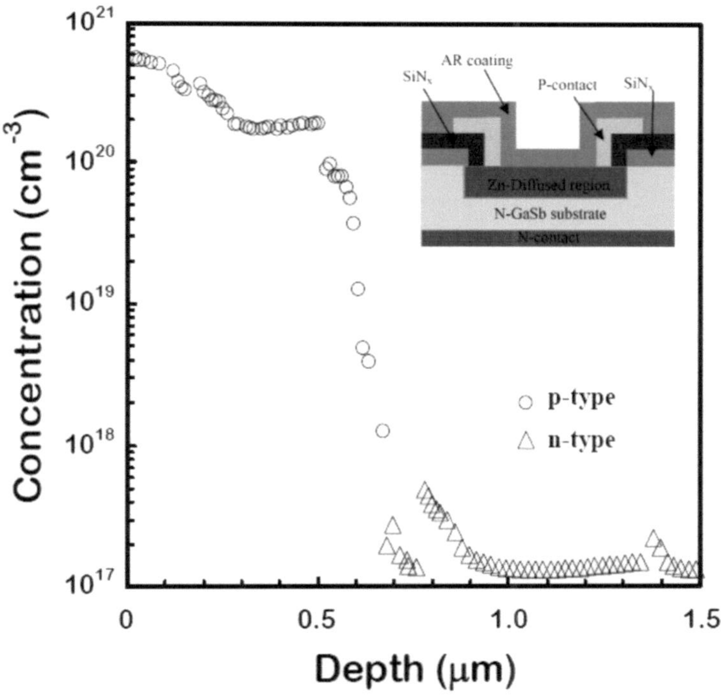

Fig. 3. The diffusion depth measured by electrochemical C-V profiler. The inset of Fig. 3 shows the fabricated p-n junction structure.

3. Fabrication of InGaAs/InP p-i-n photodiodes [10]

The InP/InGaAs/InP p-i-n PD is constructed to be capable of speedily and efficiently detecting light signals of wavelengths ranging from 0.7 μm to 1.65 μm. This range covers all the wavelengths of interest nowadays in fiberoptic communications: 0.85 μm, 1.3 μm, and 1.55 μm.

The InP/InGaAs/InP epitaxial device structure was grown by metal-organic chemical vapor deposition (MOCVD) on the n^+-InP substrate. A first layer of 0.5-μm undoped InP was grown for buffering the growth process. A second layer of 2.5-μm undoped indium gallium arsenide (InGaAs) was grown for light absorption layer. A third layer of 1.0-μm undoped InP was grown as the wide-bandgap cap layer. Highly reliable SAD planar device process, either by sealed-ampoule diffusion or spin-on diffusion, was utilized for device fabrication. Silicon nitride (SiN_x) film with 1500-Å thickness was deposited onto the entire wafer by plasma-enhanced chemical vapor deposition (PECVD). Through conventional photolithographic process and reactive ion etching (RIE), diffusion windows with 50-μm diameter were opened on the dielectric film. Afterwards, wafer was loaded into semi-closed diffusion system and zinc (Zn)–diffusion process was performed at 550℃ for 10 min. Such a temperature and period produced a 10^{17} cm^{-3} acceptor front at 1.2-μm deep below the surface. Due to a rather slow diffusion of Zn in InGaAs (3 times slower than that in InP), the Zn protrusion depth into the InGaAs can be well controlled to be about 0.1-0.2 μm, which was designed for reliability, wide spectral range, and high-speed operation considerations. After impurity activation by RTA and conventionally photolithographic process, ring-shaped p-contact metallization chromium (Cr)/gold (Au)/AuZn/Cr was deposited on heavily doped p-type InP cap layer. The contact adhesion was enhanced by heat treatment. Then the InP cap layer inside the 30-μm-diameter coupling aperture was removed by a reactive ion etching with the CH_4/H_2 source to slightly etch and recondition the InP surface and then by chemical etching based on the $3H_3PO_4$: HCl solution to selectively clean the InP thickness left. Afterwards, double-layer SiN_x/SiO_x antireflection (AR) coating and Cr/Au for bondpad metallizations were deposited in sequence. Wafers were then lapped and polished down to about 150 μm and the polished backside was coated with Ti/Pt/Au n-contact metallizations. Lastly, samples were annealed at 420℃ for 20 sec to reduce the contact resistance. The cross-sectional view of a finished device is schematically drawn in Fig. 4. Then the wafer was separated into chips, and several were packaged for characterizations.

The essence of this PD is to remove the InP cap layer inside the illumination aperture (directly above the light-absorption layer) while leave a shallow p-n junction in the narrow-bandgap InGaAs absorption layer. Also, the contact is made on the thick-enough InP cap layer for preventing the alloy spike from shortening device layers with opposite conducting types. There are several critical considerations in performing this art.

First, a well-controlled diffusion process needs to be established. The junction protrusion depth into the InGaAs absorption layer is consequential. Too shallow junction depth would result in the following two detrimental effects: surface depletion and large sheet resistance. The former leads to large dark leakage even at low bias, while the latter results in non-uniform spatial response. To the other extreme, deep junction results in excessive absorption in the quasi-neutral p-type region. Electrons generated in this low-field region slowly diffuse (as compared to drift process) out of the region or recombine with holes. As a consequence, low-efficiency and low-speed device performance can be expected for certain

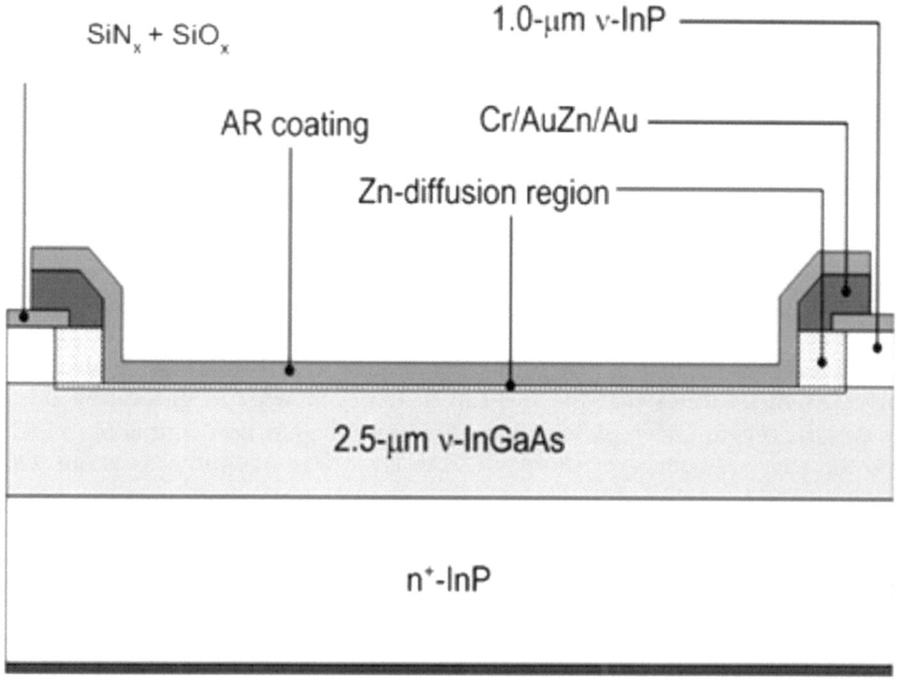

Fig. 4. Schematic drawing of device cross section. Note the absence of the InP cap inside the aperture.

wavelength operation. Shorter wavelength typically has a larger absorption coefficient that is equivalent to a shallower absorption depth, and vice versa. How deep the junction can protrude into the InGaAs absorption layer is thus quite dependent on the operating wavelength for avoiding slow carrier diffusion process. Typically the depth is designed about the reciprocal of the absorption coefficient corresponding to the minimum wavelength the detector operates.

Second, a well-controlled etching process is required to remove the InP cap layer inside the aperture region. Due to the target region is surrounded by contact metallizations, the metal films should adhere to the cap layer well enough for inhibiting non-uniform and excessive localized undercut during wet etching process. The wet etching time should also be controlled. Excessive undercut could expose the junction periphery in the InGaAs absorption layer and consequently result in severe surface leakage, which could go further if there would be process-induced damages. It would be advantageous to have the metal/semiconductor interface slightly alloyed before etching process. By using a wider p-contact span "s" or adopting a second dielectric passivation, the wet etching process can be more tolerable. Of course, this benefit is at the price of reduced coupling aperture. Besides adhesion consideration, dry etching process can be performed as an assist before wet process to minimize undercut. Nevertheless, to avoid impact damages, some thickness of cap layer should be left for wet process. It is essential that the wet process has desired selectivity; that is, the etching process stops automatically once the etching solution sees the absorption layer. If the etching solution or the material system has poor selectivity, the etching time should be accurately controlled, otherwise the shallow p-n junction can be also removed and, as a consequence, no biasing field for generated carriers.

Third, for high efficiency in all the operating spectral range, device has an anti-reflection dielectric coating designed for the wavelengths of interest. It is impossible for single-layer dielectric film to serve a broad spectral range. Therefore, at least two layers of dielectric films are required.

4. Device characteristics [10]

Fig. 5 shows both current-voltage (I-V) and capacitance-voltage (C-V) characteristics of the devices measured at room temperature. Typically, the InGaAs PD exhibits a 50-pA dark current and a 0.3-pF capacitance at –5 V. Obviously from this figure, the device leakage behaves just as of those conventional planar InGaAs p-i-n PDs, which keeps a slightly increasing leakage as the bias increases till the tunneling-type breakdown is reached. Such a low dark current indicates that the InP cap is removed without generating the surface damage and the severe undercut, which might expose the junction periphery in InGaAs. In addition, the sheet resistance of the p^+-InGaAs layer was measured as about 720 Ω/□, which is comparable to that of the p^+-InGaAs base of the InP-based heterojunction bipolar transistors (HBTs) [15]. The measured series resistance of the InGaAs PD is about 18 Ω, which indicates that the 0.1-0.2 μm p^+-InGaAs layer has been doped high enough (~10^{19} cm^{-3}) for high-speed applications. The low capacitance indicates a well-controlled junction depth and a significantly reduced parasites, which results in a 0.1-pF junction capacitance and a 0.2-pF parasitic capacitance, respectively. The estimated frequency response deduced from the series resistance and the measured capacitance is about 10.1 GHz.

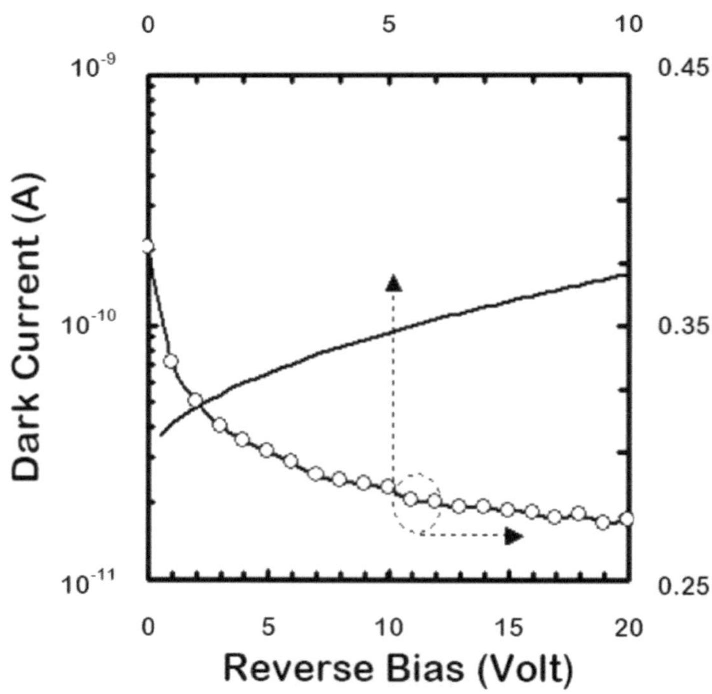

Fig. 5. Characteristics of dark current and capacitance versus reverse bias at room temperature.

For obtaining the responsivity spectrum, we utilized a tungsten lamp/monochromator/multi-mode fiber (MMF) combination as the optical source for measurement. Fig. 6 shows the measurement results of the InGaAs pin PD with the InP cap removed. The device exhibits a quantum efficiency higher than 80% in the 0.85-1.65 μm wavelength range and higher than 70% in the 0.55-1.65 μm wavelength range.

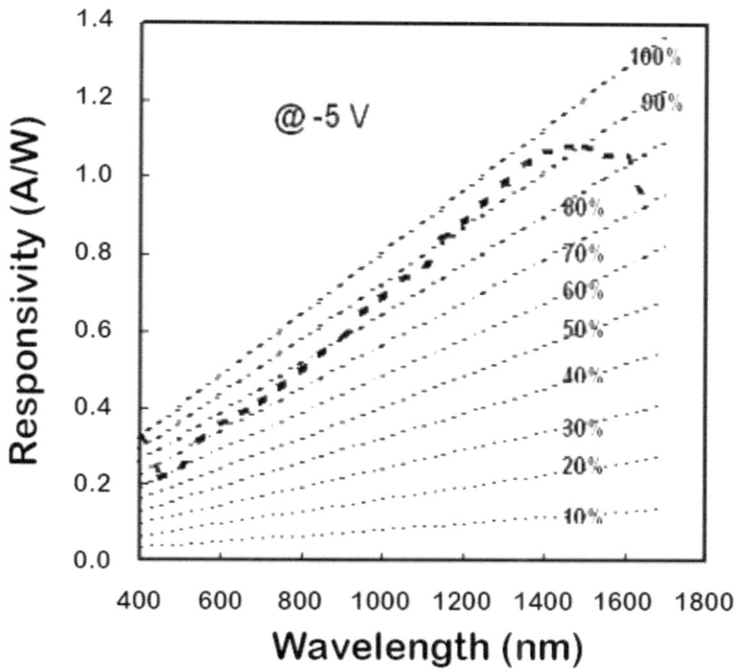

Fig. 6. Responsivity spectra measured at -5 V.

To see if the device with the InP cap removed still retains its high-frequency operation capabilities, the device was mounted onto a SMA-connector for dynamic characterizations. For the 3-dB bandwidth measurements, the packaged device was characterized at 1.3-μm wavelength using HP8703 lightwave component analyzer. As shown in Fig. 7, the device operating at -5 V achieves a 3-dB bandwidth of about 10.3 GHz. Furthermore, to see the transmission characteristics, the non-return-to-zero (NRZ) pseudorandom codes of length 2^{31}-1 at 10.3 Gbps data rate using the 0.85-μm multimode and 1.3-μm singlemode fibers were fed into the photodiode, respectively. Fig. 8 shows the back-to-back eye diagrams. It is observed that both the eye diagrams of 0.85-μm (Fig. 8(a)) and 1.3-μm (Fig. 8(b)) wavelengths are distinguishably open and free of intersymbol interference and noise. These characteristics prove that the InGaAs p-i-n photodiode is well qualified for high-speed fiber communication

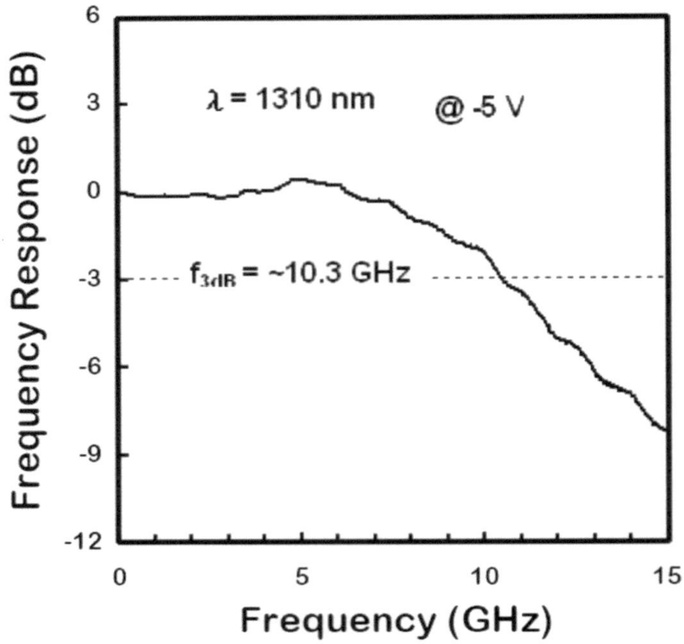

Fig. 7. Device characteristics in frequency response at the 1.3-μm wavelength.

5. 10-GBPS InGaP-GaAs p-i-n photodiodes with wide spectral range [11]

The epitaxial structure of InGaP-GaAs p-i-n PD was grown by MOCVD on the n^+-GaAs substrate. A 2.5-μm non-intentional doped GaAs absorption layer was grown on a 200 nm GaAs buffer layer. This was followed by a 10 nm $Al_{0.3}Ga_{0.7}As$ grading layer which was doped p type with a carrier concentration of approximately 1×10^{18} cm^{-3}. Here, a 10 nm p-$Al_{0.3}Ga_{0.7}As$ intermediate layer was inserted to reduce the band off-set at the interface between the absorption layer and the window layer to eliminate the hole trapping problem. An $In_{0.5}Ga_{0.5}P$ etching stop layer was doped p type and its thickness was 20 nm. The wafer was finally capped with a 200 nm thick p^+-GaAs contact layer with a hole concentration higher than 1×10^{18} cm^{-3}.

The process started with depositing a 2000 Å SiN_x film and then creating the 50-μm-in-diameter windows for the following chemical wet etching process. A circular mesa structure of a 50-μm diameter was formed by $1H_3PO_4$: $1H_2O_2$: $20H_2O$ solution for etching GaAs and AlGaAs, and 1HCl: $3H_3PO_4$ solution for etching InGaP. In order to attain a low dark current, the mesa etching was stopped at the middle of absorption layer so the current goes through the bulk region. To reduce the parasitic capacitance, a double-layer passivation of 1500 Å SiN_x and 5000 Å SiO_2 was deposited by PECVD. After a ring-shaped Cr/AuZn/Au p-contact metal deposition, the GaAs cap layer inside the 30-μm-in-diameter coupling aperture was removed by selective etching process. Afterwards, the double-layer SiN_x/SiO_x antireflection (AR) coating and Cr/Au for bondpad metallizations were deposited in sequence. Wafers were then lapped and polished down to about 300 μm and the polished backside was coated with Cu/AuGeNi/Au n-contact metallizations. Lastly, the samples were annealed at 400°C for 20 sec to reduce the contact resistance. The cross-sectional view of a finished device is schematically drawn in Fig. 9.

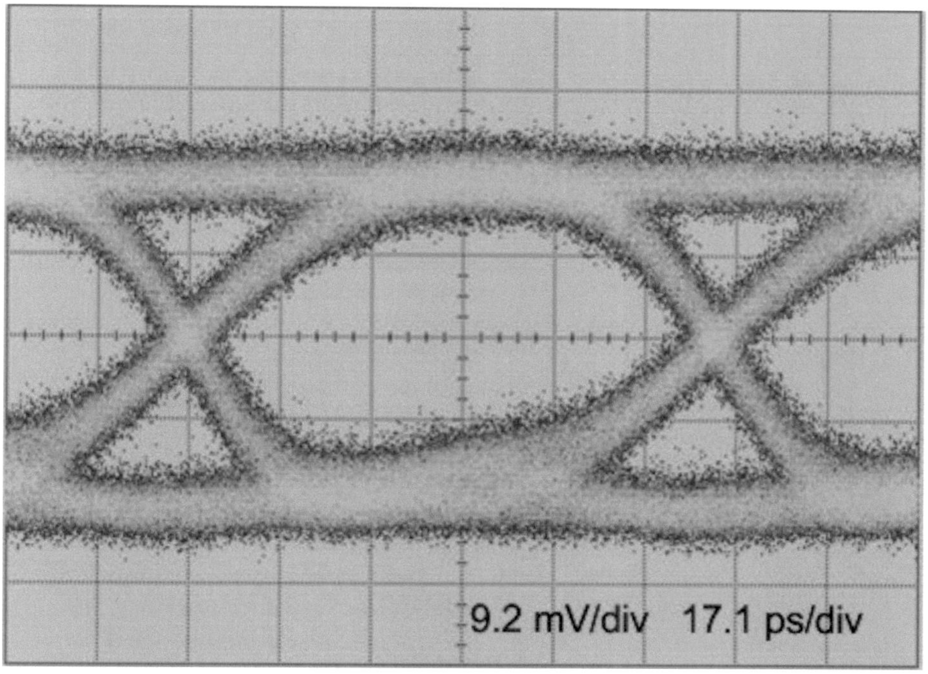

(a) Huang et al.

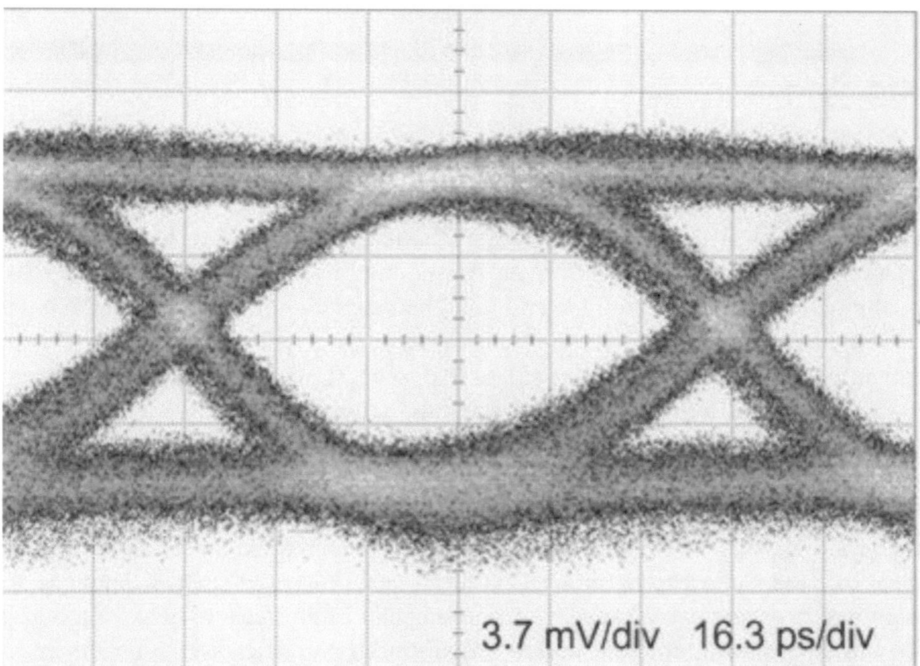

(b) Huang et al.

Fig. 8. Eye diagrams of back-to-back test for a SMA packaged device operating at –5 V and 10.3 Gb/s with PRBS of 2^{31}-1 word length at (a) 1.3-μm and (b) 0.85-μm wavelengths.

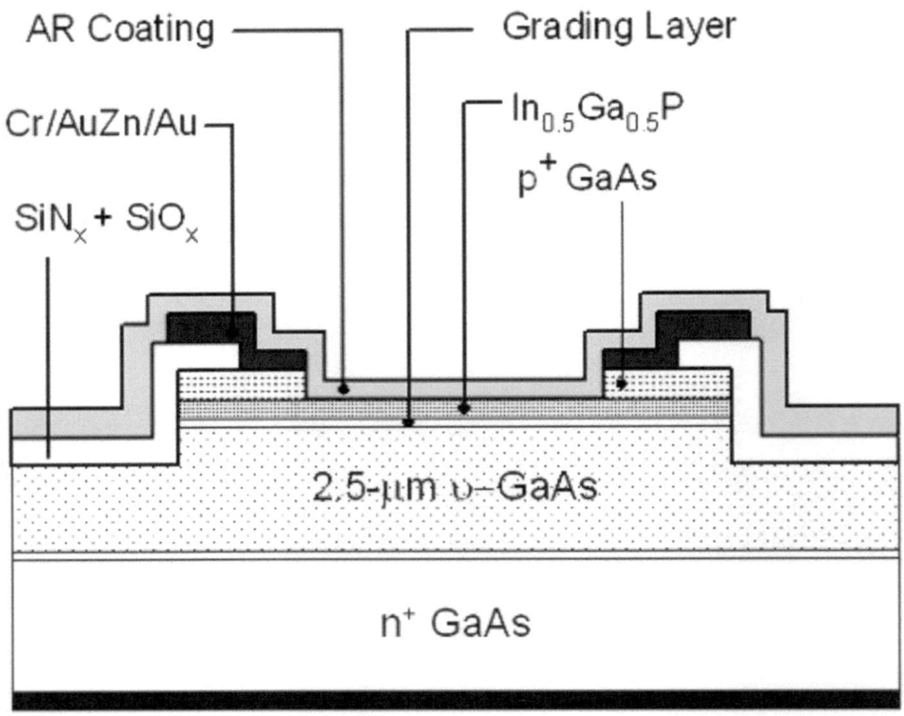

Fig. 9. Schematic drawing of device cross section. Note the absence of the GaAs cap inside the aperture.

The dark current of an InGaP/GaAs p-i-n PD is usually too low to have any significant influence on receiver sensitivity. However, it is an important parameter for process control and reliability. Fig. 10 shows both I-V and C-V characteristics of the devices with a window of 50 μm in diameter measured at room temperature. The fabricated InGaP-GaAs p-i-n PDs exhibit a sufficiently low dark current of less than several pA and a small capacitance of 0.3 pF at -5 V. All the tested p-i-n PDs show a breakdown voltage over 40 V. These characteristics indicate the high crystalline quality of the epitaxial layers grown by MOCVD and without generating the surface damage after removing the GaAs cap layer. Inspection of this figure reveals that the device leakage behaves just as of those conventional p-i-n PDs, which keeps a slightly increasing leakage as the bias increases. Such a low dark current illustrates that the GaAs cap is removed without generating the surface damages and the severe undercut. A low capacitance is of fundamental importance to achieve a high-speed PD. The low capacitance indicates significantly reduced parasitics, which results in a 0.1-pF junction capacitance and a 0.2-pF parasitic capacitance. To minimize the noise and maximize the bandwidth, the series resistance R_S should be as low as possible. The derived series resistance is about 5 Ω from the estimation of series resistance as $R_S \approx dV/dI$ at a relatively large forward current of 50 mA.

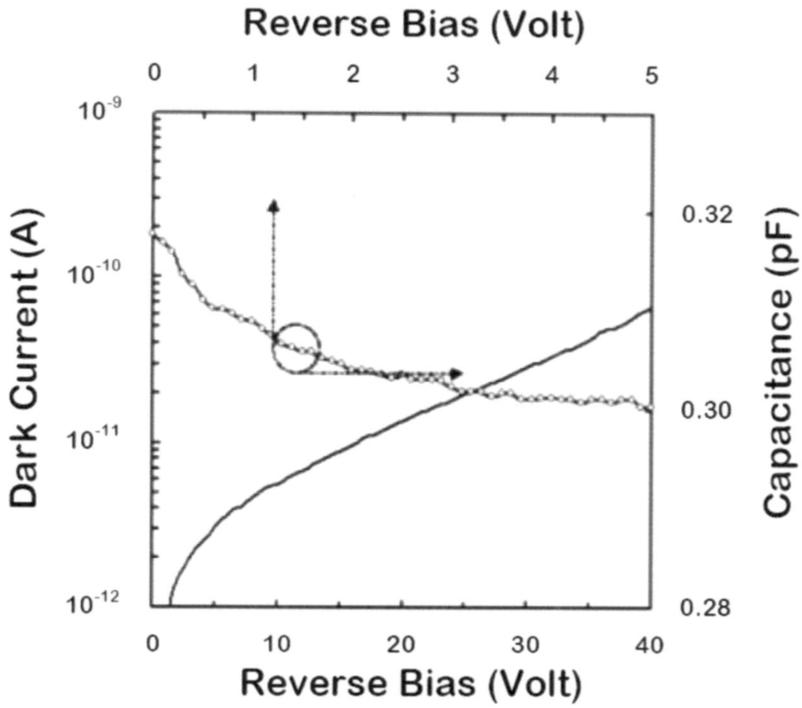

Fig. 10.Characteristics of dark current and capacitance versus reverse bias at room temperature.

For obtaining the responsivity spectrum, we utilized a tungsten lamp/monochromator/multi-mode fiber (MMF) combination as the optical source for measurements. Fig. 11 shows the measured responsivity spectra of the InGaP-GaAs p-i-n PD with the GaAs cap layer removed and a commercial Si PD. Our device exhibits a quantum efficiency higher than 90% in the 420-850 nm wavelength range and higher than 70% in 360-870 nm range, which is obviously superior to the Si PD in this wavelength range.

Fig. 12 is the simple equivalent circuit of InGaP-GaAs pin PD. The calculated frequency response deduced from the series resistance, junction capacitance, bondpad capacitance, and the transit time is approximate 8 GHz. To see if the device with the GaAs cap layer removed still retains its high-frequency operation capabilities, the device was mounted onto a SMA-connector for dynamic characterizations. For the 3-dB bandwidth measurements of 850 nm wavelength, we have established a high frequency measurement system which includes an 850 nm laser source, a 0-20 GHz modulator, a signal generator (Agilent E8257D), and a spectrum analyzer (Agilent E4448A). The influence of used cables and bias tee on the measured frequency responses has been amended carefully. The 3-dB bandwidth of this device is expected as about 8 GHz, which is dominated by RC time constant. The thickness of the absorption layer is only 2.5 μm, which is expected to have a 3-dB bandwidth larger than 11 GHz, when we only consider the transit time factor. As shown in Fig. 13, the measured result of device operating at -5 V achieves a 3-dB bandwidth of about 9.7 GHz, which is a combination result of carrier transit, RC discharge, and inductance of bonding wire. The measured 3-dB bandwidth of packaged PD is enhanced due to inductance peaking. Furthermore, to see the transmission characteristics, the non-return-to-zero (NRZ)

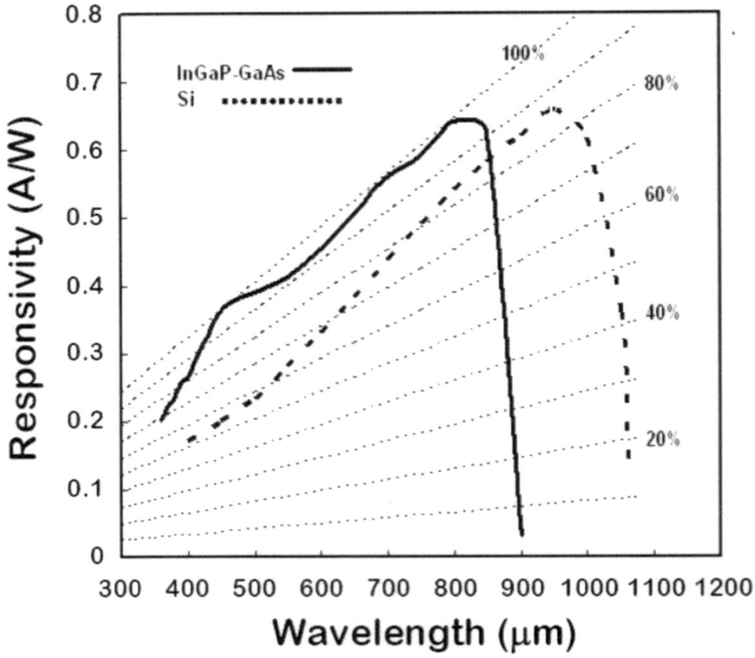

Fig. 11. Responsivity spectra measured at -10 V

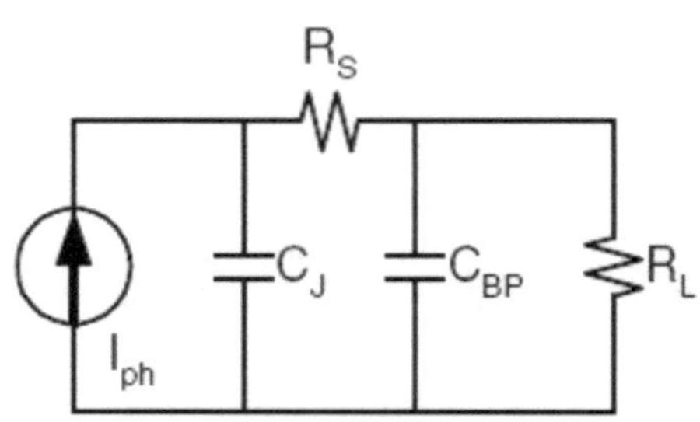

Fig. 12. Equivalent circuit of InGaP-GaAs p-i-n photodiode.

pseudorandom codes of length 2^{31}-1 at 10.4 Gbps data rate using the 850-nm multimode fibers was fed into the PD. Fig. 14 shows the back-to-back eye diagram. It is observed that the eye diagram at 850-nm wavelength is distinguishably open and free of intersymbol interference and noise. These characteristics prove that the InGaP-GaAs p-i-n PD is well qualified for high-speed fiber communications.

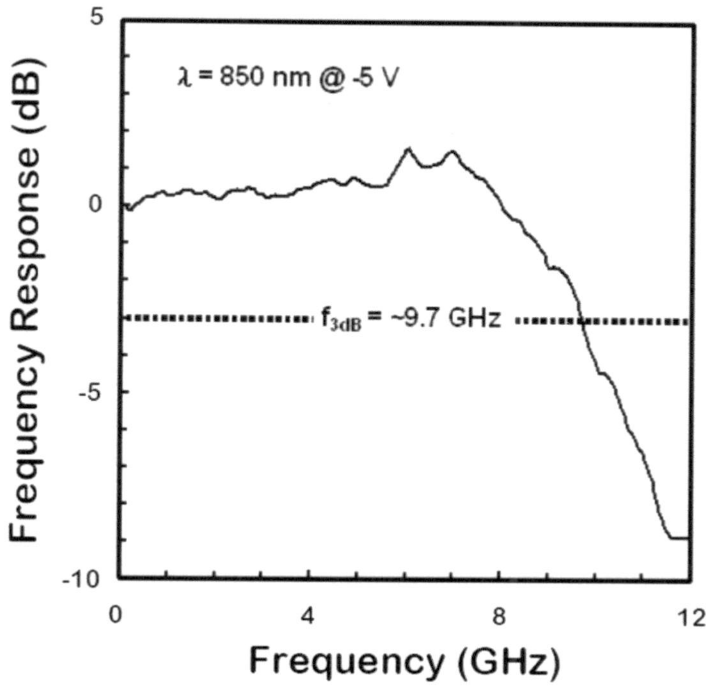

Fig. 13. Device characteristics in frequency response at the 850 nm wavelength.

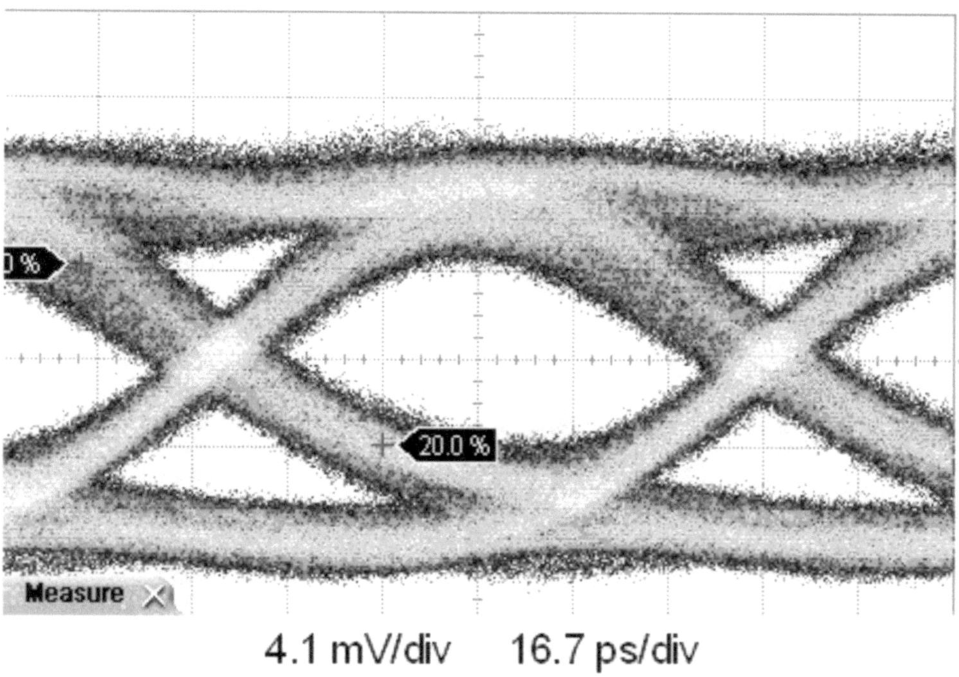

Fig. 14. Eye diagrams of back-to-back test for a SMA packaged device operating at –5 V and 10.4 Gb/s with PRBS of 2^{31-1} word length at 850 nm wavelength.

6. Alignment-tolerance enlargement of a high-speed photodiode by a self-positioned micro-ball lens

To widen the alignment tolerance of a 10-Gb/s InGaAs p-i-n PD, which typically has an optical coupling aperture of only 30 μm in diameter; we propose a self-positioning ball-lens-on-chip scheme for enlarging the effective coupling aperture of the device [16]. A Monte-Carlo ray trace simulation, which is suitable for either on-axis or off-axis simulation of various optical or optoelectronic systems in the three-dimensional (3D) space [17]-[19], is utilized to optimize the conditions of this micro-ball-lens (MBL) integrated high speed p-i-n PD [20]. The effectiveness of the MBL and the Monte-Carlo ray trace modeling demonstrates through the measurements of the spatial response uniformity of the MBL-integrated InGaAs p-i-n PD.

We shall report the detailed analyses of ϕ = 250 μm ruby ball-lens integrated photodiode. With a single-mode fiber light source, the optimal spatial response uniformity and alignment tolerance are demonstrated through the ray trace simulation and the practical measurements. The dynamic response of the MBL-integrated high speed InGaAs p-i-n PD is also characterized.

6.1 Fabrication

The photolithographic process is to define and develop the MBL-socket made of SU-8 in concentric with the coupling aperture; therefore the optical axis of the photodiode will be automatically aligned to the MBL. The inner diameter D and the height H of the socket, which was controlled by the patterned conditions and the spin-coating speed, respectively, are designed to accommodate a commercially available ruby micro-ball-lens.

After the photodiode chip was die- and wire-bonded onto a modified subminiature-version-A (SMA) connector, a sufficient UV-cured epoxy was filled into the socket and then the MBL was placed over. The MBL fell into the socket to find an equilibrium position automatically, as shown in Fig. 15. Then, the chip was fully cured by UV light to secure the ball-lens on the socket. Such a lens-on-socket scheme is inherently a self-positioning process.

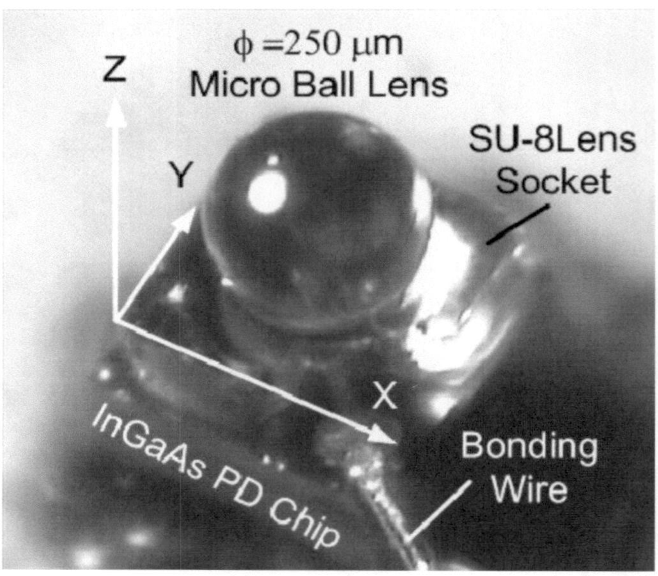

Fig. 15 Schematic diagrams of a ϕ = 250-μm ruby MBL on the lens socket.

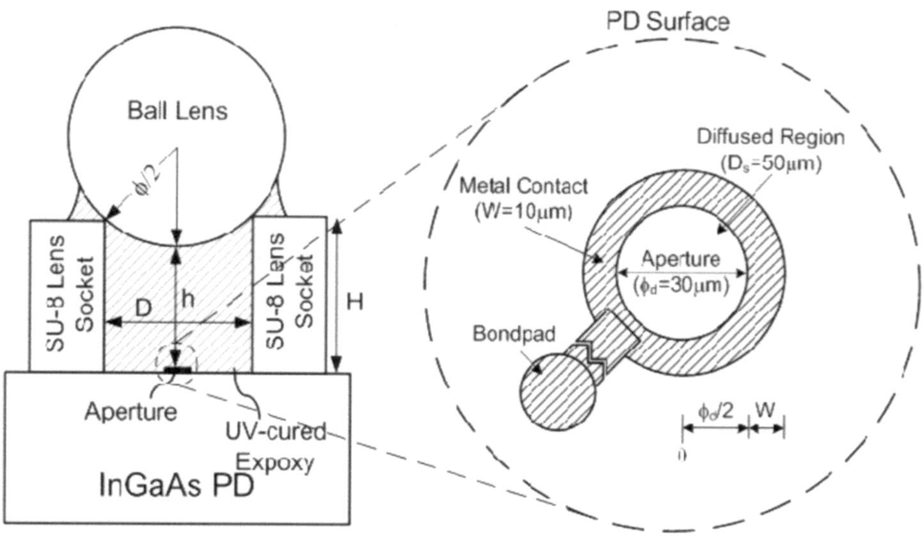

Fig. 16. Structure drawing of the MBL integrated chip and the InGaAs photodiode surface.

The detailed structural drawing of the MBL-integrated photodiode is illustrated in Fig. 16. For an ideal situation, the distance between the bottom of the MBL and the aperture, h, at that equilibrium position can be calculated by

$$h = H - \left(\frac{\phi}{2} - \sqrt{\left(\frac{\phi}{2}\right)^2 - \left(\frac{D}{2}\right)^2}\right) \tag{1}$$

where H is the height of the lens-socket, ϕ is the diameter of the MBL, and D is the inner diameter of the socket.

The pattern on the chip surface, including a metal contact ring (W = 10 μm), a bondpad, and a connection metal line, is also illustrated in Fig. 16. The area within the metal contact ring (ϕ_d = 30 μm) is the detection region wherein the selective diffusion region is wider (D_s = 50 μm).

In this study, a SU-8 ball-lens socket with a 130-μm inner diameter (D) on the InGaAs photodiode has been fabricated to sustain a ϕ = 250 μm ruby MBL. The height of lens-socket is a parameter to find an optimal condition.

6.2 Results

To evaluate the effectiveness of the integrated MBL, the response (coupling) uniformity of a photodiode with a micro-ball-lens is characterized and is compared to a bare chip. By transversely scanning (i.e., parallel to the X-Y plane defined in Fig. 15) a single-mode fiber (SMF) across the center of the entire chip, we are able to evaluate the X (Y)-axis response (coupling) uniformity. On the other hand, the axial scan (along the optical axis) provides the Z-axis response (coupling) uniformity. As a reference coordinate, ΔX and ΔY are used to represent the SMF's output facet position with respect to the optical axis ($\Delta X = \Delta Y = 0$), and ΔZ represents the distance between the SMF's output facet and the nearest coupling plane along the optical axis. The nearest coupling plane herein means the plane of aperture (without MBL) or the vertex of the ball-lens (with MBL) normal to the optical axis.

A Monte-Carlo ray trace simulation has been constructed to imitate this optical system in Ref. 20. It is a useful tool to analyze the MBL integrated photodiode. The simulated data for the ruby MBL integrated photodiode, whose lens diameter is 250 μm, are shown in Fig. 17. In the figure, the dash lines represent the responsivities that only accumulate the rays detected within the metal contact ring on the photodiode surface. The solid lines additionally include the rays that are incident at the effective detection regions outside the metal ring. It is therefore greater than the dash lines under the same conditions. However, the deviation between the solid and dash lines is undesired. The out slow diffusing carriers can degrade the dynamic performance of a high speed InGaAs photodiode.

Fig. 17(a) shows the Z-axis response uniformity along optical axis (X = 0 μm). The variation of curves caused by H from 150 to 30 μm (ΔH = -20 μm) is quite obvious. By defining the 1-dB optical loss (responsivity = 0.83) as the alignment limit, we can obtain the Z-axis alignment tolerances. These data extracted from the curves are listed in Table 1. As compared to the narrow 170-μm tolerance of a bare chip from measurements, the improvements can be at least 3.65 fold (H = 150 μm), except the case of H = 30 μm which is hard to define. Moreover, the maximum value (1150 μm) derived from the curve of H = 50 μm amazingly achieves 6.76 times the alignment tolerance of a bare chip.

In order to prove the modeling results, various MBL-integrated photodiodes with H from 50 to 110 μm were fabricated and were characterized by a single-mode fiber light source (λ = 1.3 μm). The alignment tolerances extracted from the measurements are also listed in Table 1. According to the results, they are 1120 μm (H = 50 μm), 1020 μm (H = 70 μm), 920 μm (H = 90 μm), and 850 μm (H = 110 μm), respectively. The practical alignment tolerances quite match the simulated results. In addition, the responsivities with the conditions of H = 110 μm (triangle) and H = 50 μm (circle) are chosen to be plotted in the same figure for comparison.

The alignment tolerance along X axis is more important practically, because it is much narrower than that in Z axis. The size of PD's active area, concerning with the dynamic response, limits the available alignment region. The X-axis alignment tolerances at the chosen position of ΔZ = 400 μm are characterized by transversely scanning across various MBL-integrated photodiodes. As shown in Fig. 17(b), as the H decreases, the central main peak becomes wider and hence the alignment tolerance is larger. Nevertheless, the central responsivity (ΔX = 0) starts to degrade as the H < 70 μm. The reduction of the central responsivity is attributed to the bigger beam size focused on the PD surface by the micro-ball-lens as compared to the aperture within the metal contact for the narrower distance between the micro-ball-lens and the photodiode surface.

According to the Monte-Carlo simulation, the X-axis alignment tolerances, respectively, are 140 μm for H = 50 μm, 116 μm for H = 70 μm, 96 μm for H = 90 μm, 78 μm for H = 110 μm, 64 μm for H = 130 μm, and 56 μm for H = 150 μm, as listed in Table 1, except the condition of H = 30 μm which is also hard to define. The maximum improvement can be 7 times the alignment tolerance of a bare chip.

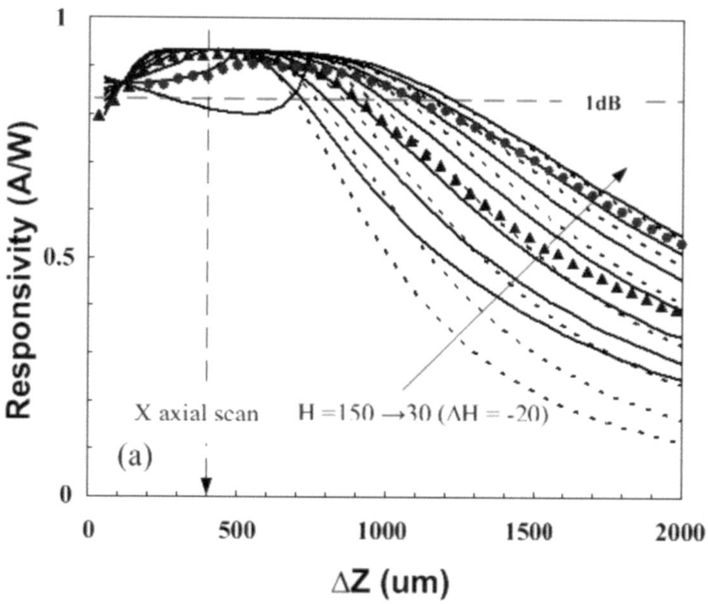

(a)

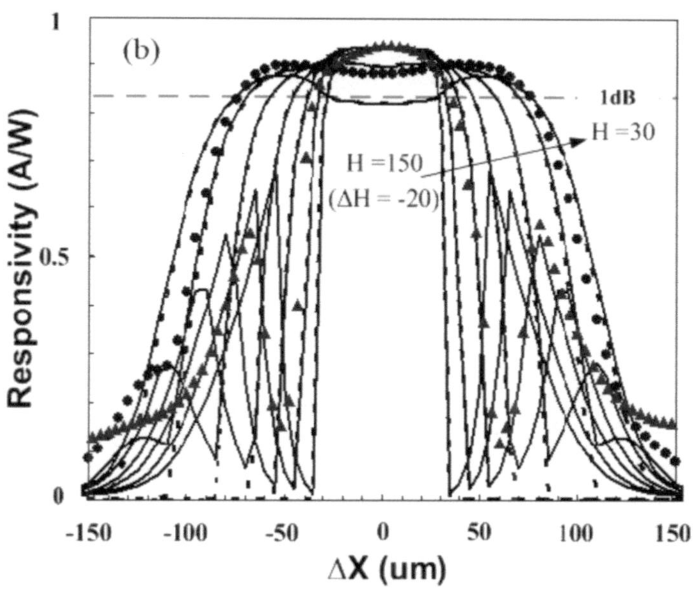

(b)

Fig. 17. Simulated (lines) responsivity curves along (a) Z axis (ΔX = 0 μm) (b) X axis (ΔZ = 400 μm) of the ϕ = 250-μm ruby MBL-integrated PD, in which the triangles and the circles are the practical measured data of H = 110 μm (triangles) and H = 50 μm, respectively. The difference between the solid line and dash line at the same condition is the former to count the effective sensitive area outside the metal ring but the latter doesn't.

		Bare Chip	H = 30 μm	H = 50 μm	H = 70 μm	H = 90 μm	H = 110 μm	H = 130 μm	H = 150 μm
Experiment	1-dB Tolerance (Z-axis)	~170 μm	-	1120	1020	920	850	-	-
	(Improvement)	(×1)	-	(×6.56)	(×6)	(×5.51)	(×5)	-	-
	1-dB Tolerance (X-axis)	~20 μm	-	150	110	86	62	-	-
	(Improvement)	(×1)	-	(×7.5)	(×5.5)	(×4.3)	(×3.1)	-	-
Simulatiom	1-dB Tolerance (Z-axis)	-	*	1150	1050	980	840	700	620
	(Improvement)	-	-	(×6.76)	(×6.18)	(×5.76)	(×4.94)	(×4.12)	(×3.65)
	1-dB Tolerance (X-axis)	-	*	140	116	96	78	64	56
	(Improvement)	-	-	(×7)	(×5.8)	(×4.8)	(×3.9)	(×3.2)	(×2.8)

* the responsivity curve are hard to define

Table 1. Alignment tolerances of ϕ = 250 μm MBL-integrated InGaAs photodiode

The effectiveness of the ϕ = 250 μm MBL-integrated photodiode is also demonstrated by the practical device fabrication and measurements. As the results of H = 110 μm (triangles) and H = 50 μm (circles) are plotted in Fig. 17(b), the measured alignment tolerances along X axis are 62 and 150 μm, respectively. Thus, the improvements are 3.1 and 7.5 folds, respectively. These data still match the simulated results. Such the wide alignment tolerances are enough for the conventional passive scheme for photodiode package.

As revealed from the results above, the optimal condition for this ϕ = 250 μm ruby ball-lens integrated photodiode would be H = 50 μm and the theoretical alignment tolerance is 1150 μm × 140 μm. It means that we can have a MBL integrated photodiode with a 6.76- × 7-fold improvement. The practical device based on this condition has been fabricated and has an alignment tolerance of 1120 μm × 150 μm which matches the simulated results.

The one-dimensional simulation provides a promising way to quickly search the optimal condition for this MBL integrated photodiode. Furthermore, the two-dimensional simulation can show us more clear information for widening the alignment tolerance. The two-dimensional responsivity for the optimal case is drawn in Fig. 18(a). According to the simulation, the alignment tolerance is 1150 μm × 180 μm. The result in X axial here is wider than that from one-dimensional simulation. It represents that the optimal X-axis alignment tolerance is not at ΔZ = 400 μm but at ΔZ = 150 μm.

For ray trace analyses, we recorded the ray incident location on the photodiode surface. The four positions we chose to locate a SMF light source are b: (ΔX, ΔZ in μm) = (0, 200), c: (0, 800), d: (0, 1900), and e: (-80, 300) labeled on the responsivity surface. The corresponding maps are sequentially illustrated in Figs. 18(b), 18(c), 18(d), and 18(e). Inspection of this figure reveals that only if the SMF is operated within the alignment tolerance region, most rays are incident at the active area, not the area outside which may degrade the high-speed performance of the InGaAs photodiode. The reason why the responsivity is reduced as observed in Figs. 18(b), 18(d), and 18(e) is attributed to some of the rays to be incident at the metal contact and then can be reflected.

Such evidence indicates that no detrimental effects but alignment tolerance enhancements are brought by the MBL integration. Fig. 19 shows the experimental results by two-dimensional scanning across the device. The practical alignment tolerance for this ϕ = 250 μm ruby ball-lens integrated photodiode is 1120 μm × 174 μm marked in this figure. Within the 1-dB alignment tolerance region, we try to measure its dynamic characteristic. Fig. 20 shows the back-to-back eye diagram at the 10.3-Gb/s data rate. It is observed that the eye diagram of 1.3-μm wavelength is distinguishably open and free of intersymbol interference and noise. These characteristics prove that the MBL integrated InGaAs p-i-n PD is indeed well qualified for high-speed fiber communication.

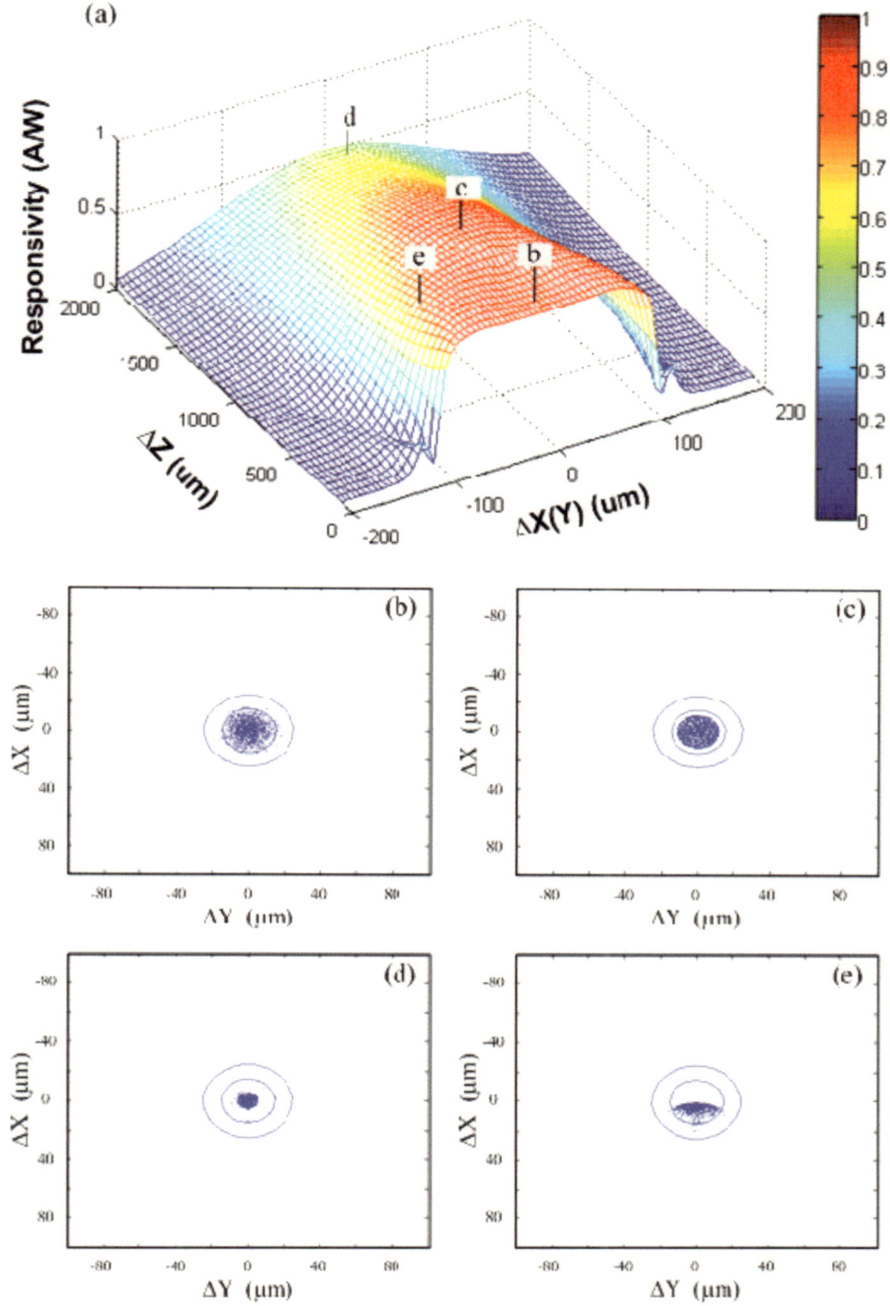

Fig. 18. (a) Modeling two-dimensional response uniformity of the ϕ = 250-μm ruby MBL-integrated PD (H = 50 μm) across the X-Z plane. The ray trace maps are derived from the positions (ΔX and ΔZ in μm), (b) (0, 200), (c) (0, 800), (d) (0, 1900), and (e) (-80, 300) labeled on the responsivity surface.

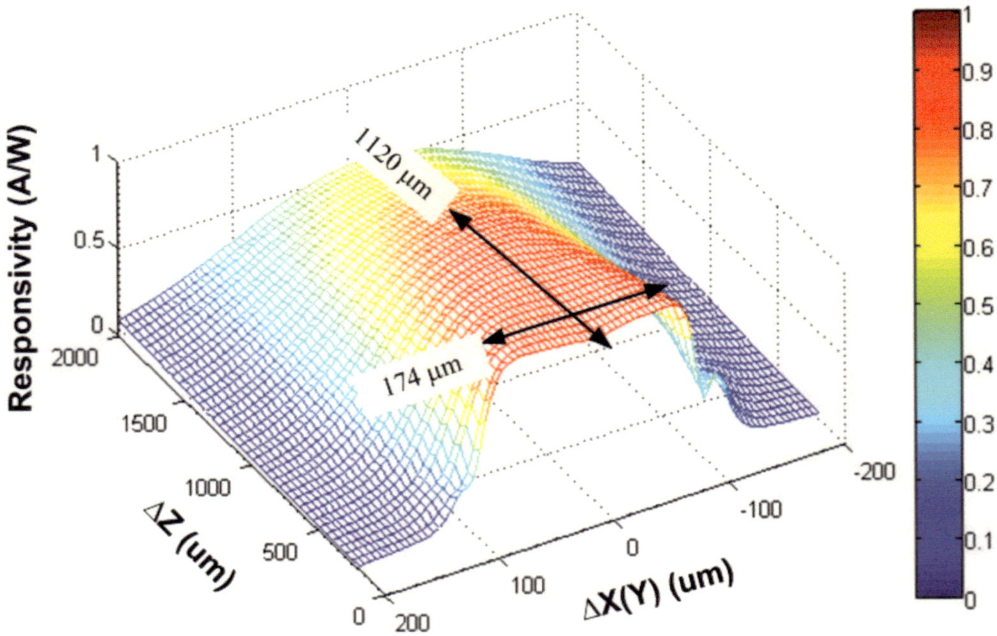

Fig. 19. Measurements of two-dimensional response uniformity of the ϕ = 250-μm ruby MBL-integrated PD across the X-Z plane. The numbers over the surface are the 1-dB alignment tolerances.

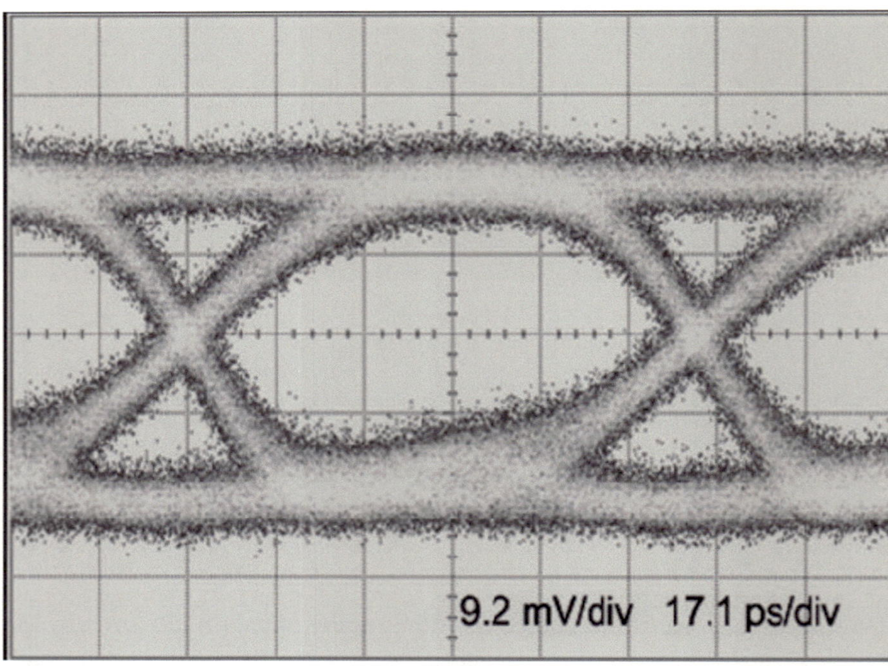

Fig. 20. Eye diagram of back-to-back test for an SMA-packaged device operated at -5 V and 10.3 Gb/s with PRBS of 2^{31}-1 word length at 1.3 μm wavelength.

7. Conclusions

We have demonstrated the PIN PDs whose configuration is suitable for broad spectral range operation. The spin-on diffusion has several excellent advantages for junction formation, such as good uniformity, simple process, high yield and low cost. Here we present the process and condition for creating shallow p-n junctions of InP/InGaAs and GaSb wafer applied SOD technique. By selectively removing the wide-bandgap cap layer on top of the conventional InP/InGaAs or InGaP/GaAs p-i-n PDs, such a wide-wavelength photodiode can be achieved. This spectral range covers all the wavelengths of interest nowadays in fiberoptic communications: 0.65, 0.85, 1.3, and 1.55 μm. With the optimized design of antireflection coating, both the PDs exhibit a dark current smaller than several pA, the photodiode has a 3-dB bandwidth of ~10 GHz. In addition, InP/InGaAs and InGaP/GaAs p-i-n PDs show high quantum efficiency in the 300-850 nm and 0.85-1.65 μm spectral range, respectively. Since both high-efficiency and high-speed operation can be achieved, receivers based on such devices are suitable for both 850- and 650-nm fiber communication systems.
By selectively removing the InP cap layer and integrating with a micro-ball-lens, we have demonstrated the alignment tolerance enhancement of 10-Gb/s InGaAs p-i-n photodiode by integrating a ϕ = 250 μm ruby micro ball lens on a chip. According to the Monte-Carlo ray trace simulation, the available alignment tolerance for positioning a 1.3-μm single-mode fiber light source can be found by varying the lens socket height. The maximum alignment tolerances are 1150 and 180 μm along the longitudinal and transverse axes, respectively, which are 6.76- and 7.5-fold improvements to the chip without micro-ball-lens integration. The modeling results are proven through practical measurements. Such ball-lens-on-chip scheme for enlarging the effective coupling aperture is efficient and cost-reductive process for the small-aperture photodiode package.

8. References

[1] C. L. Ho, M. C. Wu, W. J. Ho, J. W. Liaw, and H. H. Wang, "Effectiveness of the pseudowindow for edge-coupled InP-InGaAs-InP PIN photodiodes," *IEEE J. Quantum Electron.*, vol. 36, no. 3, pp. 333-339, 2000.

[2] S. Kagawa, K. Inoue, I. Ogawa, Y. Takada, and T. Shibata, "Wide-wavelength InGaAs/InP PIN photodiodes sensitive from 0.7 to 1.6□μm," *Jpn. J. Appl. Phys.*, vol. 28, no. 10, pp. 1843-1846, 1989.

[3] J. B. Williamson, K. W. Carey, F. G. Kellert, D. M. Braum, L. A. Hodge, and D. W. Loncasty, "High-density, planar Zn-diffused InGaAs/InP photodetector arrays with extended short-wavelength response", *IEEE Trans. Electron Devices*, vol. 38, no. 12, p. 2707, 1991.

[4] D. Wake, R. H. Walling, I. D. Henning, and D. G. Parker, "Planar-junction, top-illuminated GaInAs/InP pin photodiode with bandwidth of 25 GHz," *Electron. Lett.*, vol. 25, no. 15, pp. 967-968, 1989.

[5] M. Makiuchi, O. Wada, T. Kumai, H. Hamaguchi, O. Aoki, and Y. Oikawa, "Small-junction-area GaInAs/InP pin photodiode with monolithic microlens," *Electron. Lett.*, vol. 24, no. 2, pp. 109-110, 1989.

[6] S. R. Cho, J. Kim, K. S. Oh, S. K. Yang, J. M. Baek, D. H. Jang, T. I. Kim, and H. Jeon, "Enhanced optical coupling performance in an In-GaAs photodiode integrated

with wet-etched microlens," *IEEE Photon. Technol. Lett.*, vol. 14, no. 3, pp. 378-380, 2002.

[7] C. R. King, L. Y. Lin, and M. C. Wu, "Out-of-plane refractive microlens fabricated by surface micromachining," *IEEE Photon. Technol. Lett.*, vol. 8, no. 10, pp. 1349-1351, 1996.

[8] Z. L. Liau, D. E. Mull, C. L. Dennis, and R. C. Williamson, "Largenumerical-aperture microlens fabrication by one-step etching and masstransport smoothing," *Appl. Phys. Lett.*, vol. 64, pp. 1484-1486, 1994.

[9] M. Hutley, R. Stevens, and D. Daly, "Microlens arrays," *Phys. World*, vol. 4, no. 7, p. 27, 1991.

[10] Y. H. Huang, C. C. Yang, T. C. Peng, F. Y. Cheng, M. C. Wu, Y. T. Tsai, and C. L. Ho, "10-Gbps InGaAs p-i-n photodiodes with wide spectral range and enhanced visible spectral response" *IEEE Photon. Technol. Lett.* vol. 19, no. 5, pp. 339-341, 2007.

[11] M. C. Wu, Y. H. Huang, and C. L. Ho, "High-speed InGaP-GaAs p-i-n photodiodes with wide spectral range", *IEEE Electron Device Lett.* vol. 28, no. 9, pp. 797-799, 2007.

[12] N. Arnold, R. Schmitt, and H. Heime, "Diffusion in III-V semiconductors from spin-on film sources" *J. Phys. D: Appl. Phys.* vol. 17, no. 3, pp. 443-474, 1984.

[13] U. Schade and P. Enders "Rapid thermal processing of zinc diffusion in indium phosphide," *Semicond. Sci. Technol.*, vol. 7, pp. 752-757, 1992.

[14] H. Yoon, Y. H. Lee, D. H. Yeo, and S. J. Kim, "The characteristics of Zn-doped InP using spin-on dopant as a diffusion source" *J. Electron. Mater.*, vol. 31, no. 4, pp. 244-247, 2002.

[15] D. Huber, R. Bauknecht, C. Bergamaschi, M. Bitter, A. Huber, T. Morf, A. Neiger, M. Rohner, I. Schnyder, V. Schwarz, and H. Jackel, "A InP-InGaAs single HBT technology for photoreceiver OEIC's at 40 Gb/s and beyond," *J. Lightwave Technol.*, vol. 18, pp. 992-1000, 2000.

[16] Y. H. Huang, C. C. Yang, T. C. Peng, M. C. Wu, C. L. Ho, and W. J. Ho, "Alignment tolerance enlargement of a high-speed photodiode by a self-positioned microball lens," *IEEE Photon. Technol. Lett.*, vol. 18, pp. 112-114, 2006.

[17] N. Lindlein, "Simulation of micro-optical systems including microlens arrays," *J. Opt. A: Pure Appl. Opt.*, vol. 4, pp. S1-S9, 2002.

[18] B. Iske, B. Jäger, and U. Rückert, "A ray-tracing approach for simulating recognition abilities of active infrared sensor arrays," *IEEE Sensors J.*, vol. 4, pp. 237-247, 2004.

[19] Y. M. Govaerts and M. M. Verstraete, "Raytran: A Monte Carlo ray-tracing model to compute light scattering in three-dimensional heterogeneous media," *IEEE Trans. Geosci. Remote Sensing*, vol. 36, pp. 493-505, 1998.

[20] C. C. Yang, Y. H. Huang, T. C. Peng, M. C. Wu, C. L. Ho, C. C. Hong, I. M. Liu, and Y. T. Tsai, "Monte-Carlo ray trace simulation for micro-ball-lens-integrated high-speed InGaAs p-i-n photodiodes," *J. Appl. Phys.*, vol. 101, pp. 33107-33115, 2007.

2

Evaluation of Uni-Traveling Carrier Photodiode Performance at Low Temperatures and Applications to Superconducting Electronics

Hideo Suzuki
International Superconductivity Technology Center
Japan

1. Introduction

High-speed photodiodes are useful devices for optical-telecommunication systems and scientific applications. A uni-traveling carrier photodiode (UTC-PD), has extremely wide band performance of over 300 GHz and used for many high-frequency or high-speed applications. Signal transmission using optical fibers, which has several advantages such as its wide band transmission and low transmission loss, is an indispensable technology that forms the foundation of the Internet. Optical fibers also exhibit low thermal conductance and are capable of electrical isolation. These features are useful for interfacing between low-temperature and room-temperature electronics. Superconducting devices and circuits are attractive for high-speed, low-power, and quantum mechanical operations.

However, such devices and circuits have to be cooled below the critical temperatures of superconducting materials, Tc. For high-temperature superconducting materials such as YBCO, the operating temperature is around that of liquid nitrogen, 77 K, and for low-temperature metal-based superconducting materials, such as Nb and NbN, the operating temperature is around that of liquid helium, 4 K. Input/output links are one of the bottle necks preventing practical application of superconducting devices and circuits. In particular, devices and circuits using low-temperature superconductors exhibit serious problems because the high-frequency electrical I/O cables consume a large amount of cooling power. However, cooling power, especially at around 4 K and below, is quite small, typically less than 1 W, though the input AC power is as large as several KW. The amount of AC input power can be reduced by reducing the cooling power. Our goal is to use a compact cryocooler. Such a cryocooler has limited cooling capability; however, it is enough for most applications of superconducting devices due to their low power requirements. Optical I/O has potential to overcome the problem by using optical fibers and photo devices such as photodiodes. A UTC-PD seems to be the most attractive device because of its high-speed performance and is required to operate at low temperatures for application in superconducting systems. In this chapter, we describe UTC-PD performance at low temperatures and its applications in superconducting systems.

2. Customized structure and dc characteristics of UTC-PD module at low temperatures

We investigated the performance of a UTC-PD chip and modules at low temperature, which had not been done previously. The response at temperatures as low as 4 K was characterized for a commercially available standard UTC-PD module and a customized one we developed for superconducting devices. To apply the UTC-PD modules into various superconducting analog devices and systems using superconducting microchips (ICs) of digital and analog/digital circuits, the UTC-PD modules should be located near the superconducting ICs to maintain signal integrity. Ferromagnetic materials, which are widely used in many optical components, are used in the standard UTC-PD module. In general, superconducting devices and microchips, such as single flux quantum (SFQ) circuits and Josephson voltage standards (JVS), are strongly affected by the remnant magnetic field. Therefore, these materials must not be used near the chips. Hence, we developed a UTC-PD module using a customized package and a fiber lens technique for superconducting devices.

2.1 Band diagram and gap energy at cryogenic temperature

We studied the characteristics of a UTC-PD chip at low temperatures. The energy band diagram of a UTC-PD chip is illustrated in Fig. 1. The electrons generated by incoming optical irradiation in the InGaAs absorption layer are transported at high-speed to the InP wideband-depleted and n^+InP layers with drift by electrical field. In principle, UTC-PD uses the electrons as minority carriers for transporting current, which determines the operating speed. On the other hand, holes are not important for operating speed because those in the InGaAs layer are majority carriers and respond with dielectric relaxation time. This situation differs from a commonly used pin photo diode (pin-PD) using electrons and holes as minority carriers in the depletion layer. The features of a UTC-PD chip enable it to respond faster than a commonly used pin-PD chip. The optical absorption layer consists of

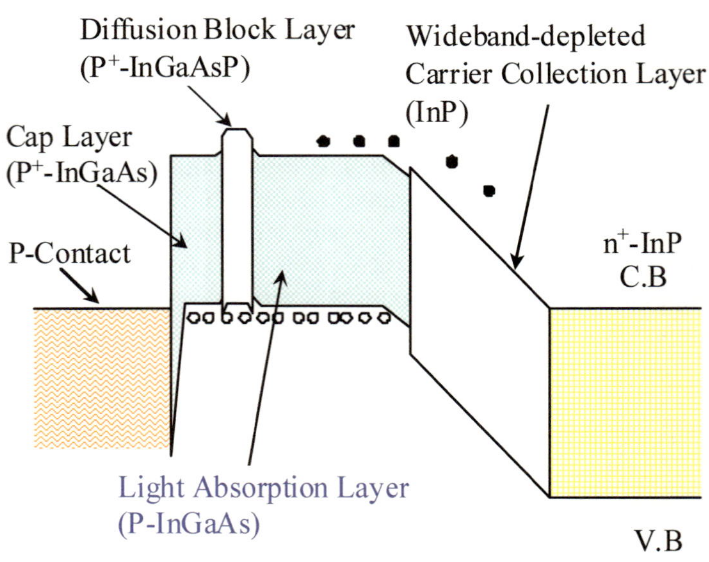

Fig. 1. Band diagram of uni-traveling carrier photodiode (UTC-PD).

$In_{1-x}Ga_xAs$ (x=0.47). The temperature dependence of the absorption coefficient vs. photon energy of the InGaAs can be seen in the handbook series on semiconductor parameters editted by Goldberg Yu.A. and N.M. Schmidt. The photon energy, at which the absorption is decreased, is critical for low-temperature performance. The transition point of the photon energy was plotted based on the handbook, as shown in Fig. 2. The wavelength, $\lambda=hc/E$, and the energy corresponding to the photon energy, E, are also plotted in this figure. Basically, a UTC-PD chip does not seem to have sensitivity at a wavelength of 1550 nm to optical irradiation at cryogenic temperature between 4 - 77 K. However, we assume that they must have sensitivity even at cryogenic temperature because the absorption layer, the InGaAs layer, is p-doped, blurring the band edge of the conduction band.

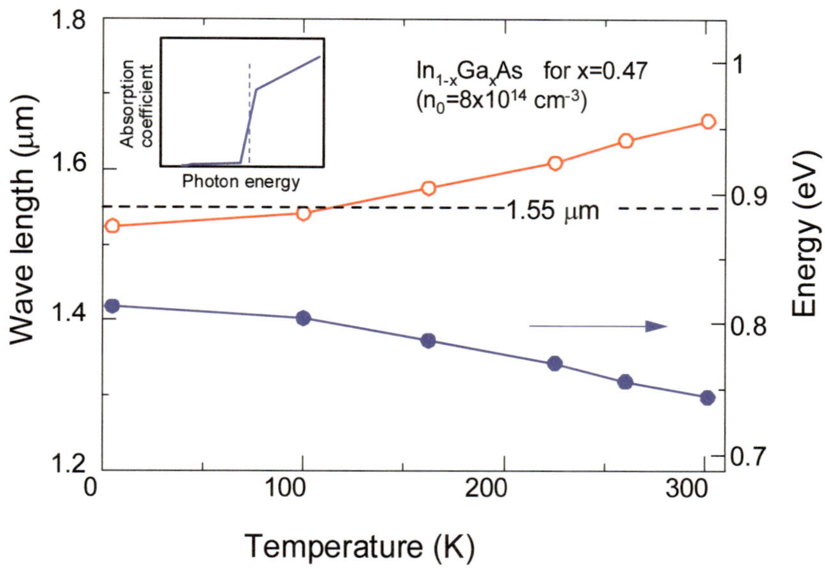

Fig. 2. Gap energy and its corresponding wavelength dependence as function of temperature for $In_{1-x}Ga_xAs$ (x=0.47) used as absorption layer in UTC-PD.

2.2 Structure and optical dc sensitivity at low temperature

Figure 3 shows an illustration of two types of UTC-PD modules, standard and customized. The photo diode chips have the same specifications as follows, over 60-GHz band width, negative type output, optical acceptance area of 100 μm^2, incident light irradiated to the edge of the chip, which is chemically etched along the facet of the InP substrate, and facet angle of 55 degrees, making the incident angle 35 degrees to the facet. Hence, the incident light comes from the InP substrate to the absorption layer. The standard module has two lenses, collimation and focus, between the optical fiber and the UTC-PD chip to effectively introduce the light, as shown in Fig. 3(a). In this structure, ferromagnetic cobalt material is commonly used to fix the lenses in the package. For most applications of superconducting electronics, however, remnant magnetism must be avoided for use near superconducting ICs. Therefore, an optical fiber lens technique, in which the optical fiber is rounded at the edge, is used in the customized UTC-PD module instead of normal optical lenses, as shown in Fig. 3(b). The working distance between the fiber lens and chip is around 80 μm in the customized module.

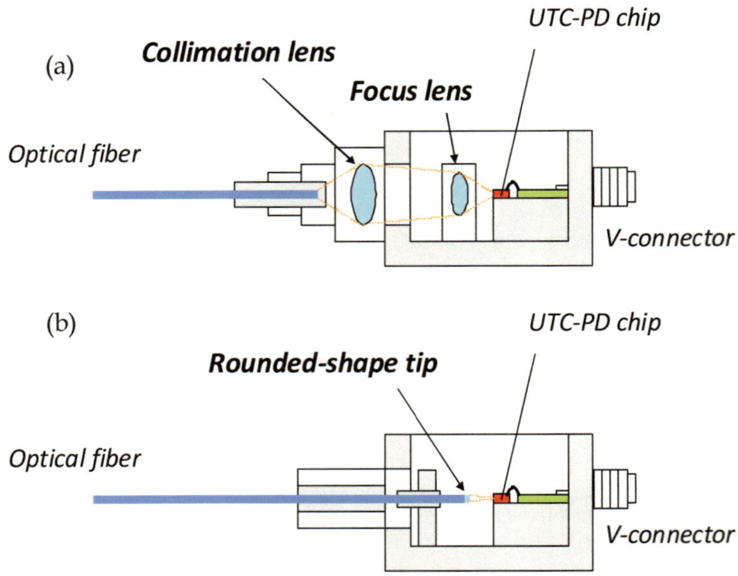

Fig. 3. Structures of (a) standard UTC-PD module, and (b) non-magnetic UTC-PD module using fiber lens specialized for superconducting device.

The beam size is 8 μm in $1/e^2$ reduction of the intensity, and the tolerance of the beam position for optical coupling is shown in Fig.4. Optical output reduction was 50% for a beam position movement of 3 μm from the ideal central position. The optical beam was irradiated from the edge of the UTC-PD chip, which corresponds to an incident angle of 35 degrees to the facet of the InP substrate. One of the important problems with our customized module is that the optical axis is misaligned, when the module is cooled to cryogenic temperature. The optical sensitivity of our customized UTC-PD module at around 4 K was reduced to less than one tenth of that at 300 K in the initial version, which was developed in the beginning

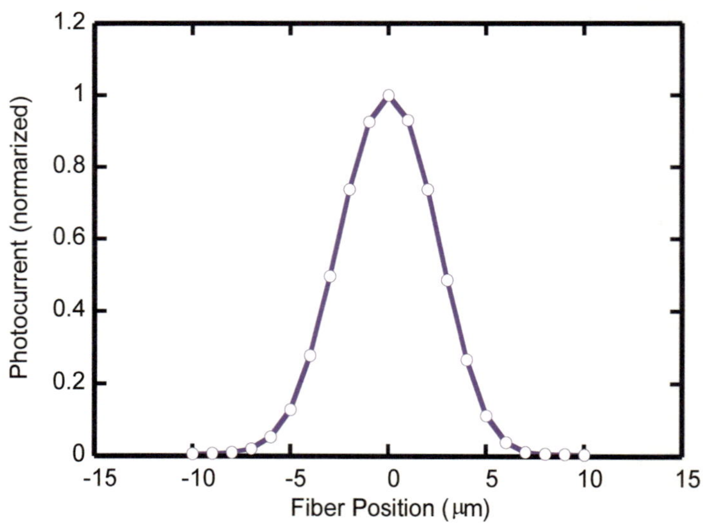

Fig. 4. Intensity dependence of beam on offset distance from ideal central position.

of the customized modules. On the other hand, misalignment did not occur for the standard one. The cause of the misalignment was due to the bending of the optical fiber. The problem was finally resolved by shortening the free space of the fiber without ferrule and by uniformly gluing the fiber to the ferrule with epoxy resin, as shown in Fig. 5(a). Figure 5(b) is a photograph of the entire module, which has a coaxial V-connector for a wide-band electrical output and DC terminals.

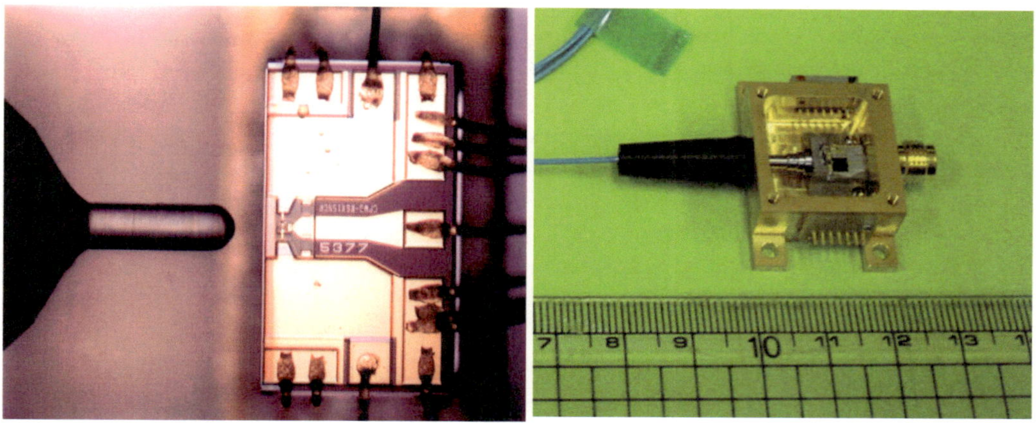

(a) (b)

Fig. 5. Photographs of customized UTC-PD; (a) UTC-PD chip and fiber lens and (b) entire module.

The equivalent circuit of a negative type UTC-PD module is shown in Fig. 6. In the negative type, the UTC-PD module is usually negatively biased to accelerate electron drift in the depletion layer, increasing the operating speed. The output signal is inverted to the input signal. A termination resistor of 50 Ω for impedance matching is integrated at the output of the chip.

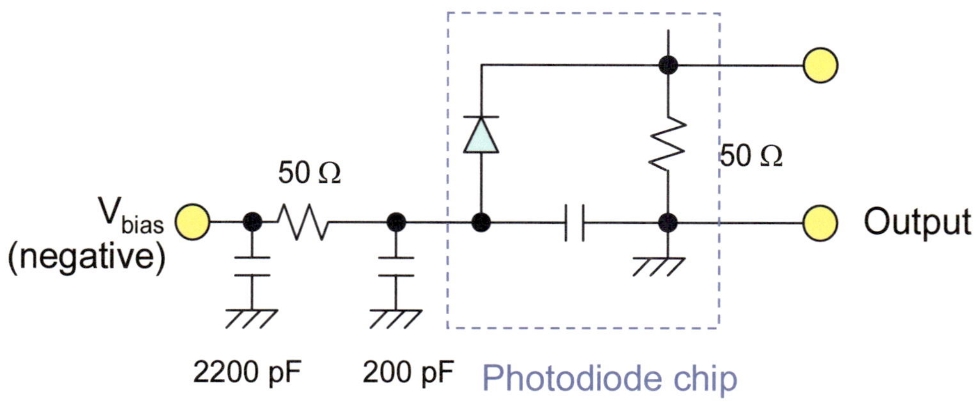

Fig. 6. Equivalent circuit of negative-type UTC-PD module.

2.3 DC characteristics at low temperature

The current versus voltage (I-V) characteristics of our customized UTC-PD module was measured at operating temperatures from 4 to 300 K, as shown in Fig. 7. No electrical and

mechanical damage was observed from the I-V characteristics in our experiments when the UTC-PD module was cooled using a cryocooler at a cooling rate of around 1 degree/minute. Since the gap energy of the InGaAs increased and thermal energy decreased, the forward voltage, at which the current rapidly increased, somewhat increased. The forward voltage increased around 0.16 V by cooling from 300 K to 4 K. The forward current increased sharply at this forward voltage as the operating temperature decreased.
Dependence of optical sensitivity on temperature was measured for both modules, as shown in Fig. 8. The optical wave length was 1550 nm and the input optical power was 0.7 W. Both the UTC-PD modules were biased at -2 V, and the output voltage was measured with a digital voltmeter. The output voltage decreased as the temperature decreased. The output voltage of the standard UTC-PD module was larger than that of customized UTC-PD module over the entire temperature range. The temperature dependences, however, showed relatively similar changes between the two modules. The difference in the results for the two modules was probably due to the difference in the coupling efficiency between the lens and the chip. The output voltage of the customized module is still large enough. We can, therefore, conclude that the customized module using a fiber lens is useful for most applications that require a non-magnetic environment, such as those for superconducting devices.

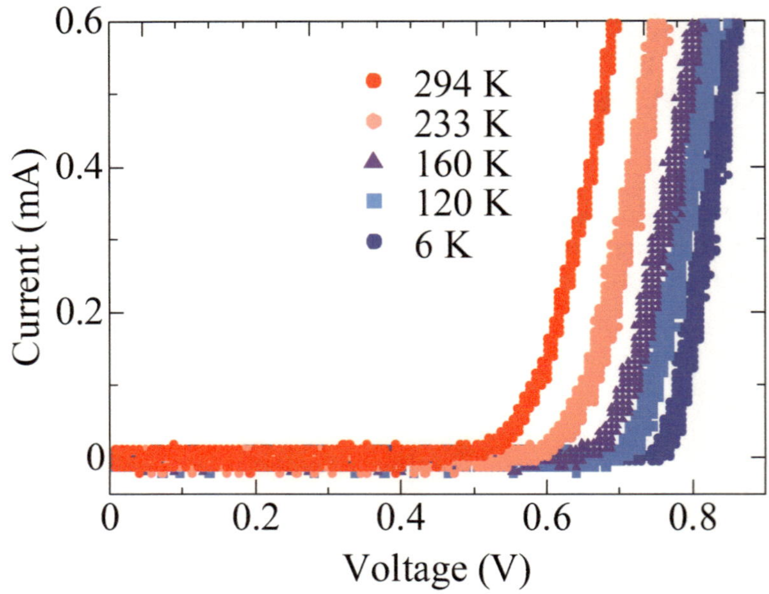

Fig. 7. Current versus voltage (I-V) curves at temperatures between 6 and 294 K.

3. High-frequency and high-speed operation

The high-frequency response of a UTC-PD module at low temperature is important. We evaluated this response using a high-speed optical measurement system. We needed several electronic and optical instruments to produce an optical signal modulated with various high-speed bit pattern signals. The measurement system and the high-speed response of our customized UTC-PD module are discussed in this section. The cryocooling system for cooling the customized UTC-PD module and superconducting devices is discussed in the next section.

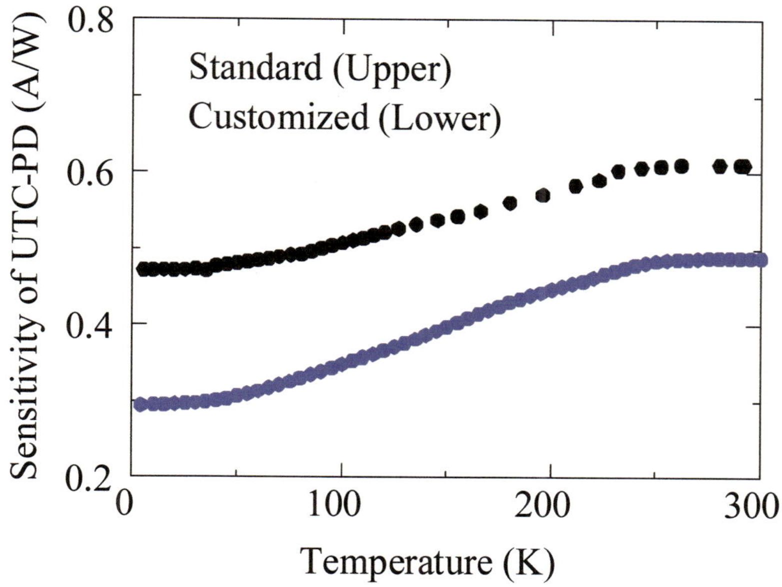

Fig. 8. Temperature dependence of sensitivity of standard and customized UTC-PD modules.

3.1 Optical input measurement system

Figure 9 shows a block diagram of the optical measurement system, which can output 47-Gbps high-speed optical signals. The main clock signal is generated with a signal generator (Anritsu MG3695B: 2 - 50 GHz), and the pulse pattern is generated with a 4-channel pulse pattern generator (Anritsu MP1758A: 10 MHz - 12.5GHz) and serialized with a multiplexer (MUX), which enables us to generate a non-return-to-zero (NRZ) pulse pattern of up to 47 GHz. The MUX and pulse pattern generator (PPG) were synchronized and the timing of the digital data from the PPG to the clock signal in the MUX was adjusted with delay lines. An electrical/optical (E/O) converter with a MUX (Anritsu MP1806A), which includes a laser diode, an optical modulator with an automatic bias controller (ABC), generated arbitrary optical digital pattern signals with a modulation depth of almost 100%. The optical signal was amplified with an erbium-doped fiber amplifier (EDFA) and the output power was adjusted with a power controller and attenuator (Agilent 8163B). The controlled output signal was applied to the customized UTC-PD module, which converted the optical signal to an electrical signal at around 4 K. The electrical output was connected to a cryoprobe, which was also cooled at around 4 K, through a 1.19-mmϕ copper coaxial cable of 230 mm in length.

3.2 High-frequency performance

The high-speed performance of the customized UTC-PD module cooled around 4 K was measured and confirmed for up to a 40-Gbps NRZ signal. The customized UTC-PD module was set on the 2nd stage in the cryocooling system, which is discussed in Section 4.1. Figures 10(a) and (b) show typical eye diagrams of the input optical signal and the output electrical signal observed with a sampling oscilloscope (Agilent 86100C). The modulation depth was automatically adjusted to almost 100%. The input signal was a pseudo random bit stream (PRBS) signal with a data length of 2^{31}-1. A block diagram of the measurement system is shown in Fig. 9. The output line includes a loss of 2.8 dB at 40 GHz in a 510-mm-long coaxial cable in the cooling system.

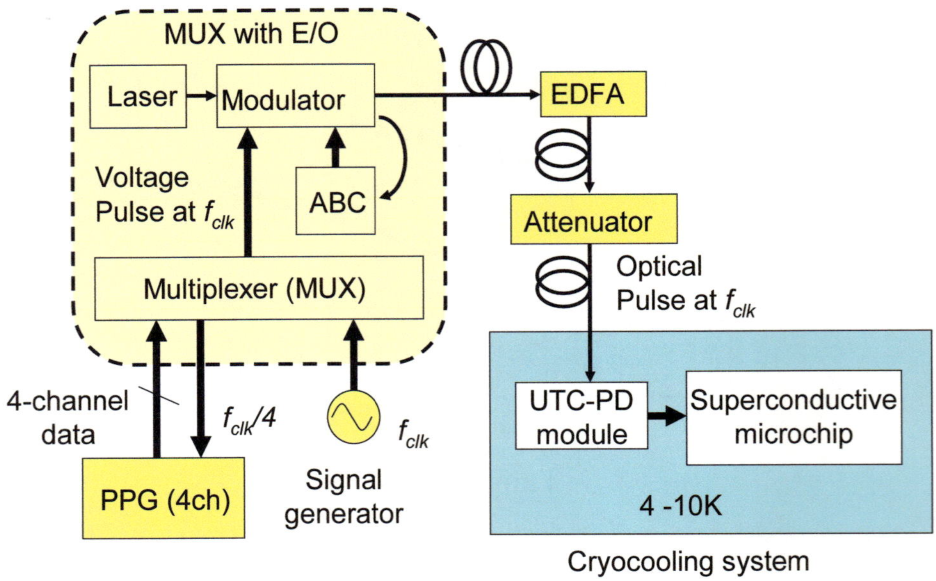

Fig. 9. Setup of optical measurement system that can produce optical digital signal at data rate of up to 47 Gbps

The amplitude of the output signal was 90 mV in a peak-to-peak voltage for an input optical signal power of 10 mW at a wavelength of 1550 nm. We evaluated the linearity for the amplitude of the output voltage to the optical input signal power. Since there was no difference observed for the data length between 2^{31}-1 and 2^{7}-1 of the PRBS signals, a data length of 2^{7}-1 was used to save time. Figure 11 shows the optical input power versus the output voltage for 10, 20, and 40-Gbps PRBS data input, resulting in good linearity over the input optical power of 10 mW. In the above evaluation, the customized UTC-PD module

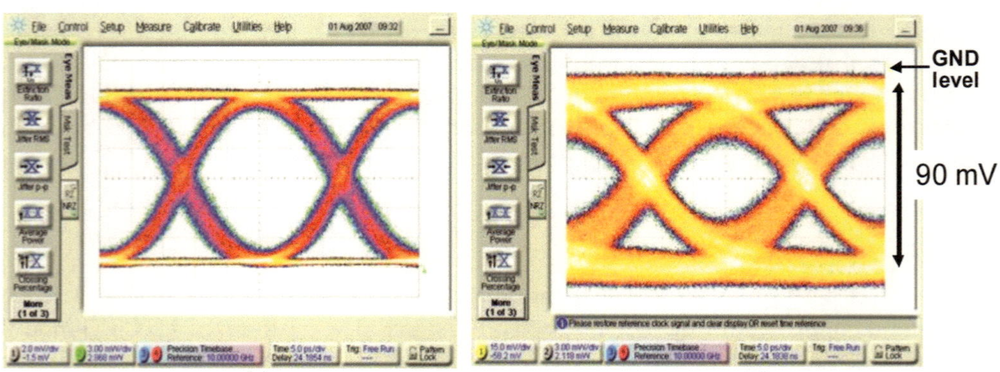

Fig. 10. Eye patterns of (a) optical output signal of optical measurement system for 31-stage pseudo random bit stream (PRBS) digital signal and (b) electrical output signal of customized UTC-PD module cooled at 5 K.

was DC biased at -2 V, which is definitely required for high-speed performance at room temperature. It should be noted that the customized UTC-PD module operated at high speed even at zero DC bias voltage, which may be due to the increment of the built-in electric field in the absorption and depletion layers.

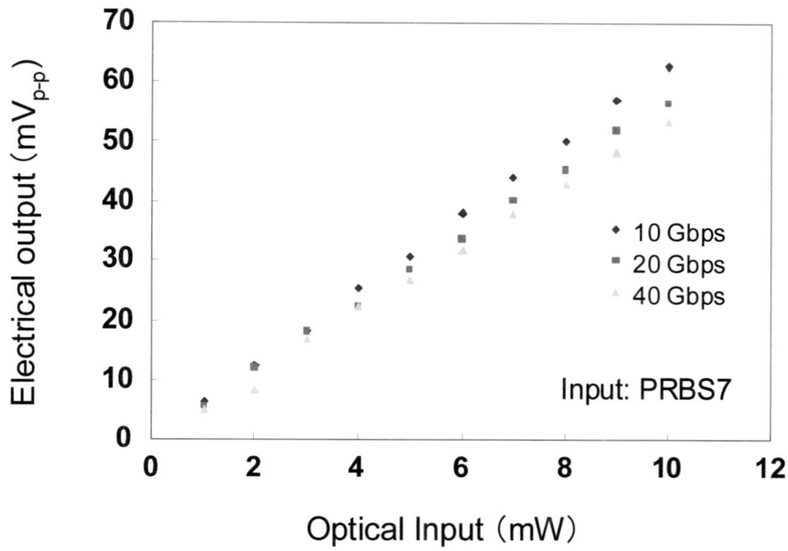

Fig. 11. Electrical output voltages as function of optical input power of customized UTC-PD module cooled at 5 K for 10, 20, and 40-Gbps PRBS data input.

4. Applications of UTC-PD module operating at cryogenic temperature to superconducting electronics

The optical link of the input signal between semiconducting devices operating at room temperature and superconducting devices at cryogenic temperature has several advantages. The thermal conductivity of optical fibers is extreamly small compared with metal-based electric links, such as coaxial and flexible film cables. The themal conductivity of quatz, which is a base material in a single-mode opitical fiber, is 1.4 W/m/K; therefore, the thermal conductivity of a single-mode optical fiber having a crad diameter of 125 μm and a length of 1 m is as small as 5.2 x 10^{-6} W. The signal loss is also extremely small, e.g., < 0.2 dB/km for a wavelength of 1550 nm and < 0.4 dB/km for 1310 nm. The signal loss of the optical fiber is negligible for our applications such as analogue to digital converters (ADC) using SFQ circuits, which require short distance transmission. It is small enough even if we use a longer, e.g., 1 km, optical fiber. The signal loss seems to be rather large at optical connectors and other parts.

4.1 Cryocooling system for superconducting electronics system

Single flux quantum circuits have been investigated for superconducting digital and analog/digital applications. In most of these investigations, superconducting IC chips were cooled by directly immersing them in liquid helium. It is convenient to cool IC chips to cryogenic temperature for laboratory use due to the immediate cooling time. Many superconducting systems, however, require a cryocooler for practical applications. Even for laboratory use, a cooling system using a cryocooler is desirable for system-level tests and high-speed or high-frequency tests because the signal loss and distortion between room temperature and cryogenic temperature may especially cause problems and restrict experiments. A cryocooling system using a two-stage 4-K Gifford MacMahon (GM) cryocooler was developed at the international Superconductivity Technology Center (ISTEC) for demonstrating superconducting digital and analog ICs based on the Nb/AlOx/Nb Josephson junctions. A photograph and illustration of the system is shown in Fig. 12. The 2nd cold stage, 4-K stage, including a superconducting chip, a cryoprobe, and our customized UTC-PD module is surrounded with a thermal shield with a temperature of 50 K using the 1st cold stage of the cooler. Cryogenic amplifiers are attached to the thermal shied. The cryocooler (RDK-408D) and the compressor (CSA-71A) are from Sumitomo heavy industries Ltd. The cooling capacity is 1 W at 4.2 K for the 2nd cold stage and 60 W at 50 K for the 1st cold stage. The total input AC power of the cooler is 6.5 kW. The system has twenty-four high-frequency I/O terminals with V-connectors and two optical input ports using the customized UTC-PD module. The 1st cold stage of the cooler, the 50-K stage, can effectively be used for cooling the cryogenic amplifiers, thermal shied, and thermal anchor.

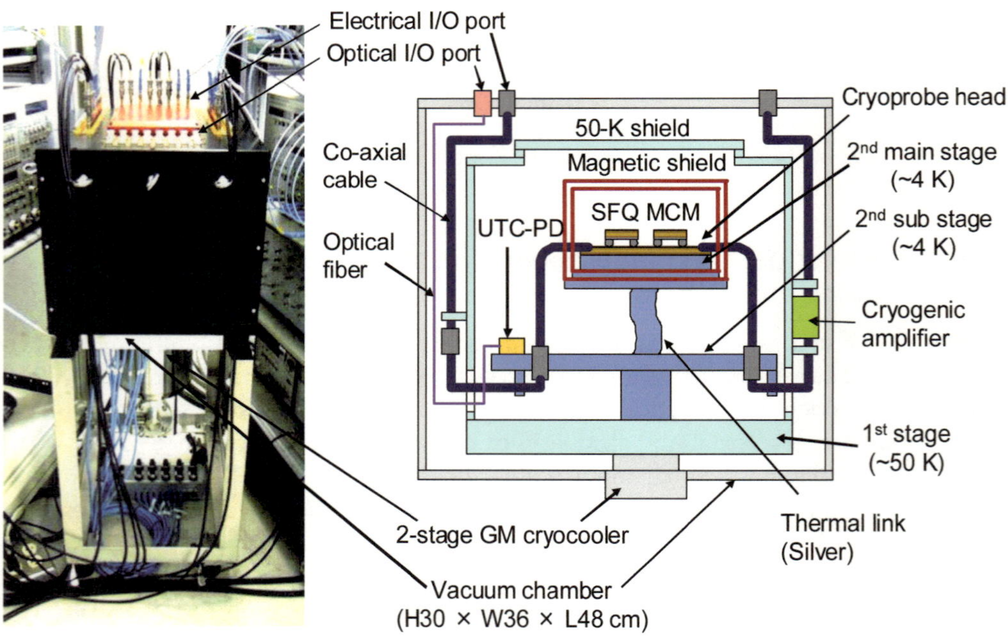

Fig. 12. Cryocooling system for supeconducting devices. Left is photograph of system and right is cross-sectional illustration.

Figure 13 shows a photograph of the 2nd stage arrangement with a cryoprobe and two customized UTC-PD modules placed on the sub 2nd cold stage located in a short distance around 100 mm from the SFQ multi-chip module (MCM) on the main 2nd stage, as shown in Figs. 12 and 13; therefore, the temperature was a little high, between 5-6 K. We developed MCM technology with flip-chip bonding and a cryoprobe for superconducting systems, which enable us to conduct high-speed measurements of superconducting circuits. The SFQ chips mounted on the MCM substrate including the cryoprobe was attached to the main 2nd stage, which was magnetically shielded with a two-folded permalloy enclosure. However, the customized UTC-PD module was placed outside the magnetic shield. The main 4-K stage was cooled with thermal conduction through a thermal link made of silver and the magnetic shield from the 2nd cold head of the cryocooler. The vibration of the temperature at the main 4-K stage was then stabilized to as low as 10 mK, which ensured the stable operation of SFQ circuits.

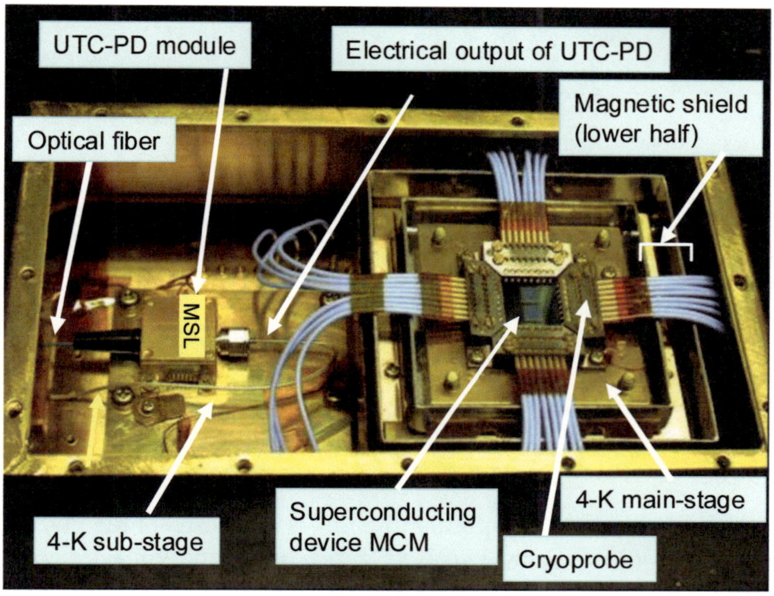

Fig. 13. Arrangement of 4-K cold stages in cooling system; superconducting IC chip with multi-chip module (MCM) and cryoprobe surrounded by double magnetic shield (right side; the lids are removed to show the contents) on main cold stage, and customized UTC-PD module operating at 4 K for introducing high-frequency optical signal into cryostat through optical fiber was placed on sub-cold stage.

4.2 Superconducting single flux quantum (SFQ) digital circuits

We designed an SFQ circuit chip, which includes an input interface between the customized UTC-PD module and SFQ circuit. Figures 14 (a) and (b) show an equivalent circuit and a microphotograph of the PD/SFQ converter. The chip was fabricated with the ISTEC standard process 3 (STP3) using Nb/AlOx/Nb Josephson junctions with a current density of 10 kA/cm^2. The input signal was magnetically coupled to the SFQ circuit, making it possible to accept both polarities of the input signal by changing the direction of the coupling in the transformer. The negative polarity signal from the customized UTC-PD module was then able to be received directly without any offset current and inverter by the PD/SFQ converter shown in Fig. 14. Josephson junctions, J1 and J2, and inductances, L1 and L2, construct a superconducting quantum interference device (SQUID). When the input signal, data "1", is applied, the SQUID stores the single flux quantum in the superconducting loop, producing clockwise circulating current. By applying the clock pulse, the SFQ pulse is output by switching J2 and J3. When data "0" is applied, no SFQ pulse is output. In this case, the SFQ pulse produced by the clock pulse is escaped from J5. The converter can then produce SFQ pulses from the normal NRZ signal from the customized UTC-PD module, where the SFQ pulse

$$\int V(t)dt = \Phi_0 = h / 2e \sim 2.07 \times 10^{-15}[Wb] \tag{1}$$

acts as the quantized information medium in SFQ circuits.

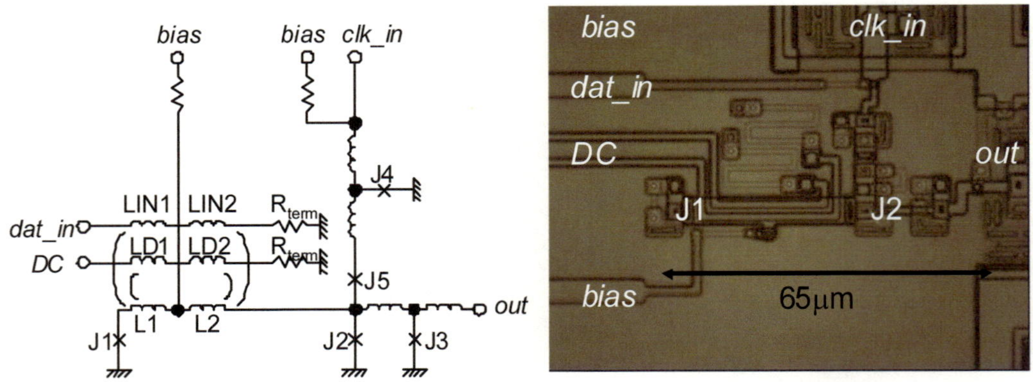

Fig. 14. UTC-PD to single flux quantum (SFQ) converter; (a) equivalent circuit and (b) microphotograph.

The SFQ circuit chip for testing the optical input link is composed of the PD/SFQ convertor, a 1-2 demultiplexer (DEMUX), and two NRZ superconducting voltage drivers (SVDs), as shown in Fig. 15. Signal flux quantum pulses have a narrow width (~2 ps) and a low signal level (~1 mV), and the circuit can be operated faster than that in semiconductor devices. The SFQ output data of the PD/SFQ is alternately output to the two outputs with the 1:2 DEMUX in parallel to reduce the output data rate to half the input data rate. Then, the SFQ pulse signal is converted to an NRZ signal by the SVDs.

Figure 16 shows an NRZ SQUID voltage driver (NRZ SVD). This NRZ SVD consists of a splitter (SPL), which divides a single SFQ signal into 16 splitter outputs, RS flip-flops (RSFFs), each of which stores an SFQ signal, and 16 serially connected SQUIDs, which amplify the SFQ signal stored in the RSFF to 2-mV NRZ data streams up to 23.5 GHz. There are a total of 318 junctions, and the bias current is 43 mA. The 5 x 5 mm SFQ chip was flip-chip bonded on a 16 mm x 16 mm MCM carrier with InSn bumps, as shown in Fig. 17(a).

Both the chip and carrier are made of the same Si substrate, which prevents stress due to the difference in thermal expansion coefficients when they are cooled. Figure 17 (b) shows InSn bumps for the signal and ground, in which the signal bump was connected to a 50 Ω micro-strip line (MSL) in the chip. The height of the bump was as small as 8 µm, as shown in Fig. 17(c), which enabled us to transmit high-frequency signals over 100 GHz. The MCM carrier was mounted on the 4-K main base plate of the cryoprobe, as shown in Fig. 13. Copper-molybdenum alloy was chosen as the base plate material to decrease the difference in the thermal expansion coefficient. The cryoprobe was adjusted to ensure contact of the chip pads. The optical link was tested using the test circuit at a high-speed data rate.

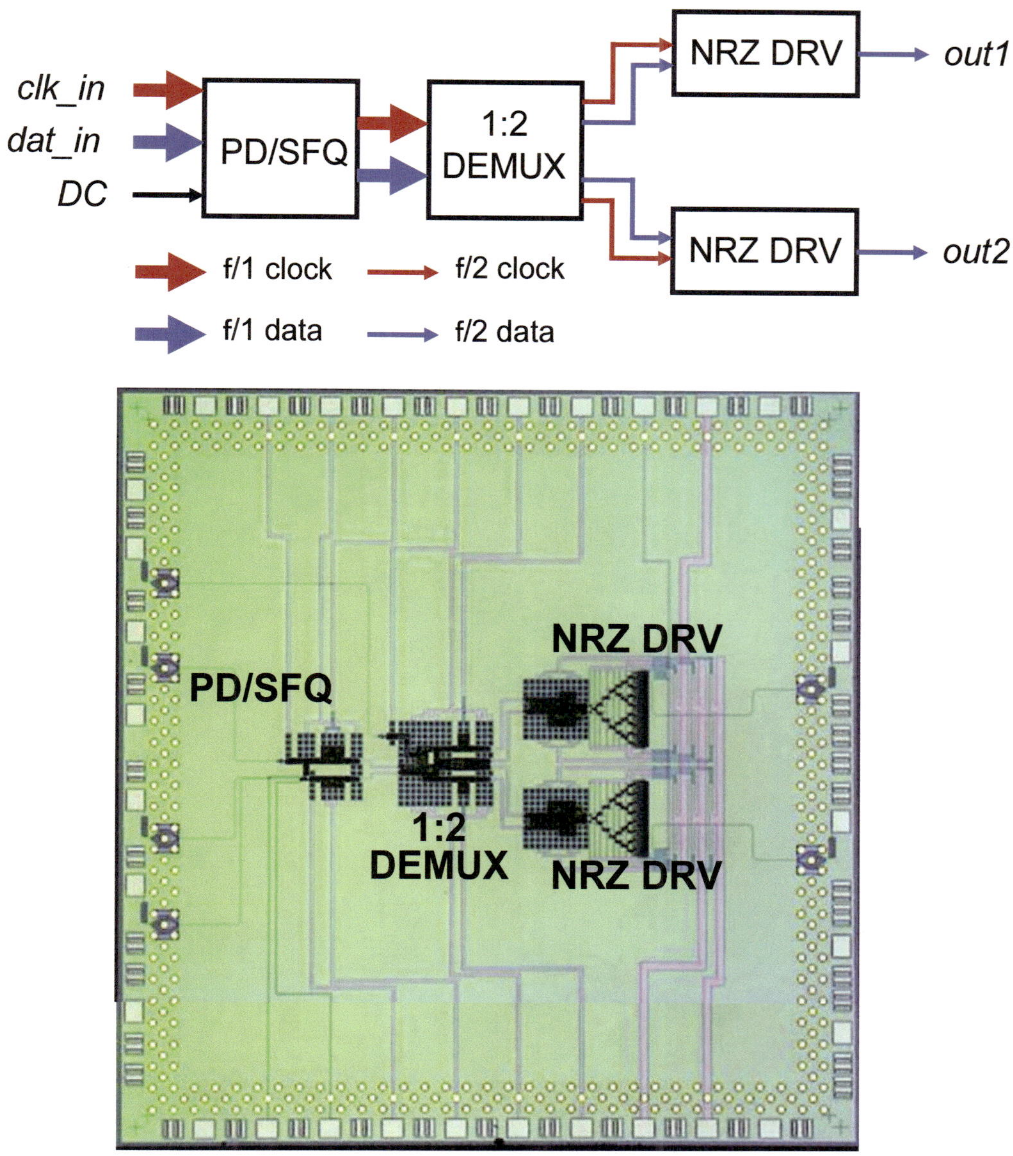

Fig. 15. Block diagram and microphotograph of SFQ test chip for optical input.

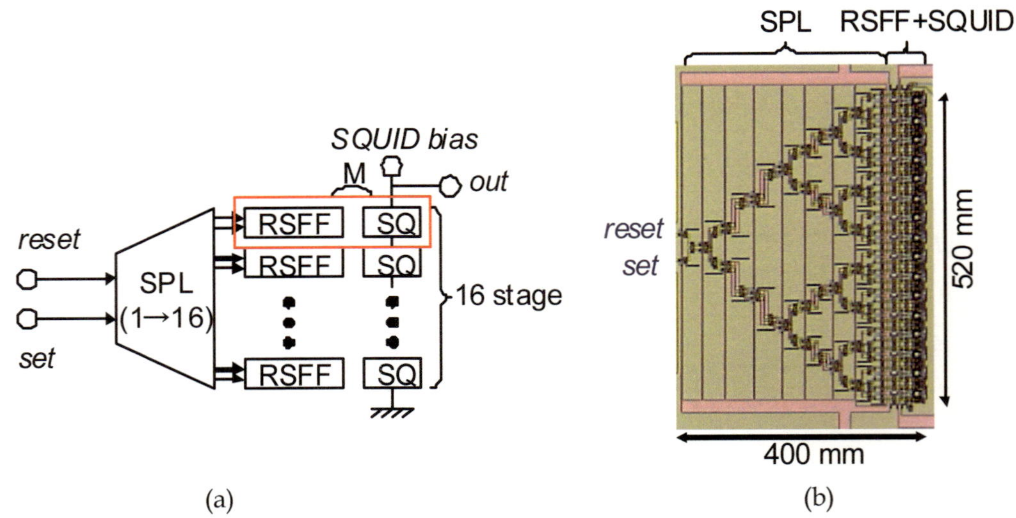

Fig. 16. Non-return-to-zero (NRZ) superconducting quantum interference device (SQUID) voltage driver; (a) block diagram and (b) microphotograph.

Fig. 17. Photographs of, (a) flip-chip bonded MCM carrier and superconducting micro-chip, (b) flip-chip bumps on chip, and (c) cross sectional view of flip-chip bonded bump.

The output signals of the SVDs are further amplified by GaAs cryogenic amplifiers mounted on the 1st stage of the cryocooling system, as shown in Fig. 12. The cryogenic amplifier, SHF105C, was developed by SHF communication Technology AG originally for SFQ circuits in collaboration with ISTEC. The output voltage of the SVDs was amplified to around 50 mV with the cryogenic amplifiers, which have a gain of around 30 dB at 23 K and a typical bandwidth of 30 GHz. The optical digital data of up to 47 Gbps was applied to the customized UTC-PD module, and the converted electrical signal was applied to the test chip through a Cu coaxial flexible cable of 1.19 mm in diameter and length of 230 mm. Figure 18 shows the experimental results for the input data rate of 47-Gbps data; (a) the outputs of the two SVDs for patterned digital data and (b) eye pattern for PRBS of 2^{31} -1. We can clearly see an open eye pattern. The bit error rate (BER) was measured with an error detector (Advantest D3286). Figure 19 shows the dependences of the BER for PRBS of 2^7-1, (a) on the bias current of the PD/SFQ converter and (b) on the input optical power. Sufficiently small BER of less than 10^{-12} at 40 Gbit/s in the output was obtained with the test circuit for the optical input signal through the customized UTC-PD module as an O/E converter.

(a)

(b)

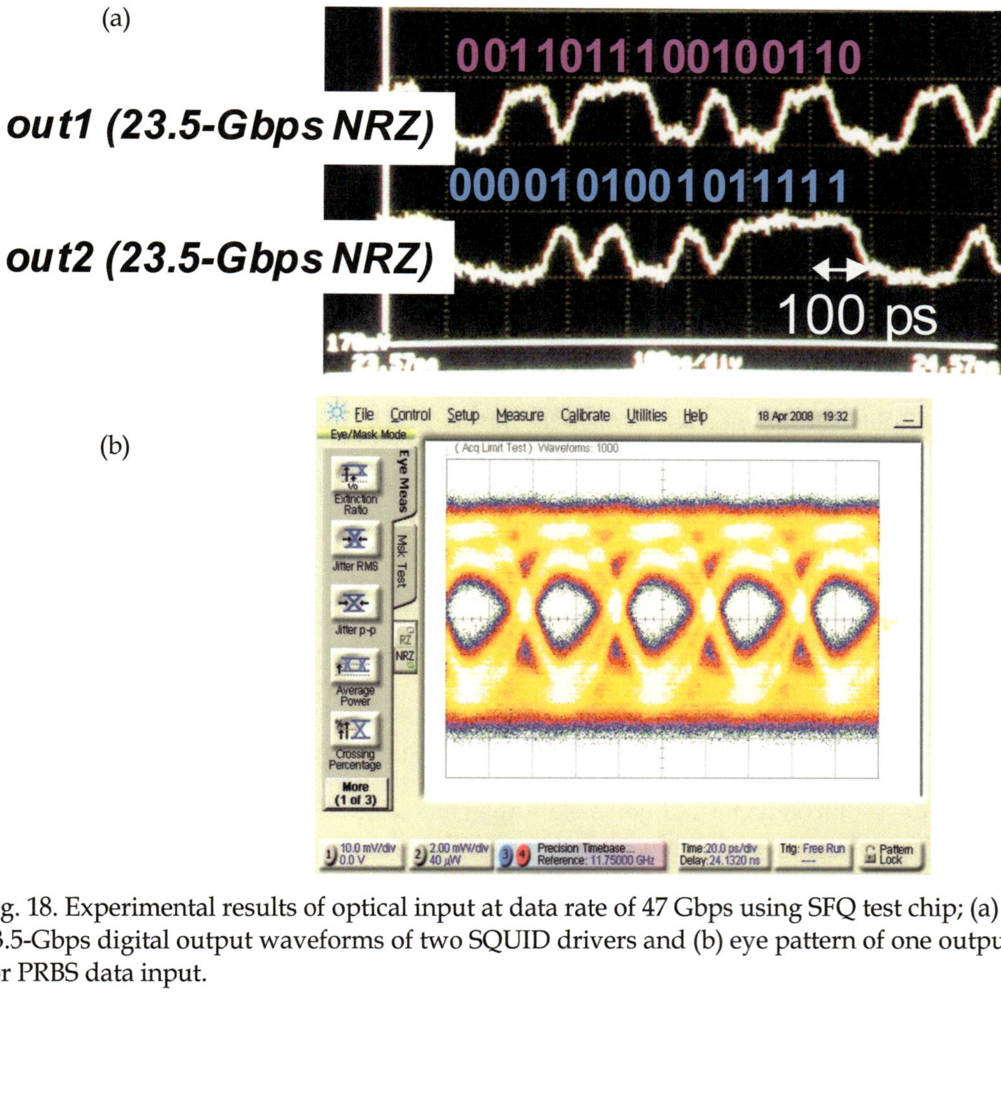

Fig. 18. Experimental results of optical input at data rate of 47 Gbps using SFQ test chip; (a) 23.5-Gbps digital output waveforms of two SQUID drivers and (b) eye pattern of one output for PRBS data input.

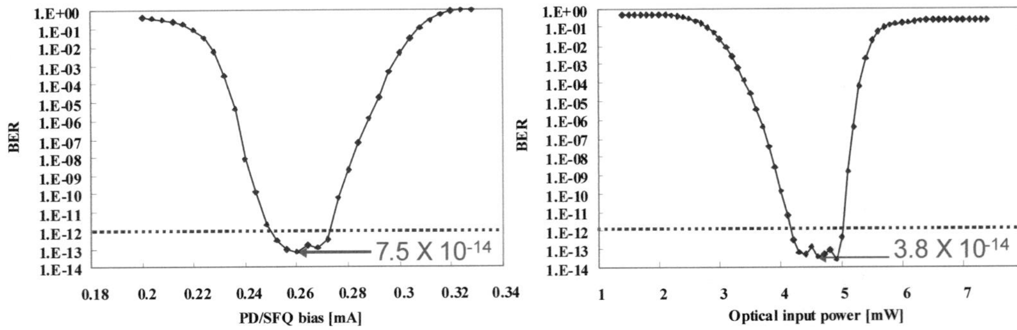

Fig. 19. Bit error rate (BER) as function of (a) bias current of PD-SFQ converter and (b) optical input power for UTC-PD module.

4.3 Josephson voltage standards

Josephson voltage standards (JVS) have been used as a DC voltage standard since 1990 because of their quantum mechanical accuracy. These standards consist of an under-damped superconductor-insulator-superconductor (SIS) junction array, which is DC biased and radiated with microwave. The voltage is determined with the microwave frequency and physical constant, which ensure its quantum mechanical accuracy. Although, JVS are suitable for DC voltage standards, they cannot be applicable to AC voltage standards. Because JVS use the hysteresis of SIS junctions, a proper procedure for applying the DC bias and microwaves and time to fix to the desired voltage is required.

The pulse-driven Josephson arbitrary waveform synthesizer (PD-JVS) is a device for producing AC voltage standard, which is one of AC JVS. This device is also called as Josephson arbitrary waveform synthesizer (JAWS). The principle is based on a 1-bit sigma delta digital-to-analog converter. The basic idea is that the amplitude of a signal waveform is represented as a pulse density. The pulse pattern is properly calculated for desired waveform and generated with a pulse pattern generator, which is applied to a JAWS chip. The JAWS chip consists of over-damped Josephson junction arrays (JJAs), which are capable of producing quantized voltage pulses. High-speed pulses, of which a pattern is calculated for producing the desired waveform, is generated in room-temperature electronics, and the optical signal is transferred to an electrical signal with the customized UTC-PD module at cryogenic temperature, which enables us to apply the high-speed signal to the SFQ chip with extremely low noise as well as low signal losses and distortions. The operation of the synthesizer was demonstrated by the National Institute of Advanced Industrial Science and Technology (AIST) and ISTEC using the cooling system with the customized UTC-PD module. We have to use junctions without hysteresis for the JAWS. The JAWS chips were fabricated in two superconducting microchip processes; one with Nb/AlOx/Al/AlOx/Nb Josephson junctions, which are superconductor-insulator-normal metal-superconductor (SINS) junctions, developed by ISTEC, and the other with NbN/TiNx/NbN junctions, which are superconductor-normal metal-superconductor (SNS) junctions, developed by AIST. Figure 20 shows an IC chip fabricated with the Nb/AlOx/Al/AlOx/Nb junctions.

The chip consists of an array of 100 serially connected junctions, which can increase the output voltage. The array was arranged in the center of a 50 Ω coplanar waveguide input line in the chip. The 5 x 5 mm chip was flip-chip bonded on the MCM carrier, in the same manner as SFQ chips. PD-JVS chips were also fabricated with the NbN/TiNx/NbN junctions, in which 480 junctions were serially connected to increase the output voltage. The chip using the NbN junctions can operate at higher temperatures than that using the Nb junctions, which enable us to use a 10-K cryocooler.

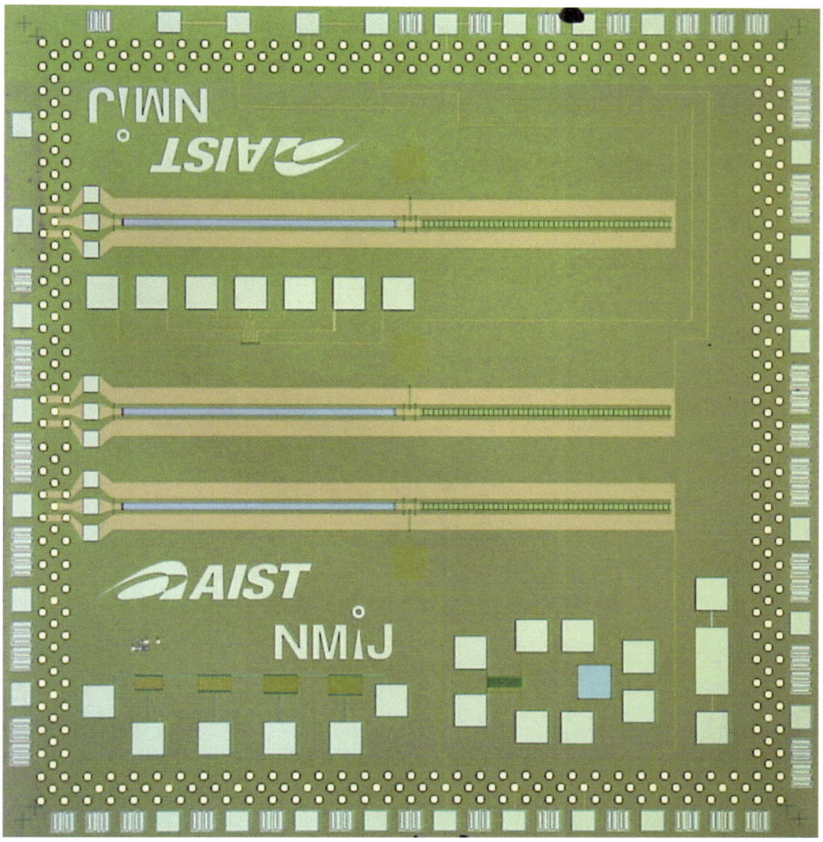

Fig. 20. Microphotograph of pulse-driven Josephson arbitrary waveform synthesizer (PD-JVS) chip fabricated with Nb/AlOx/Al/AlOx/Nb junction technology.

The JAWS can produce any waveform by applying a properly calculated pulse pattern. Figure 21 shows examples of synthesized waveforms; (a) triangular, (b) rectangular, and (c) saw-tooth. The left charts show the frequency spectrum and the right ones show generated waveforms. A high-precision sine wave was generated with a JAWS chips fabricated with both Nb/AlOx/Al/AlOx/Nb and NbN/TiNx/NbN Josephson junctions. Figure 22 shows the frequency spectrum of a 152.6-kHz sine wave with the PD-JVS using the Nb junctions. The sampling frequency was 10 GHz and the output voltage of 1.24 mV with spurious free dynamic range (SFDR) of -75 dBc was obtained from the chip. Figure 23 shows the frequency spectrum of a 59.6-Hz sine wave generated with the 480 NbN-SNS junctions, of which the frequency is important because it is around the commercial (mains) frequencies of 50 and 60 Hz. The sampling frequency was 8 GHz and a 134,217,728-bit-long (=2^{27} bit) binary pulse pattern was used for generating the 59.6-Hz sine wave. A sine wave was clearly observed with both PD-JVS chips. However, the SFDR was limited to -67 dBc due to odd harmonics of 50 Hz. The SFDR omitting these harmonics was as low as -80 dBc. The reduction of signal-to-noise ratio (SNR) due to the odd harmonics of 50 Hz seemed to be affected by noise from the ground loops. The ground noise could be avoided by isolating the grounds in the I/Os.

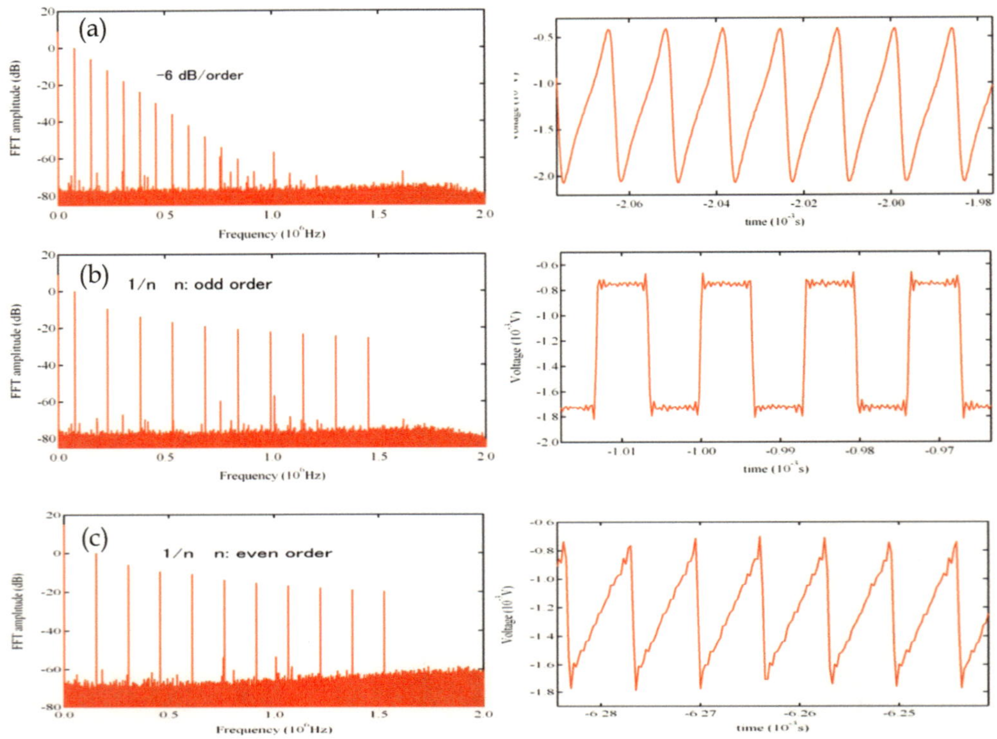

Fig. 21. Examples of frequency spectrum and waveforms synthesized using PD-JVS; (a) triangular, (b) rectangular, and (c) saw-tooth.

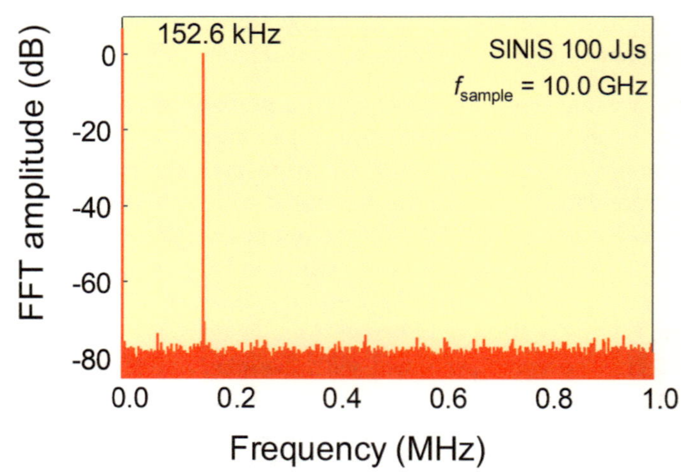

Fig. 22. Frequency spectrum of synthesized sine wave of 152.6 KHz with the PD-JVS using Nb/AlOx/Al/AlOx/Nb junctions.

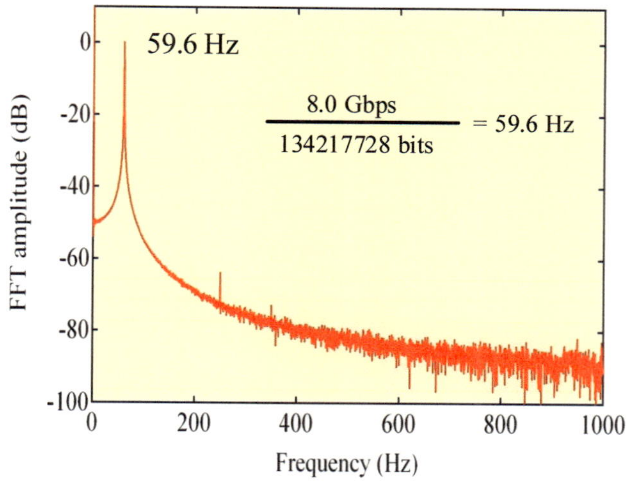

Fig. 23. Frequency spectrum of synthesized sine wave of 59.6 KHz with the PD-JVS using NbN/TiNx/NbN Josephson junctions.

5. Conclusion

We studied the performance of a standard UTC-PD module at low temperature and developed a customized module for superconducting devices. In the customized module, an optical fiber lens was used to avoid using ferromagnetic material for fixing the optical lens. The performance of the customized UTC-PD modules at cryogenic temperature as low as 4 K was confirmed experimentally for the first time. High-speed operation of up to 40 Gbps was confirmed using a cryocooling system we developed for superconducting circuits, especially SFQ circuits. This cryocooling system uses a 4-K GM cryocooler and worked well for evaluating our customized UTC-PD module and for demonstrating superconducting circuits with high-speed data rate using an optical input link with our customized UTC-PD module and optical fibers. First, a basic SFQ digital circuit, which has a PD-SFQ converter with the output signal from the UTC-PD module for the input link, a 1-2 DEMUX, two sets of driver circuits for the output links, operated at a data rate of up to 47 GHz. Second, the performance of the PD-JVS with an optical input link was successfully demonstrated using the same cryocooling system at AIST in collaboration with ISTEC.

6. Acknowledgments

We would like to thank Tadao Ishibashi of NTT Electronics Ltd., and Takeshi Konno, Koichiro Uekusa, and Masayuki Kawabata of Advantest Lab. Ltd. for their contributions to the development of the UTC-PD for superconducting devices and their useful comments, and express our gratitude to Nobuhisa Kaneko, Chiharu Urano, Michitaka Maruyama for giving the result of a pulse-driven AC voltage standard. We also would like to thank Yoshiji Hashimoto for his many of contributions to this work, Michiyo Isaka and the members of ISTEC-SRL for fabricating the IC chips, and Mayumi Katsume for assembling the MCMs. We also express our gratitude to Seizo Akasaka of Kawashima Manufacturing Co, Ltd. for developing the MCM package and connector. The National Institute of Advanced Industrial Science and Technology partially contributed to the circuit fabrication. This work was partially supported by the New Energy and Industrial Technology Development

Organization (NEDO) as Development of Next-Generation High-Efficiency Network Device Project. The National Institute of Advanced Industrial Science and Technology (AIST) partially contributed to the circuit fabrication.

7. References

E.Zielinski, H.Schweizer, K.Streubel, H.Eisele, G.Weimann, *J. Appl. Phys.*, 59, no.6, pp.2196-2204(1986)

http://www.ioffe.ru/SVA/NSM//Semicond/GaInAs/optic.html

Goldberg Yu.A. and N.M. Schmidt *Handbook Series on Semiconductor Parameters*, vol.2, M. Levinshtein, S. Rumyantsev and M. Shur, ed., World Scientific, London, 1999, pp. 62-88

K. Likharev and V. K. Semenov, "RSFQ logic/memory family : A new Josephson-junction technology for sub-terahertz-clock frequency digital systems, " *IEEE Trans.Appl. Superconductivity*, vol. 1, no. 1, pp. 3–28, Mar. 1991

Y. Hshimoto, S. Yorozu, T. Satoh, and T. Miyazaki, "Demonstration of chip-to-chip transmission of single-flux-quantum pulses at throughputs beyond 100 Gbps, " Appl. Phys. Lett., 2005, 022502

Y. Hashimoto, S. Yorozu, T. Miyazaki, Y. Kameda, H. Suzuki, and N. Yoshikawa, "Implementation and experimental evaluation of a cryocooled system prototype for high-throughput SFQ digital applications," *IEEE Trans.Appl. Superconductivity*, vol. 17, no. 2, pp. 546–551, Jun. 2007

Y. Hashimoto, H. Suzuki, S. Nagasawa, M. Maruyama, K. Fujiwara, and M. Hidaka, "Measurement of superconductive voltage drivers up to 25 Gb/s/ch," *IEEE Trans.Appl. Superconductivity*, vol. 19, no. 3, pp. 1022–1025, Jun. 2009

M. Maruyama, K. Uekusa, T. Konno, N. Sato, M. Kawabata, T. Hato, H. Suzuki, and K. Tanabe, "HTS sampler with optical signal input," *IEEE Trans.Appl. Superconductivity*, vol. 17, no. 2, pp. 573–576, Jun. 2007

H. Ito, S. Kodama, Y. Muramoto, T. Furuta, T. Nagatsuma, and T. Ishibashi, "High-speed and High-output InP-InGaAs unitraveling-carrier photodiodes, " *IEEE J. Selected Topics in Quantum Electronics,* vol. 10, no. 4, pp. 709–727, July/Aug. 2004

H. Ito, T. Furuta, T. Nagatsuma, F. Nakajima, K. Yoshino, and T. Ishibashi, "Photonic generation of continuous THz wave using Uni-Traveling-carrier photodiode, " *IEEE J. Lightwave Technology,* vol. 23, no. 12, pp. 4016–4021, Dec. 2005

H. Suzuki, T. Hato, M. Maruyama, H. Wakana, K. Nakayama, Y. Ishimaru, O. Horibe, S. Adachi, A. Kamitani, K. Suzuki, Y. Oshikubo, Y. Tarutani, K. Tanabe, T. Konno, K. Uekusa, N. Sato, and H. Miyamoto, "Progress in HTS sampler development," Physica C 426-431, pp. 1643-1649, 2005

H. Suzuki, M. Maruyama, Y. Hashimoto, K. Fujiwara, and M. Hidaka, "Possible application of flash-type SFQ A/D converter to optical communication systems and their measuring instruments,", *IEEE Trans. Appl. Supercon..* vol. 19, pp. 611-616, Jun. 2009

H. Suzuki, M. Oikawa, K. Nishii, K. Ishihara, K. Fujiwara, M. Maruyama, and M. Hidaka, "Design and demonstration of a 5-bit flash-type SFQ A/D converter integrated with error correction and interleaving circuits," to be published in *IEEE Trans. Appl. Supercon..*, Jun. 2011

T. Ishibashi, and N. Shimizu "Uni-traveling-carrier photodiode as an optoelectronic driver, " *OSA TOPS,* vol. 28, Ultrafast Electronics and Optoelectronics, John Bowers and Wayne Knox (eds.)

http://www.shf.de/fileadmin/download/Amp/Datasheet_SHF105C_V001.pdf

3

The Optimum Link Design Using a Linear PIN-PD for WiMAX RoF Communication

Koyu Chinen
Okinawa National College of Technologies
Japan

1. Introduction

Worldwide Interoperability for Microwave Access (WiMAX) is a new standard for high-speed wireless communication that covers wider area than that of Wireless Local Area Network (WLAN). In the WiMAX, the original data are first mapped on the symbols of Quadrature Phase Shift Keying (QPSK), 16 Quadrature Amplitude Modulation (QAM), or 64 QAM, depending on the data speed. The complex numbers of the symbols are allocated to subcarriers and the subcarriers are converted to time-domain I and Q data by Inverse Fast Fourier Transform (IFFT) algorithm. The parallel data of I and Q are converted to serial data by using a parallel to serial converter. The digital serial data are converted to analog data by using digital to analog converter. The analog I and Q data are orthogonally modulated by a carrier frequency and multiplexed to generate a time domain waveform. Therefore, in the Orthogonal Frequency Division Multiplexing (OFDM), the Peak-to-Average Power Ratio (PAPR) of the waveform becomes higher when the number of sub-carriers increases. When the waveform is converted by Electrical-Optical converter (E/O) to optical signal and transmitted over a fiber and is converted by Optical-Electrical converter (O/E) to electrical signal, the larger PAPR causes larger distortion in those optical components. Therefore it is strictly important to design the WiMAX communication link by using highly linear optical signal converters. Since the linearity in the actual optical components is insufficient to cover all modulation conditions in the WiMAX communications, the optimum design of the E/O, the O/E, the modulation, and the demodulation is necessity, based on the specific condition of the communication systems. But it is obvious to use the linear PIN photodiode (PIN-PD) for all of the WiMAX Radio-over-Fiber (RoF) links. Because, the structure and the performance are stable and simple, in comparison with that of other active optical components, such as Avalanche Photo diode (APD) and Distribute feedback Laser diode (DFB-LD).

2. An RCE calculation model for RoF of WiMAX

Relative Constellation Error (RCE) is an important standard for evaluation of the transmission quality in the WiMAX. Since the modulation of the WiMAX consists of QPSK, QAM, and OFDM, the RCE is sensitive to the change in the phase and the amplitude of the signals. The phase and the amplitude of the signals are influenced with optical components.

Therefore in the Radio over Fiber (RoF) link, the RCE is determined with many component factors, such as the modulation power, the type of optical transmitters, optical fiber length, optical receiver, and the type of antennas.
An RCE calculation model was theoretically and experimentally derived, for the RoF system of the WiMAX, when the system was configured with linear characteristic components and the Polarization Mode Dispersion (PMD) was suppressed by an optimum modulation condition. In hybrid optical links, the influence of the WiMAX signal on the digital baseband was also investigated [1]. It is also important to characterize the WiMAX signal behavior in the digital optical links.

2.1 RCE degradation due to PMD and PML

The Polarization Mode Dispersion (PMD) and the Polarization Mode coupling Loss (PML) were suspected in the RoF link. A measurement setup shown in Fig.2.1 was used to investigate its influence on the RCE. The baseband signal generated at a vector signal generator (VSG) complies with the IEEE 802.16-2004 downlink standard. The sub-frame structure of the WiMAX signal generated included three types of the burst data of QPSK, 16QAM, and 64QAM. Three different types of DFB lasers were evaluated as the optical transmitter [2].
The receiver was an 80micron diameter InGaAs PIN-PD packaged in a coaxial pigtail module. The responsivity is 0.85A/W in the typical, the cutoff frequency is 2GHz, the capacitance is 1.3pF or less.
Since this PIN-PD is designed for analog transmissions, the secondary order Intermodulation Distortion (IMD2) is less than -75dBc, the third order Intermodulation Distortion (IMD3) is less than -90dBc, with two tones of 244 MHz and 250MHz, at an Optical Modulation Depth (OMD) of 70%. The load resistance is 50Ω. This low gain circuit is sensitive to the degradation of the carrier to noise ratio (CNR) of the receiving signal. When the fiber length, the signal frequency, and the PMD and PML change, the digradation in the CNR affects the RCE in the WiMAX transmission.
The RCE was measured at a Vector Signal Analyzer (VSA). To investigate the influence of the carrier frequency and the fiber length, the frequency changed from 1 to 2.5GHz, and the standard single-mode-fiber (SMF) length changed from 0 to 10km. The measured RCE was influenced strongly by the modulation carrier frequency and the fiber length, as shown in Fig.2.2. At higher frequency the RCE measurement results were unstable. This is due to the PMD caused in the SMF. In case of the lower RCE, the constellation map showed small size dots, as shown in Fig.2.3 (a). But, the RCE for all sub-frames measured after the 10km SMF, showed large deviations in the magnitude and the phase, as shown in Fig.2.3 (b). When increasing the reflected light intensity into the fused-type optical coupler, the RCE showed increase of the magnitude, as shown in Fig.2.3(c). This was due to the PML caused in the fused-type optical coupler. The returned light did not affect the transmitter noise, since there was a 60dB optical isolator in front of the DFB laser. In addition to the fact, it was also confirmed that the returned light did not cause any instability in the RCE, when the isolation at the DFB laser decreased to 30dB. These results were the same for the 1550nm Multi Quantum Well (MQW) laser, 1310nm MQW laser, and 1310nm Electro-Absorption-Modulator integrated DFB laser. The optimization of the carrier frequency and the fiber length has to be first carried out to achieve the lower RCE.

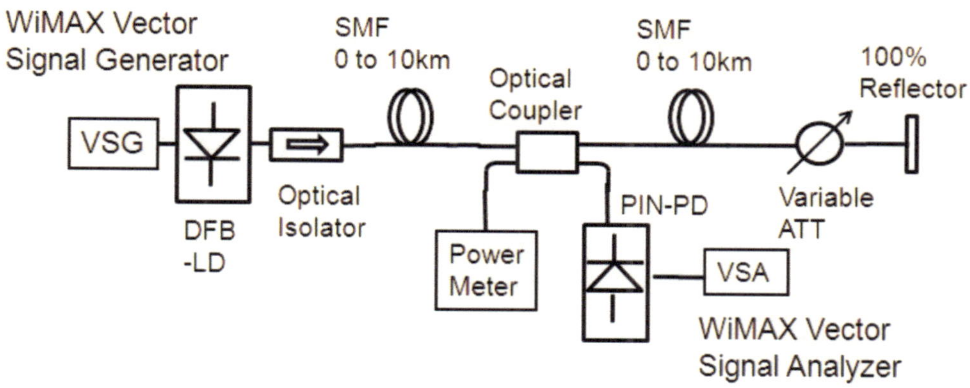

Fig. 2.1. The RCE measurement setup configured with DFB laser, single mode fiber, optical coupler, and optical reflector.

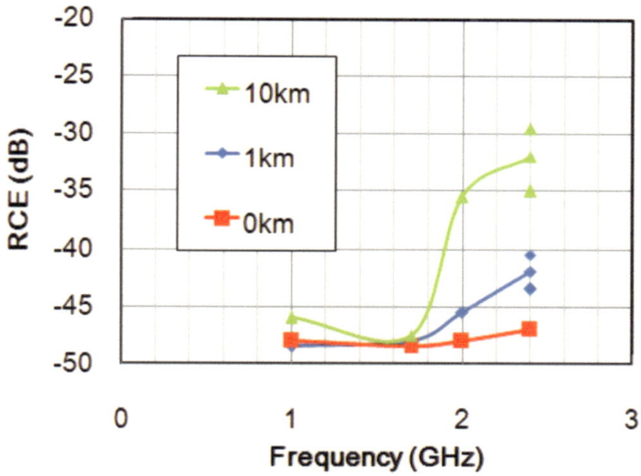

Fig. 2.2. The RCE was measured with different fiber lengths and modulation frequencies.

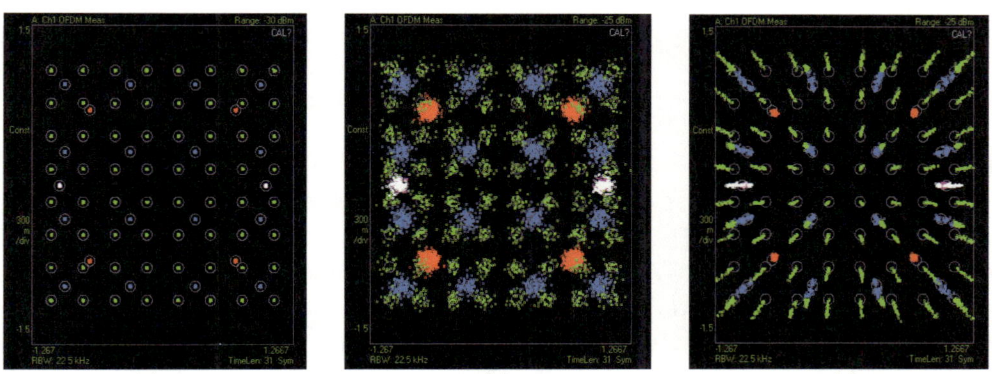

(a) (b) (c)

Fig. 2.3. (a): Constellation on normal condition, (b): with degradation by the PMD, and (c): with degradation by the PML.

2.2 An RoF and wireless link system configuration

It was found that the RCE of the WiMAX was determined with the burst signal waveform, and that the PAPR at a Complementary Cumulative Distribution Function (CCDF) was not changed by the carrier modulation bandwidth for the burst signal waveform, and was close to the Gaussian curve. Therefore the RCE was not varied with the carrier modulation bandwidth between 5 to 20 MHz. When the RoF link was constructed with electrically and optically linear characteristic components, the RCE was determined with theCNR of the received burst signal power. Figure 2.4. shows an RoF and wireless system used for the RCE measurements. The system was configured with a transmitter of the 1550nm DFB laser, a standard 10km SMF, a PIN-PD receiver, and parabolic grid antennas.

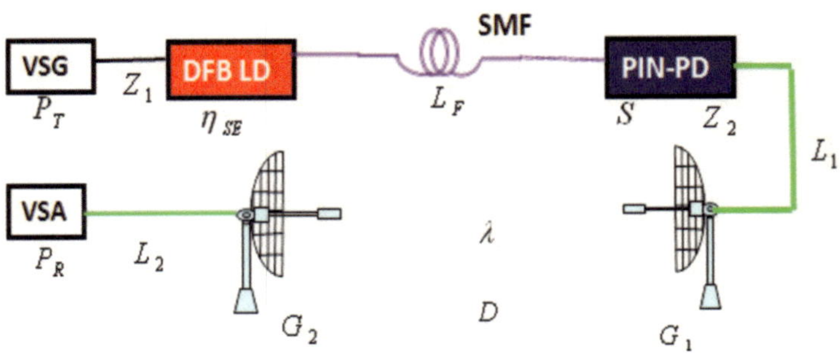

Fig. 2.4. The RoF and wireless system configured with DFB laser, optical single mode fiber, and 1.7GHz parabolic grid antennas.

2.3 Received power in RoF

By using the power level at each component of the signal generator (VSG), the DFB laser, the single mode fiber (SMF), the PIN-PD, the coaxial cables, the transmitter (Tx), the transmitter (Tx) parabolic antenna, space, and the receiver (Rx) parabolic antenna, the received power at the signal analyzer (VSA) can be expressed as

$$P_R = P_T - 2\,L_F + 20\log\left(\sqrt{\frac{Z_2}{Z_1}}\cdot\eta_{SE}\cdot S\right) - L_1 - L_2$$

$$+10\log\left(\left(\frac{\lambda}{4\,\pi\,D}\right)^2 + G_1 + G_2\right) \tag{2.1}$$

where, P_R [dBm] is the received power at the VSA, P_T [dBm] is the DFB laser modulation power generated at the VSG, L_F [dB] is the optical fiber loss, Z_1 [Ω] is the DFB laser input impedance, η_{SE} [mW/mA] is the DFB laser slope efficiency, S [A/W] is the PIN-PD responsivity, Z_2 [Ω] is the PIN-PD output impedance, L_1 [dB]and L_2 [dB]are the cable loss, D [m] is the antennas distance, λ [m] is the carrier wavelength, G_1 [dBi] is the Tx antenna gain, and G_2 [dBi] is the Rx antenna gain. In the calculation the following parameters are used; L_F =2dB, Z_1 =25Ω, η_{SE} =0.17 mW/mA, S =0.85A/W, Z_2 =50Ω, L_1 and L_2 =5dB in total, D =16m, λ =0.17647m, G_1 =20dBi, and G_2 =19dBi.

The received power was measured with different system configurations. The first configuration used coaxial cables only between the VSG and VSA. The second one used the coaxial cable and the 10km optical fiber between the VSG and VSA. The third one used the coaxial cable, the 10km optical fiber, and the parabolic grid antennas between the VSG and the VSA. Figure2.5 shows the experimental data and the theoretical data that are calculated with the equation (2.1). The experimental data of the received power coincided within 2dB with the theoretical data.

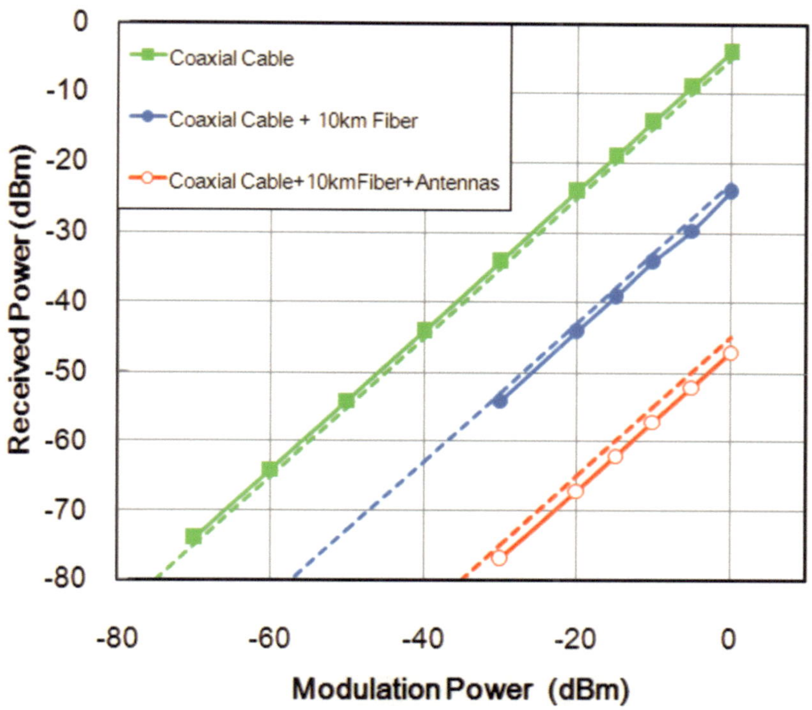

Fig. 2.5. The received powers were measured with modulation powers of driving the DFB laser for three different types of system configurations. The solid lines showed the experimental data and the dashed lines showed theoretical data calculated with the equation

2.4 RCE in RoF of WiMAX

It was considered that the RCE was determined with the received power and the noise power ratio when the RoF system was configured with linear characteristic components and the optimized modulation condition that were not influenced by the PMD and PML. However it is not realistic to completely eliminate the influence by the PMD and PML in the actual link system. A compensation factor to the received power has to be taken into account. Hence, the RCE is expressed using a compensation factor and a noise power as

$$RCE = P_N - A \cdot P_R \tag{2.2}$$

where, RCE [dB] is the Relative Constellation Error, P_R [dBm] is the received power, A is the compensation factor to the received power, and P_N is the system noise power. In the calculation the following parameters are used; A= 0.857 and P_N =-84.3[dBm].

The RCE was also measured with those three different system configurations. Figure 2.6 shows the experimental data and the theoretical data that are calculated with the equation (2.2). In the full components configuration system, the RCE experimental data coincide within 2 dB with the theoretical data. In the coaxial cable components configuration system, there was a little discrepancy between the experimental and the theoretical data in the lower RCE. This was due to that the noise power P_N used in the calculation was derived for the full components configuration system. It is obvious that the noise power P_N has to be used for unique value for each system configuration. But, by using the noise power P_N of full components configuration system, the RCE for the different type of the configuration system can be estimated. This means that the equation (2.1) and (2.2) are valid for any components configuration of the RoF system for the WiMAX, and are useful for the RoF system design and analysis.

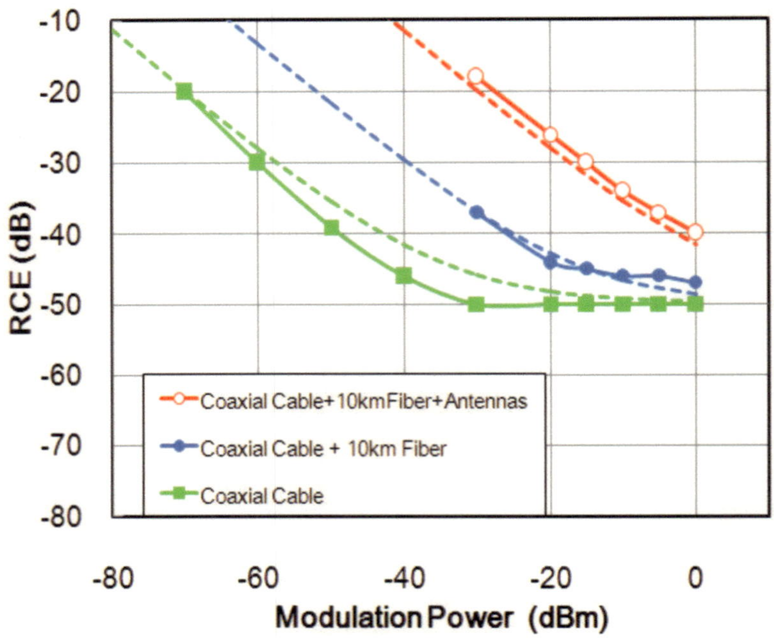

Fig. 2.6. The RCE were measured with modulation powers of DFB laser for the different type of system configurations. The solid lines show the experimental data and the dashed lines show theoretical data calculated with the equation (2.2).

It is possible to minimize the influence of the PMD and the PML by optimizing the carrier frequency, the fiber length, the type of the fiber, and the type of the coupler. The burst signal received power of the RoF of WiMAX was determined by the transmitter modulation, when the RoF link was configured with optically and electrically linear characteristic components. Since the RCE in the RoF of the WiMAX was related to the burst signal receiver power, the RCE was expressed in the linear relation with the burst signal received power. The experimental data and theoretical data mostly coincided within 2dB for the received power and the RCE.

3. E/O and O/E in WiMAX RoF

The WiMAX is a new standard for high-speed wireless communication that covers wider area than that of WLAN. For the field service, many access points are required, and it is important to design them with small size, low power consumption, and high reliability. Therefore, the complicated RF modem and signal processing functions are transferred from the access points to a central control office [3]. To extend the distance between the access points and the central office, the use of RoF is suitable for the WiMAX. There have been several studies of the lower cost and the high performance solutions for the RoF of WLAN [4]. The use of Vertical Cavity Surface Emitting Laser (VCSEL) or Fabry-Perot Laser Diode (FP-LD) was suggested for a low cost solution, and Mach-Zender Modulator (MZM) and Electro-Absorption Modulator (EAM) were used to achieve a high performance. However, there have been few studies for the RoF of WiMAX. A cost effective design was investigated for the E/O and the O/E that satisfy both the low cost and the high performance for the WiMAX RoF.

3.1 WiMAX RoF access points

In the WiMAX field service, a lot of access points are required, as shown in Fig.3.1. Therefore, it is important to design the access points with low power consumption, small mechanical size, high reliability, long distance installation, and low cost. A solution to those requirements is that the signal processing function such as the frequency up-down converter, modulator and demodulator, and A/D converter shall be transferred to the central office, and the access point is modified to the remote antenna unit (RAU) that has a transceiver antenna, and E/O and O/E converters [5], [6]. The RoF link enables installation of the access units in a long distance from the central office.

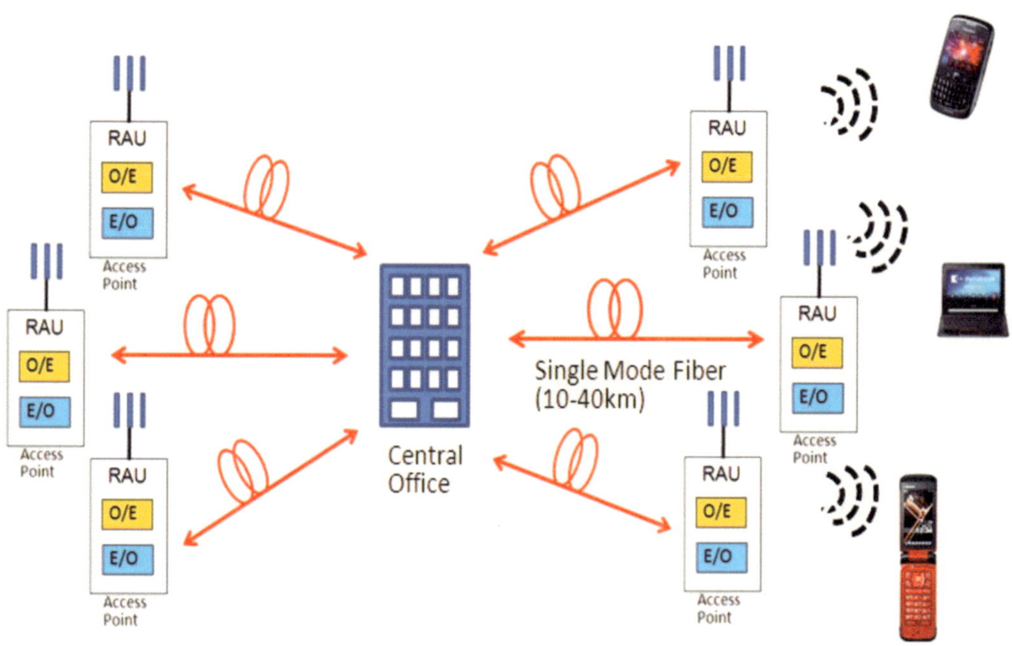

Fig. 3.1. The WiMAX RoF links connect between the central office and the access points.

3.2 Input Impedance adjustments in E/O converters

Two different types of 1310nm InGaAsP lasers of DFB and FP type lasers were used for the E/O converter. The package of the laser was Transmitter Optical Sub-Assembly (TOSA) that was a low cost design with a receptacle optical interface. We investigated if the electrical and optical conversion efficiency was increased with adjusting the input impedance of the E/O converter. The input impedance of the lasers at the 2.4GHz was adjusted on a network analyzer by changing the R, C, and L elements at the Radio-Frequency (RF) line of the Bias-Tee. In case of adjusting the input impedance to 50Ω, the input impedance was moved from the initial value of 15-j55Ωto the matching point of 49.5-j33Ω by adding 33Ω resistor, 1nH inductor, and 4.7pF capacitor to the RF line, as shown in Fig.3.2. In case of adjusting the input impedance to a low value, the impedance was moved from the initial value of 15-j50Ωto the low value point of 7-j2.07Ω by adding 22nH inductor to the RF line, as shown in Fig.3.3. In case of FP-LD, in addition to the inductor, a capacitor was used. After the input impedance adjustment, the E/O conversion efficiency was measured by a light-wave optical component analyzer (N4373A+N5230A, Agilent). The low-impedance input laser modulation showed about 5dB higher E/O conversion efficiency than that of 50Ω input impedance laser, as shown in Table 3.1. Although the length of the RF coaxial cable was about 50cm, the electrical reflectance did not affect the E/O conversion efficiency.

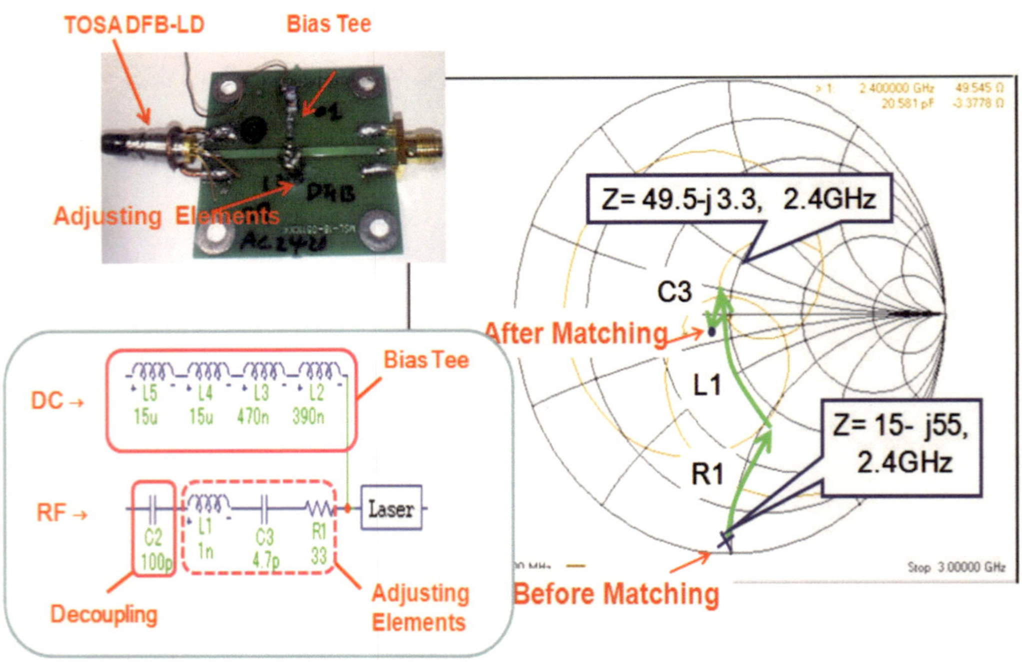

Fig. 3.2. TOSA type E/O with Bias-Tee, Bias-Tee circuit for 50Ω input- impedance adjusting , Input-impedance adjusting with R,C, and L elements, on a network analyzer.

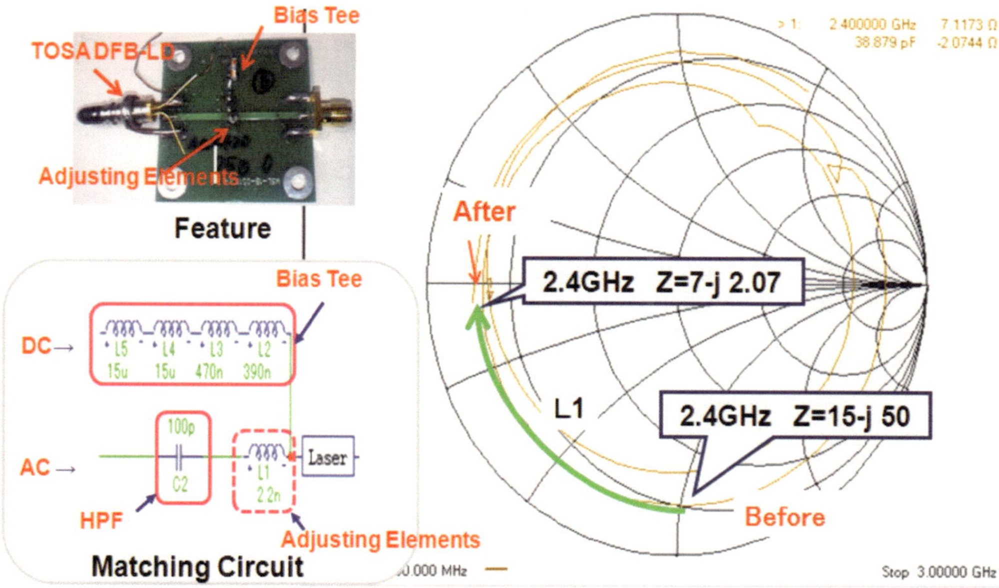

Fig. 3.3. TOSA type E/O with Bias-Tee, Bias-Tee circuit for input- low impedance adjusting , Input-impedance adjusting with L element, on a network analyzer.

Symbol	Input Impedance Adjusting		E/O Conversion
	Before (Ω)	After (Ω)	(2.4GHz) (dB)
DFB-Low	15-j50	7-j2.07	-8.3
DFB-50Ω	15-j55	49.5-j3.3	-12
FP-Low	12-j19	11-j3.95	-13
FP-50Ω	12-j19	50–j0.093	-18

Table 3.1. E/O input-impedance adjusting and E/O conversion

The E/O conversion of the laser is able to be derived by the following equation;

$$\eta_{EO} = 20 \cdot \log(\eta_{SE} \cdot \sqrt{\frac{50}{Z_{in}}}) \ [dB] \tag{3.1}$$

where, η_{EO} is the electrical and optical conversion efficiency [dB], η_{SE} is the laser slope efficiency [W/A], and Z_{in} is the laser input impedance [Ω] .
According to the equation (3.1), the electrical optical conversion efficiency of the low input impedance laser is higher than that of the 50Ω input impedance laser by 7 to 8.5dB, but the measured value showed a difference of 3.7 to 5dB. The difference might be due to the electrical reflectance between the laser and the signal generator. The theoretical calculation and the experimental result showed the low input impedance laser had higher electrical and optical conversion efficiency.

3.3 O/E converters configured with PIN-PD and various type amplifiers

There are several different types of InGaAs PIN-PDs, front-ends, and packages for the candidate of the O/E converters, since the high speed digital communication and multichannel wideband analog communication have developed the high performance PIN-PDs, the low impedance front-end, the transimpedance front end, and the reliable packages[8]. Four different types of the O/Es were configured as a cost effective solution of the design of the O/E converter, A linear PIN-PD designed for analog modulation, and a high speed PIN-PD designed for the 2.5Gb/s or 10Gb/s speeds were used. Two different types of packages of the coaxial pigtailed type and the ROSA type were used. Four different types of the pre-amplifiers designed with multi stages GaAs Enhancement-Mode Pseudomorphic High Electron Mobility Transistor (EP-HEMT), or with a combination of GaAs Trans-impedance Amplifier (TIA) and the EP-HEMT amplifier, were configured. The reason of the use of the EP-HEMT was to achieve low voltage single power supply, low noise figure, and high power gain. Table3.2 shows the O/E converters that were fabricated with various parameters of those components to investigate the optimum performance. The O/E conversion gain of those O/E converters was measured with the light-wave optical component analyzer (N4373A+N5230A, Agilent). The 2.5Gb/s digital PIN-PD has a 2kΩ trans-impedance amplifier in the ROSA package, and is followed by a 15dB gain EP-HEMT amplifier. The total gain measured with the optical component analyzer was 33.8 dB. The 10Gb/s digital PIN-PD has a 1.5kΩ TIA and a low gain pre-amplifier in the coplanar type package. The total gain measured with the optical component analyzer was 35 dB. The pre-amplifier used in the 10Gb/s O/E converter has a gain control function.

Symbol	PIN1-AMP1	PIN2-AMP2	PIN3-AMP3	PIN1-AMP4
Block Diagram	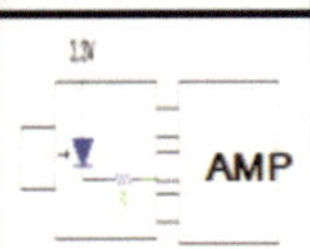	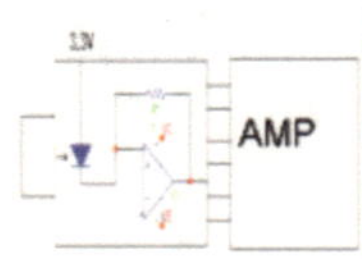	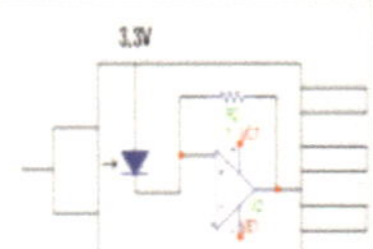	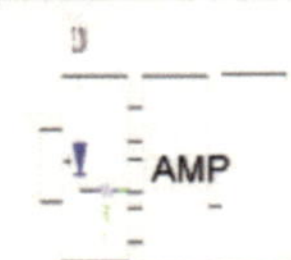
Photograph				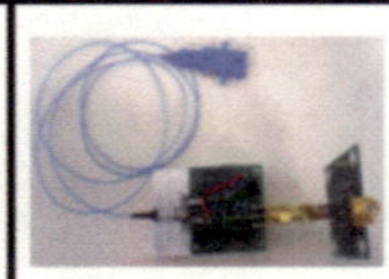
Configuration	PIN-Pigtail (2GHz-Analog) + AMP(17dB)	ROSA (PIN-TIA) (2.4GHz-Digital, Zt= 2k Ω) + AMP(15dB)	PIN-TIA (10Gb/s-Digital, Zt=1.5kΩ) + AMP(15dB)	PIN-Pigtail (2GHz-Analog)+ AMP(17dB)+ AMP(29dB)
O/E Conversion (2.5GHz)	19 dB	33.8 dB	35 dB	44.8dB

Table 3.2. O/E converters with PIN-PD and EP-HEMT Amplifiers

The O/E conversion of the laser is able to be derived by the following equation;

$$\eta_{OE} = 20 \cdot \log\left(\frac{1}{50} \cdot S_R \cdot Z_t \cdot 10^{\frac{G_v}{20}}\right) \ [dB] \tag{3.2}$$

where, η_{OE} is Optical and Electrical conversion efficiency [dB], 50 is the measurement equipment input impedance [Ω], S_R is responsivity of the PIN-PD, Z_t is the trans-impedance [Ω], G_v is the pre-amplifier gain [dB].

3.4 RCE measurements at 30 to 40km WiMAX RoF transmission

In the experiments of the evaluation of the E/O and O/E converters, a downlink WiMAX signal (IEEE 802.16) with 10MHz BW and with multi burst sub-frames of BPSK, QPSK, 16QAM, and 64QAM, was used at a 2.4GHz carrier frequency. The received signal error was evaluated by using the RCE that indicated average-error for all the sub-frames modulations. The WiMAX signal was generated by a VSG (E4438C, Agilent), and was converted to optical signal by the E/O converters. After the transmission over a SMF, the optical signal was converted to electrical signal by the O/E converter. The received signal was analysed by a VSA (89600S, Agilent).

The WiMAX downlink standard of the IEEE802.16 requires a value of -30dB for the RCE as the maximum value at the access point. The RCE measured for the FP-LD, the DFB-LD, the low and 50Ω input impedances are shown in Fig.3.4. The SMF length was 30km. The DFB-LD showed the lowest RCE. This is due to the lower relative intensity noise (RIN, about -155dB/Hz) of the DFB-LD. The low input-impedance DFB-LD showed lower RCE, this was due to the high electrical and optical conversion efficiency (see Table.3.1).

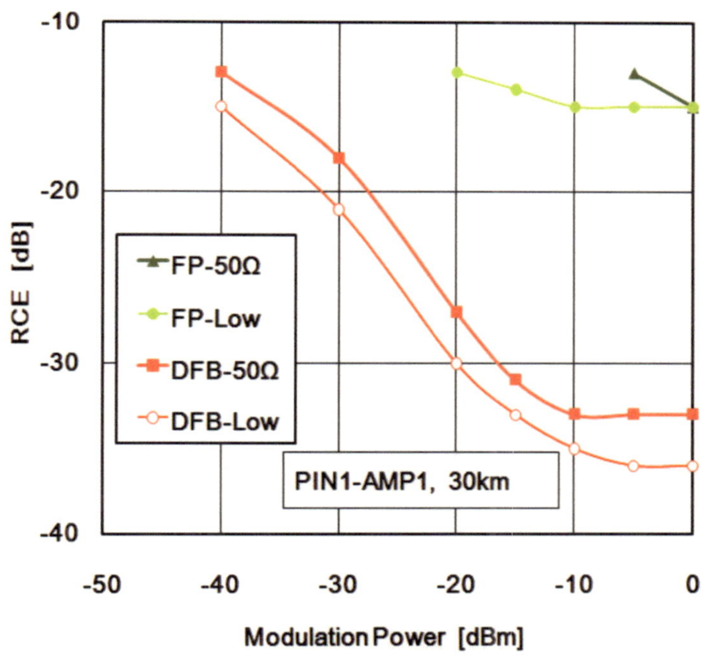

Fig. 3.4. RCE measured with FP-LD, DFB-LD, low input-impedance, and 50Ω input-impedance.

Figure3.5 shows the RCE measured for the three different types of the O/Es. The PIN1-AMP1 was configured with a linear PIN-PD and an EP-HEMT amplifier (17dB gain at 2GHz). The intrinsic layer of the linear PIN-PD was optimized for the low distortion modulation. The PIN2-AMP2 was a ROSA that was configured with a PIN-PD and a 2kΩ GaAs TIA designed for 2.5Gb/s digital transmission. The PIN1-AMP4 was configured with a linear PIN-PD and a high gain EP-HEMT amplifier (44.8dB gain at 2GHz). Each amplifier gain was confirmed on the measurements with the light-wave optical component analyzer (N4373A+N5230A, Agilent) and with the received electrical power measured at the VSA. The linear type analog PIN-PD of the type of PIN1 showed lower RCE than that of the digital PIN-PD of the type of PIN2 that was followed by relatively high gain TIA, as shown in Fig.3.5. The low gain PIN1-AMP1 showed lower RCE than that of the PIN2-AMP2. This is due to the low distortion conversion in the analog PIN-PD.

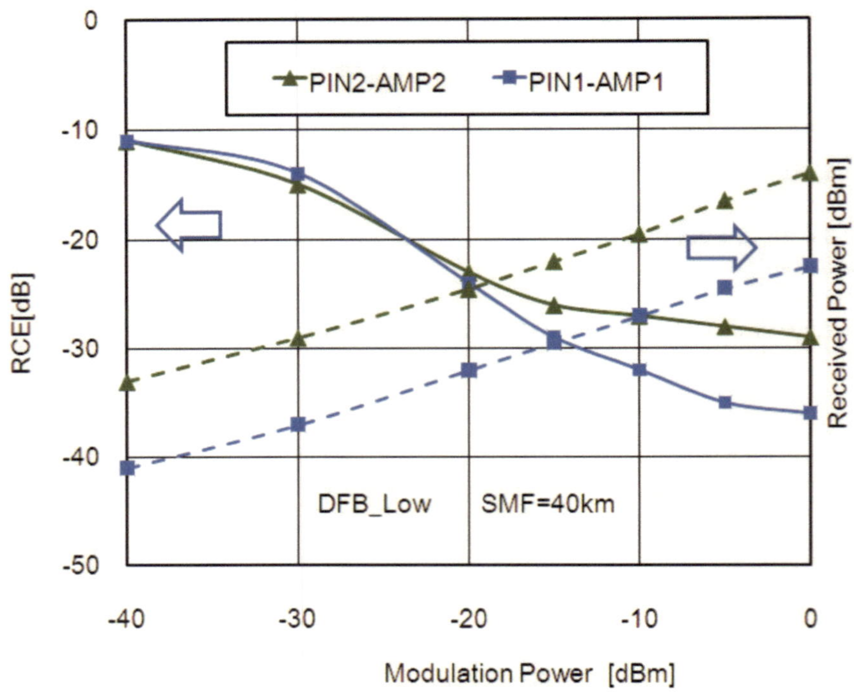

Fig. 3.5. RCE measured with analog PIN-PD (PIN1-AMP1) and 2.5Gb/s digital PIN-PD (PIN2-AMP2).

In the WiMAX RoF link, the use of the linear PIN-PD is strictly important to reduce the RCE. In general, the linear PIN-PD is achieved by a specific design of the structure [9]. The diameter of the window layer has to be determined by the received power, the focused size of the input light beam, and the capacitance. When increasing the diameter, the linearity in the higher optical power level is improved, but the cut-off frequency becomes lower. The length of the intrinsic layer determines the higher optical power level. The longer length is suitable, if the electric field is high, the distribution of the impurities is uniform, and the carrier density is low. The maximum length of the intrinsic layer is around the inverse of the optical absorption coefficient. The longer length of the intrinsic layer may cause the long carrier drift time, a low electric field, and lack of the uniformity. Since the PIN-PD designed for the high speed digital application has a small diameter and a short intrinsic layer, the optical linearity is not sufficient for the WiMAX communication.

O/E	RCE [dB]	Rec.Pwr [dBm]	O/E Conv. [dB]
PIN1-AMP1	-32	-28	19
PIN2-AMP2	-27	-39	33.8
Difference	5	11	14.8

Table. 3.3. RCE measured with analog PIN-PD (PIN1-AMP1) and 2.5Gb/s digital PIN-PD (PIN2-AMP2).

4. Conclusions

It is possible to minimize the influence of the PMD and the PML by optimizing the carrier frequency, the fiber length, the type of the fiber, and the type of the coupler. The burst signal received power of the RoF of WiMAX was determined by the transmitter modulation, when the RoF link was configured with optically and electrically linear characteristic components. Therefore it is strictly important to use the linear PIN-PD for the optical receiver. Since the RCE in the RoF of the WiMAX was related to the burst signal receiver power, the RCE was expressed in the linear relation with the burst signal received power. The experimental data and theoretical data mostly coincided within 2dB for the received power and the RCE.

An optimum design of the E/O and the O/E converters for a cost effective access point was carried out. Four different types of the E/O converters and four different types of the O/E converters were evaluated with the RCE on a WiMAX RoF link using a 2.5GHz carrier signal. At the transmission link between 30 and 40km, to satisfy the lower cost and the RCE less than -30dB, it is suitable to use the 1310nm DFB-LD with a pigtailed package, the lower input impedance than 10Ω, and an EP-HEMT multistage amplifier with the gain larger than 40dB. In this case, it is also strictly important to use the linear PIN-PD that was originally designed for analog transmission.

5. References

[1] Prasanna A. Gamage, et.al.(2008). Power Optimized Optical Links for Hybride Access Networks. Opto-Electronics and Communications Conference (OECC) and the Australian Conference on Optical Fibre Technology (ACOFT), Australia, July 7-10, 2008

[2] Koyu Chinen (2008). RCE Measurements in ROF of IEEE802.16 - 2004 (WiMAX) with Structurally Optimized DFB Lasers. The 8 th International Conference on Wireless and Optical Communications (WOC2008), Canada, May 26-28, 2008, pp.48-52

[3] H.Al-Raweshidy and S. Komaki (2002). Radio over Fiber Technologies for Mobile Communications Networks (Artech House, 2002) Chap.4 .

[4] Andrey Kobyakov. et.al.(2006). 802.11a/g WLAN Radio Transmission at 1.3um over 1.1 km Multi-mode and >30km Standard Single-mode Fiber Using InP VCSEL. *In Proc. ECOC 2006,* Cannes, France, 2006, Paper Tu1.6.1.

[5] Mohammad Shaifur Rahman, Jung Hyun Lee, Youngil Park, and Ki-Doo Kim (2009). Radio over Fiber as a Cost Effective Technology for Transmission of WiMAX Signals. *World Academy of Science, Engineering and Technology,* vol. 56, pp.424-428, (2009)

[6] Chien-Hung Yeh, Chi-WaiChow, Yen-Liang Liu, Sz-Kai Wen, Shi-Yang Chen, Chorng-Ren Sheu, Min-Chien Tseng, Jiunn-Liang Lin, Dar-Zu Hsu, and Sien Chi (2010). Theory and Technology for Standard WiMAX Over Fiber in High Speed Train Systems. *Journal of Lightwave Technology,* vol.28, No.16, Aug. 15, pp.2327-2336, (2010)

[7] Charles H.Cox, III (2004). Analog Optical Links , Cambridge University Press

[8] Eduard Sackinger(2005). Broadband Circuits for Optical Fiber Communications, John Wiley & Sons, Inc.

[9] Avigdor Brillant (2008). Digital and Analog Fiber Optic Communications for CATV and FTTx Applications, SPIE and John Wiley &Sons, Inc.

4

Single Photon Detection Using Frequency Up-Conversion with Pulse Pumping

Lijun Ma, Oliver Slattery and Xiao Tang
Information Technology laboratory, National Institute of Standards and Technology
United States of America

1. Introduction

In any quantum communication system, such as a quantum key distribution (QKD) system, data rates are mainly limited by the system clock rate and the various link losses. While the transmission clock rate is limited by the temporal resolution of the single-photon detectors, losses in a fiber-based quantum communication system can be minimized by operating in the near infrared range (NIR), at 1310 nm or 1550 nm. Commercially available InGaAs-based avalanche photo-diodes (APDs) can be operated as single-photon detectors in this wavelength range [Hadfield, 2009]. Due to the severe after-pulsing, InGaAs APDs are typically used in a gated mode and this can limit their application in high-speed quantum communications systems. Superconducting single-photon detectors (SSPDs) can work in the NIR wavelength range with good performance [Gol'tsman et al. 2001; Hadfield, 2009]. However, SSPDs require cryogenic temperatures, and are not widely available on the commercial market at present. In addition, InGaAs/InP based photomultiplier tubes (PMT) can operate in the NIR range, but its performance is limited by very low detection efficiency (1 % at 1600 nm) and large timing jitter (1.5 ns) [Hamamatsu, 2005]. Microchannel plates (MCP) are micro-capillary electron multipliers coated with an electron-emissive material and multiply photon-excited electrons from a photon cathode [Wiza, 1979]. MCPs usually have faster rise times and lower timing jitter than is achievable with PMTs. InGaAs MCPs can work in the NIR range. These MCPs, but are limited by low detection efficiency (~1 %) [Martin, J. & Hink P. 2003].

On the other hand, silicon based avalanche photo-diodes (Si APDs) are compact, relatively inexpensive, and can be operated at ambient temperatures with high detection efficiency and low noise in the visible or near-visible range. Unfortunately they do not work at wavelengths longer than 1000 nm. For those wavelengths, an up-conversion technique has been developed that uses sum-frequency generation (SFG) in a non-linear optical medium to convert the signal photons to a higher frequency (shorter wavelength) in the visible or near visible range. The up-converted photons can then be detected by a Si APD. Up-conversion detectors use commercially available components and devices, and are a practical solution for many applications in quantum communications. To date, several groups have successfully developed highly efficient up-conversion single-photon detectors in the near-infrared range using periodically poled lithium niobate (PPLN) waveguides [Diamanti et al., 2005; Langrock et al., 2005; Thew et al., 2006; Tanzilli et al., 2005; Xu et al., 2007;] and bulk crystals [Vandevender & Kwiat, 2004].

Traditionally, an up-conversion single photon detector uses continuous wave (CW) pumping at a single wavelength. For a quantum communication system, a synchronized clock signal can be obtained from classical channels. Therefore, the up-conversion detector can be operated in pulsed pump mode using the synchronized clock signal. Up-conversion detectors with a pulsed pump can provide extra advantages that CW pumped detectors do not offer. For example, because the dark count rate in an up-conversion detector is dependent on the pump power, pulsed pump can significantly reduce the noise while keeping the same conversion efficiency. Furthermore, in the CW pump mode, the temporal resolution is determined by the timing jitter of the Si APD used in the detection system. The jitter-limited temporal resolution becomes a bottleneck as the transmission rate increases in a quantum communications system. An up-conversion single-photon detector using multiple spectrally and temporally distinct pulse pumps can support transmission rates significantly higher than the jitter-limited transmission rate of a traditional Si APD detector. In this chapter, we will describe in detail an up-conversion detector using a pulsed pump and its applications in noise reduction and transmission rate increase in quantum communication systems. This detector was developed at the National Institute of Standards and Technology.

2. Up-conversion detector with one wavelength pulsed pumping

2.1 Detector configuration

The configuration of an up-conversion detector with a single wavelength pulsed pump is shown in Fig. 1.

A 1550 nm CW laser provides the pump seed. The CW seed light is modulated to an optical pulse train by a Lithium Niobate-based electro-optical modulator (EO modulator). In many quantum information systems, the photons arrive with a synchronized classical signal. One can use the synchronized signal to modulate the pump seed resulting in the up-conversion detector operating a pulsed pump mode. This feature is similar to an optical gate, which is very useful for noise reduction or high speed gating in a communications system.

The modulated pump seed is then amplified by an erbium-doped fiber amplifier (EDFA). Two 1310/1550 wavelength division multiplexer (WDM) couplers with a 25 dB extinction ratio are used to suppress noise around 1310 nm at the output of the EDFA. The amplified pump light is then combined with a weak (single photon) signal in the 1310 nm range by another WDM coupler and the combined pump and signal are then coupled into the PPLN waveguides. The input polarization state of both the signal and the pump are adjusted by the polarization controllers, PC1 and PC2 respectively, before entering the coupler. The longer the waveguide length, the lower the pump power needed to reach the maximum conversion efficiency. The PPLN waveguide for the up-conversion detector is 5-cm long, the longest length possible using current manufacturing techniques. The input of the PPLN waveguide is fiber coupled, and the output is coupled into free-space through a 710 nm anti-reflection (AR) coating on the face of the waveguide. The output light of PPLN waveguide consists of a 710 nm (SFG) up-converted weak light signal, residual 1550 nm pump light and its second harmonic generation (SHG) light at 775 nm. These beams are separated by two dispersive prisms and the 710 nm photons are detected by a Si APD. An iris and a 20 nm band-pass filter are used to reduce other noise, such as external light leaked into the system.

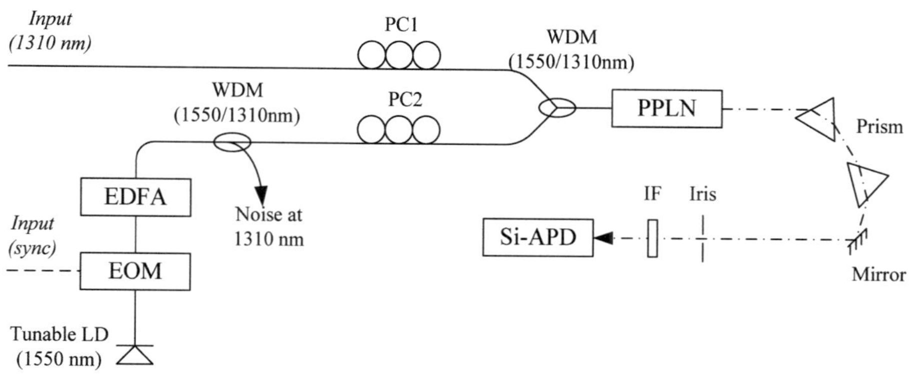

Fig. 1. Schematic diagram of an up-conversion detector with a pulsed pump. EOM: Electric-optic modulator; EDFA: Erbium-doped fiber amplifier; WDM: Wavelength-division multiplexing coupler; PC: Polarization controller; PPLN: Periodically-poled LiNbO3 waveguides; IF: Interference filter. Solid line: Optical fiber; Dash line: Free space optical transmission.

2.2 Noise reduction in quantum communication systems

The noise, or dark counts, of a single photon detector is one of its important performance parameters: a higher dark count rate can cause more errors in the quantum information system and degrade the system's fidelity.

In an up-conversion single photon detector, the total dark counts originate from the intrinsic dark counts of the Si APD and the noise from the frequency conversion process inside the crystal. The intrinsic dark count rate is very low, and is usually negligible in comparison to the noise due to the frequency conversion process. It is widely believed that the noise which arises in the frequency conversion process stems from the spontaneous Raman scattering (SRS) [Diamanti et al., 2005; Langrock et al., 2005; Thew et al., 2006; Tanzilli et al., 2005; Xu et al., 2007; Vandevender & Kwiat, 2004] and spontaneous parametric down conversion (SPDC) [Pelc et al, 2010] generated in the waveguide by the strong pump. If these SRS photons or SPDC photons are generated at wavelengths within the signal band they can be up-converted to the detection wavelength, generating noise or 'dark' counts. In our experiment, we have shown that a pump using a longer wavelength than the signal avoids the SPDC photons leaving only the noise from the anti-Stokes photons of SRS.

The significance of the Raman scattering component can be estimated by first neglecting the SFG, loss, and pump depletion, and then solving the photon number for a given Stokes mode N_s using the differential equation of the Raman scattering [Smith, 1972] in a known length PPLN waveguide. The solution is given by

$$N_s = e^{gL} - 1 , \tag{1}$$

where the Raman gain, g, is given by $\gamma P / A_{eff}$ and L represents the length. The quantity γ is the gain factor and the pump power, P, divided by the effective area, A_{eff}, gives the pump intensity.

In the above configuration, the 240-nm wavelength spacing between the pump (1550 nm) and the signal (1310 nm) is much larger than the peak Raman shift frequency of PPLN. Therefore,

$$e^{gL} - 1 \approx gL \ , \tag{2}$$

and the stimulated Raman scattering inside the PPLN is also negligible since a relatively low pump power is used. Therefore, we can assume that most of the Stokes photons are induced by the spontaneous Raman scattering inside the PPLN and this assumption enables a simple solution for the dark count.

The internal conversion efficiency (η_C) in the PPLN waveguide is a function of pump power, and is determined by the following equation [Diamanti et al., 2005; Langrock et al., 2005; Thew et al., 2006; Tanzilli et al., 2005; Xu et al., 2007; Vandevender & Kwiat, 2004]:

$$\eta_C(P) = \eta_0 \sin^2\left(k\sqrt{P / A_{eff}} \cdot L\ \right), \tag{3}$$

where η_0 is the peak conversion efficiency. η_0 is dependent on the quality of the PPLN waveguide and can reach 1 in a waveguide with high poling quality. The pump intensity is the ratio of the power, P ,to the effective area, A_{eff} . L_{PPLN} is the PPLN waveguide length. k is a constant and determined by

$$k = \left(\frac{\omega_s \omega_p d_{eff}^2}{n_s n_p c^2}\right)^{1/2}, \tag{4}$$

where ω_s and ω_p are the signal and pump wavelengths; n_s and n_p are the refractive index of lithium niobate for the signal and pump wavelengths. d_{eff} represents the effective nonlinear coefficient and c is the speed of light.

Using the above assumptions in Eq. (1), one can get a differential equation for the Stokes photon: $dN_s / dz = g = \gamma P / A_{eff}$ with z being a distance inside the PPLN waveguide from the input surface. A generated Stokes photon will be up-converted to the detection wavelength and induce a dark count. Consequently, by further applying the conversion efficiency of the PPLN, Eq. (3-4), we can write the dark count rate, DCR, as a function of pump power:

$$\begin{aligned} DCR_{Raman}(P) &= \eta_{\mathrm{APD}}\eta_{\mathrm{T}} \int_0^{L_{\mathrm{PPLN}}} \frac{\gamma \times P}{A_{\mathrm{eff}}} \sin^2\left[(L - z)k\sqrt{P / A_{eff}}\ \right] dz \\ &= \frac{\gamma}{2A_{\mathrm{eff}}} \eta_{\mathrm{APD}}\eta_{\mathrm{T}} P L_{\mathrm{PPLN}} \left[1 - \frac{\sin\left(2k\sqrt{P / A_{eff}} L\ \right)}{2k\sqrt{P / A_{eff}} L}\right], \end{aligned} \tag{5}$$

where η_{T} is the overall transmission efficiency and η_{APD} is the detection efficiency of the Si APD. The linear term of the pump power, P ,in the integral describes the generation of Stokes photons via the spontaneous Raman scattering and the sinusoidal term describes the up-conversion of the Stokes photons generated at z contributing to a dark count at the PPLN output ($z = L$). In the above derivation, we also neglected the optical loss in the PPLN waveguide. This assumption is justified by the fact that low waveguide loss is achieved with current manufacturing techniques and nearly 100 % internal conversion efficiency has been reported.

From the above analyses, one can see the conversion efficiency and noise both are dependent on the pump power. In a quantum communication system, each time bin is usually much longer than the optical pulse, because the single photon detectors have large timing jitter. When the pump pulse width is larger than the optical signal width, the photons can be converted at the highest efficiency. By turning off the pump after the signal pulse the dark counts rate can be reduced. When the pump is off, Si-APDs and counting board still operate, so the counts triggered by the signal photons delayed by the detector's jitter can still be recorded. In that case, we can reduce the noise while the efficiency is not affected.

We have experimentally demonstrated the up-conversion detector with a pulsed pump. The conversion efficiency as the function of pump power is shown in Fig. 2. The detection efficiency measured here is from a 625 MHz synchronized signal with 600 ps (FWHM) pump pulses. The optical pulse is pumped with the same synchronized signal but has a shorter (300 ps FWHM) pulse width. The detector operating in pulsed pump mode can reach the maximum conversion efficiency with a lower average pump power, which helps to reduce noise. The optimal pump powers (average) are 38 mW and 78 mW, for the pulsed and CW pump modes respectively.

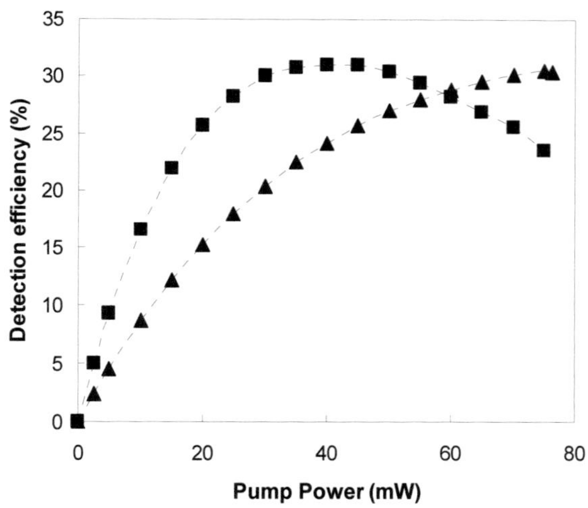

Fig. 2. The detection efficiency as a function of pump power. Two cases are studied: CW pump (triangle) and pulsed pump (square).

As shown in the Fig. 3, the pulsed pump generates more dark counts than the CW pump for a given average power since the peak power of the pulsed pump is higher than the average power. However, the pulsed pump needs less average power than the CW pump to achieve a given detection efficiency. Therefore, the pulse pump can achieve a given detection efficiency with less dark counts in comparison to the CW pump. For example, the maximum detection efficiency is reached when using the pulsed pump at 38 mW and the dark count rate is 2400 c/s. For the CW pump, a power of 78 mW is needed to achieve the maximum detection efficiency, which incurs a dark count rate of 3100 c/s. Consequently, a pulsed pump needs lower average power and effectively reduces the dark count rate compared to a CW pump.

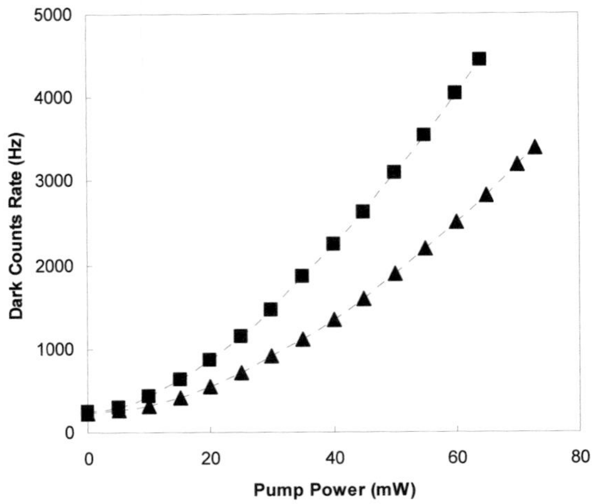

Fig. 3. The dark count rate as a function of average pump power at the PPLN input. Two cases are studied: CW pump (triangle) and pulsed pump (square).

To demonstrate the performance of the above up-conversion detector with pulsed pumping, we integrated them into one of our QKD system [Xu et al., 2007]. The quantum keys are encoded by photons in the quantum channel at 1310 nm using the B92 protocol [Bennett, 1992], as shown in Fig. 4. The QKD system uses a custom printed circuit board with a field-programmable gate array (FPGA) [Mink et al. 2006; Mink et al. 2009] to generate a random stream of quantum data for the photons and to transmit and receive the classical data. The classical data is carried by the optical signal in the classical channel at 1550 nm. To polarization-encode the quantum channel with the random quantum data, we first modulate a 1310-nm CW light into a 625 MHz pulse train, which is evenly split into two polarization channels. Each pulse train is further modulated by one of two complementary 625 Mbit/s quantum channel data streams. The two quantum channels are combined by a 45-degree polarization-maintaining combiner and attenuated to a mean photon number of 0.1 per bit, and then multiplexed with the classical channel and sent to a standard single-mode fiber. At Bob, another WDM is used to demultiplex the quantum and classical channels. The quantum channel is polarization-decoded and detected using the up-conversion single-photon detectors, and the detection events are recorded to generate raw keys. Bob's board informs Alice of the location of the detection events via the classical channel. After reconciliation and error correction, Alice and Bob obtain a common version of shared secret key bits, which are further used to encode and decode information for secure communication between Alice and Bob.

The system performance is shown in Fig. 5. During our measurements, the pump power was fixed at 40 mW. The sifted-key rate is 2.5 Mbit/s for a back-to-back connection, 1 Mbit/s at 10 km, and 60 Kbit/s at 50 km. The quantum bit error rate (QBER) is approximately 3% for the back-to-back configuration, remains below 4% up to 20 km, and reaches 8% at 50 km. The finite extinction ratio of the modulator and the system timing jitter induce a background QBER of approximately 2.5% and the rest is from dark counts generated by both the pump light and the classical channel. We also calculated the theoretical sifted-key rate and QBER and they agree well with the measured results.

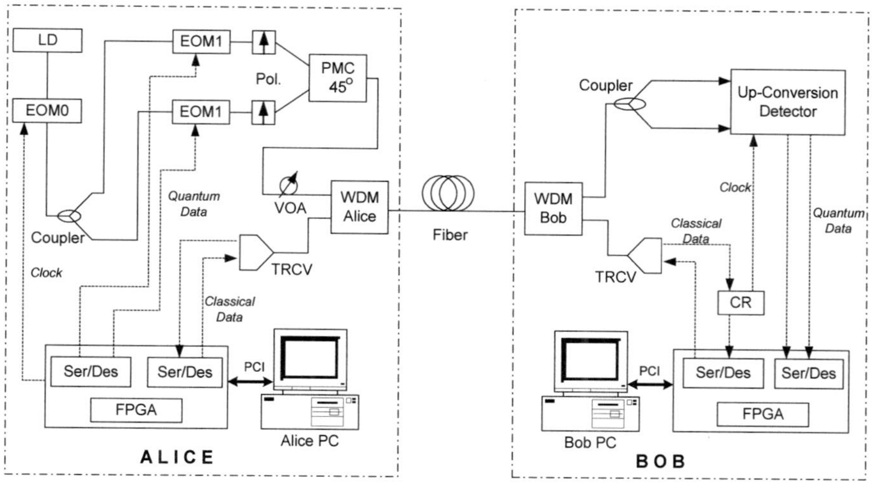

Fig. 4. The B92 polarization coding QKD system. LD: Laser diode; EOM: Electric-optic modulator (LiNbO3); PC: Polarization controller; PMC-45°: Polarization maintaining combiner that combines two light signals that are separated by 45 degrees; VOA: Variable optical attenuator; WDM: Wavelength-division multiplexer; SMF: Standard single-mode fiber; TRCV: Optical transceiver; CR: Clock recovery module; FPGA: Custom printed circuit board controlled by a field-programmable gate array; PCI: PCI connection; Dotted line: Electric cable; Solid line: Optical fiber.

Although the pump power is fixed near the optimal value for maximum up-conversion efficiency, the QBER remains small until the distance reaches close to 20 km due to the low dark count rate caused by the 1550 nm up-conversion detector. This QKD system can generate secure keys in real time for one-time-pad encryption and decryption of a continuous 200 Kbit/s video transmission stream over 10 km fiber. The system performance demonstrates that the up-conversion detectors with pulse pumping are suitable for the fiber-based polarization-encoding QKD system, realizing high speed secured data transmission.

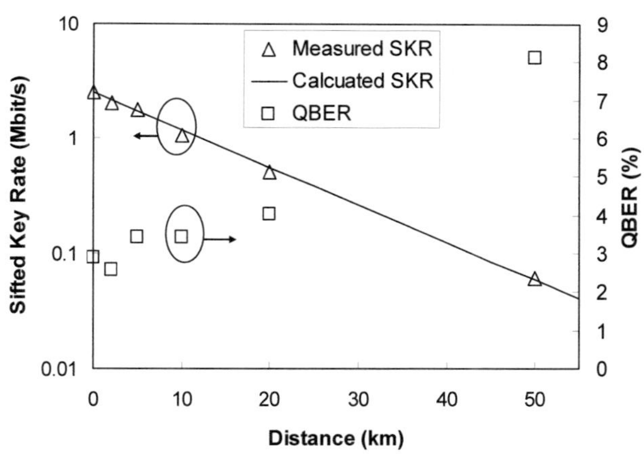

Fig. 5. The system performance of the B92 polarization-based QKD system using pulsed pump up-conversion detectors.

3. Up-conversion detector with multiple wavelength pulse pumping

3.1 The scheme and detector configuration

In many recent quantum communication systems, the photon sources can generate and temporally encode data at rates significantly higher than what single-photon detectors can resolve. For example, commercially available mode-locked lasers or optical modulators can generate sub-10 ps optical pulses, and broadband SPDC sources can readily prepare optical states of photon pairs with sub-100 fs correlation time. On the other hand, current high-resolution single-photon detectors exhibit a FWHM temporal resolution of the order of 50 ps. In a quantum communication system, insufficient temporal resolution in the detector can cause inter-symbol interference (ISI), i.e., a detection signal can be recorded at a time slot adjacent to the intended one, and this can induce a significant error rate. The transmission rate is therefore limited by the temporal resolution of the single-photon detectors. As a figure of merit, a single-photon signal can be received with an acceptable error rate when the transmission period is equal to or larger than the full width at 1% of the maximum (FW1%M) of the single-photon detector's response histogram [Restelli et al. 2009]. For most types of Si APDs the FW1%M is significantly larger than the commonly cited FWHM. At the peak of a typical Si APD's response histogram, where the FWHM is measured, the profile is approximately Gaussian. However, at lower part of the detector's response curve, the response histogram profile deviates significantly from Gaussian, often exhibiting a long exponential tail, which dramatically increases the FW1%M of the device. A typical commercially-available Si APD has a FWHM of about 350 ps, but a FW1%M of about 1100 ps that limits the transmission rate to less than 1 GHz for a quantum communication system using an up-conversion detector equipped with this type of Si APD.

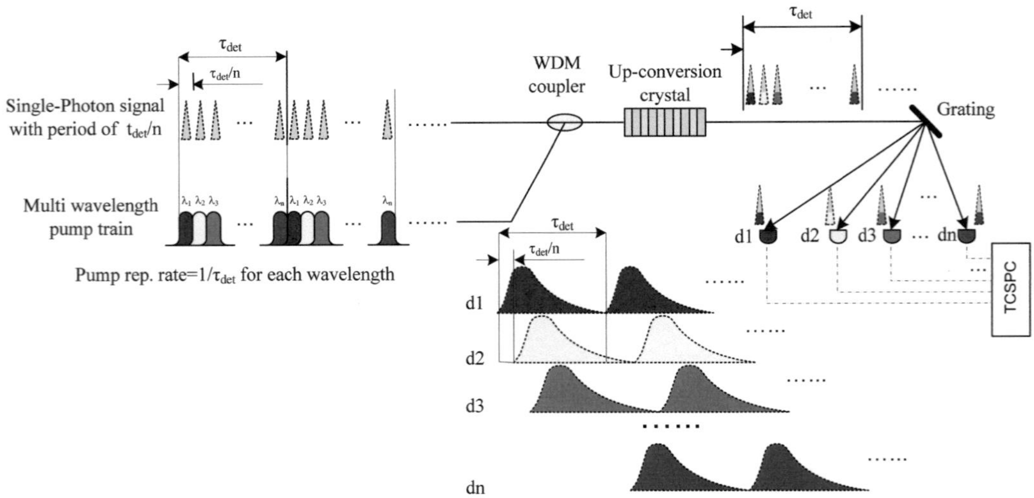

Fig. 6. Schematic diagram of up-conversion single-photon detection with multi-wavelength optical sampling. A sequence of n spectrally and temporally distinct pump pulses are used to sub-divide the minimum resolvable time bin, τ_{det}, of conventional Si APDs, increasing the temporal resolution of the overall system by a factor of n. The incident single-photon signal is combined with the sequence of pump pulses with a wavelength division multiplexer (WDM). Detection events from each of the n Si APDs are time-tagged with time-correlated single-photon counting (TCSPC).

To increase the temporal resolution of an up-conversion detection system beyond that of its constituent Si APDs, a sequence of spectrally and temporally distinct pump pulses can be used to sub-divide the minimum resolvable time period τ_{det}. This application of optical sampling is illustrated in Fig. 6, where n pump pulses with different wavelengths are used to sample the incident single-photon signal in intervals of duration τ_{det}/n. To ensure optimum detection efficiency, each pump pulse is wider than the single-photon signal pulses. The repetition rate for each particular wavelength of the pump is $1/\tau_{det}$. When a signal photon enters and co-propagates with one of the strong pump pulses in a quasi-phase matched sum-frequency device, such as PPLN, it can be up-converted to the visible range. The specific wavelength of the up-converted photon is determined by the wavelength of the pump pulse with which it interacted. A subsequent dispersive element such as a grating can separate the up-converted signal photons and distribute them to an array of Si APDs. Each Si APD in the array therefore corresponds to a particular pump wavelength, and, by extension, a particular arrival time period of duration τ_{det}/n. In such a configuration the sampling period for each Si APD is τ_{det}, allowing it to accurately resolve the signal without ambiguity due to the detector's temporal response. With this approach, the detection system is able to resolve the single-photon signal in a period as small as τ_{det}/n, representing an increase in temporal resolution by a factor of n.

The experimental configuration of an up-conversion detector with a multiple wavelength pulse pump is shown in Fig. 7. A pattern generator drives the pulse-carving systems for the two up-conversion pump sources at 1549.2 nm and 1550.0 nm. Each pump source has a period of 1.25 ns. Before the pump and the signal are combined the pulses from the first pump are aligned with the odd signal pulses, and the pulses in the second pump are aligned

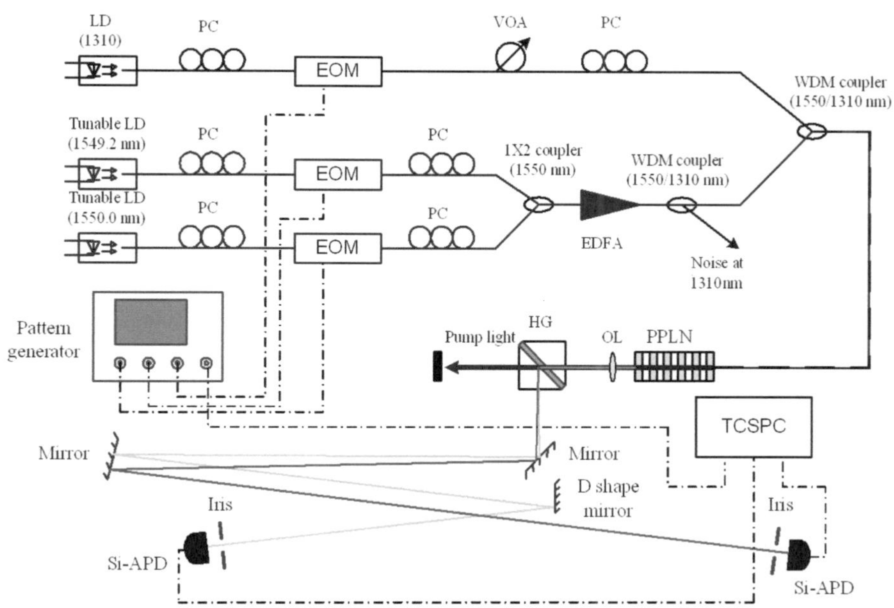

Fig. 7. Experimental setup. LD: Laser Diode, EOM: Electric-optic Modulator; EDFA: Erbium-doped fiber amplifier; WDM: Wavelength-division multiplexing coupler; PC: Polarization controller; PPLN: Periodically-poled LiNbO3 waveguides; OL, Objective Lens; HG, Holographic Grating. TCSPC: time-correlated single photon counting;

with the even signal pulses by adjusting the delays in the pattern generator, as shown in Fig. 8. The pump-pulse duration used in the experiment is 400 ps, which is wider than the 220 ps signal pulse and chosen to provide higher conversion efficiency [Xu et al., 2007]. The two pump beams are combined by a 1x2 coupler and then amplified by a 1-Watt EDFA. At the output of the EDFA, two 1310/1550 WDM couplers are used in series, giving a 50-dB extinction ratio in total, to suppress noise around 1310 nm. The amplified pump light is then combined with the 1310-nm signal by another WDM coupler, and the combined pump and signal are coupled into the up-conversion medium. Up-conversion takes place in a 1-cm PPLN waveguide that has a fiber-coupled input and a free-space output. When mixed with the slightly different pump wavelengths in the PPLN waveguide, the 1310 nm signal photons are up-converted to output photons at 710.0 nm and 709.8 nm. The output beam is filtered to remove noise and excess pump light and then diffracted by a holographic grating. After a 3-m long path the 710.0 nm and 709.8 nm photons are sufficiently separated that they can be directed onto two Si APDs. In this system an adjustable iris placed in front of the Si APD, in conjunction with the holographic grating, act as a 0.4-nm band-pass filter, which greatly reduces the dark count rate. The detected signals are then counted by a time-correlated single photon counting system.

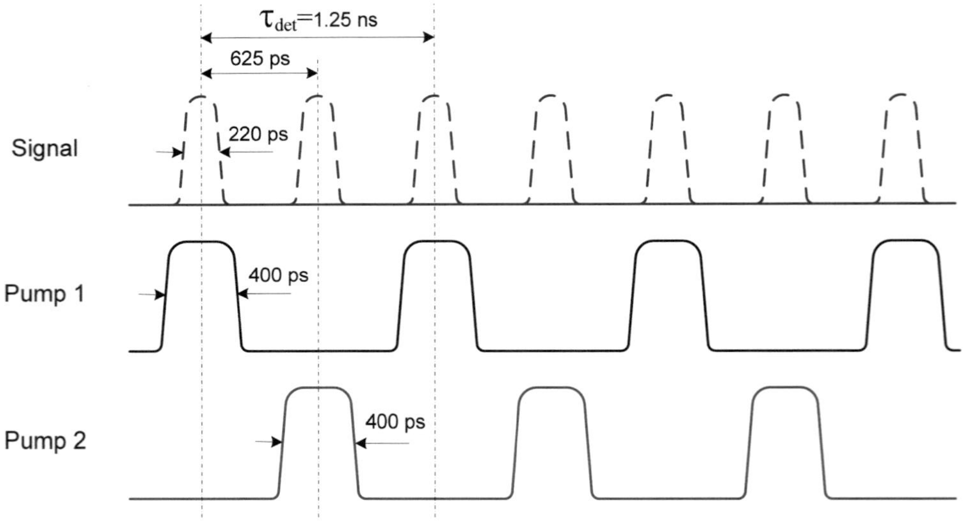

Fig. 8. Timing diagram of the signal, pump 1 and pump 2 used in the multi-wavelength optical sampling single-photon detection system.

In this experiment we use a 1-cm long PPLN waveguide. A shorter waveguide provides a broader spectral bandwidth, which is necessary to cover the spectral separation of the two output photons. In the previously described work, we have demonstrated high single-photon up-conversion efficiency with a longer 5-cm PPLN waveguide, but we find that the quasi-phase matched bandwidth for such a waveguide is too narrow for the current approach. The transfer function response of a finite-length uniform quasi-phase matching grating is [Fejer et al. 1992]:

$$I_{SFG}(\Delta k) \propto I_{pump} \cdot I_{signal} \cdot \text{sinc}^2(\Delta k \cdot L / 2), \tag{6}$$

where I_{SFG}, I_{pump}, and I_{signal} are the intensities of SFG, pump, and signal light, respectively, L is the waveguide length, and Δk is the phase-mismatching, which determines the bandwidth of the spectral response. According to Eq. (6), for a given SFG intensity, the waveguide length and the spectral response bandwidth are inversely proportional; the shorter the waveguide, the broader the spectral response bandwidth. Fig. 9 (a) shows the spectral response measured experimentally for the 1-cm PPLN waveguide. Its 3-dB bandwidth is about 1.3 nm, which is about 5 times wider than that of the 5-cm PPLN waveguide (0.25 nm) [Ma et al. 2009]. In this experiment the wider bandwidth allows two pumps, at wavelengths 1549.2 nm and 1550.0 nm, to operate with almost the same conversion efficiency, which is about 85% of the maximum conversion efficiency.

Detection efficiency is a significant trade off for a short waveguide. From Eq. (3), to compensate for the reduced conversion efficiency in a shorter waveguide the pump power must be scaled quadratically. For example, the pump power required to achieve the maximum conversion efficiency in a 1-cm waveguide is 25 times higher than that for a 5-cm waveguide. Fig. 9(b) shows the detection efficiency of the up-conversion detector as a function of the average pump power. The pump power on the x-axis is measured at the input fiber of the PPLN waveguide. Although the maximum output power of the EDFA is 1 W, the maximum power at the input fiber is approximately 510 mW due to losses in the WDM couplers and connectors between the EDFA and the waveguide. In our system the combined pulse duration of the two pumps covers 67% of each clock period, and therefore the peak power of each pulse is only 1.5 times higher than the average power. Fig. 9(b) indicates that the up-conversion efficiency of the detector does not reach its potential maximum value and is limited by the available pump power. Besides the insufficient pump power, the detection efficiency in our system is further reduced by the absence of an AR coating on the waveguide ends, causing about 26% loss, and, as stated above, the fact that the two pumps operate at wavelengths that provide 85% of the peak spectral response. Due to these factors, the overall detection efficiency is measured to be 7 %. The detection efficiency can be improved by using a higher pump power and an AR-coated waveguide. Selecting pump wavelengths closer to the center of spectral response can also improve the overall detection efficiency, but this puts more stringent demands on the spectral separation before the Si APDs.

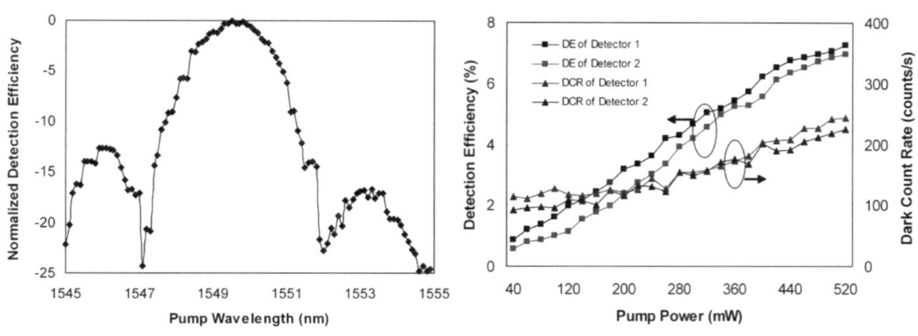

Fig. 9. (a) The spectral efficiency of the up-conversion detector. (b) The detection efficiency (DE) and dark count rate (DCR) as function of pump power. The pump power is measured in the input fiber of the PPLN waveguide.

Similar to other single photon detectors, the dark counts of this detector is caused by the anti-Stokes components of SRS in this waveguide and the intrinsic dark counts of Si APD. The SRS photons are generated over a broad spectrum, while the up-converted signal can be quite narrow. To further reduce the noise count rate, it is beneficial to use a bandpass filter with a very narrow bandwidth behind the waveguide. As stated above, in this experiment the iris in front of the Si APDs and the holographic grating constitute a band-pass filter with a bandwidth of about 0.4 nm. From Fig. 9 (b), the total dark count rate of the two Si APDs in the up-conversion detector are approximately 240 and 220 counts per second, respectively, at the maximum pump power.

3.2 Increasing transmission rate of a communication system

For a quantum communication system, inter-symbol interference (ISI) can be a significant source of errors. ISI can be caused by timing jitter of single photon detectors, and to avoid a high bit-error rate, the transmission data cycle should be equal to or larger than the FW1%M of the response histogram. For the 220-ps signal pulse used in our system, the response histogram of an up-conversion detector with a single wavelength pump is shown in Fig. 10 (black). The FW1%M of the histogram is about 1.25 ns and this detection system can therefore support a transmission rate of 800 MHz. When such a detection system is used to detect a 1.6 GHz signal, the insufficient temporal resolution of the detector results in severe ISI, as indicated by the poor pulse resolution, shown in Fig. 10 (grey). The application of optical sampling with two spectrally and temporally distinct pump pulses and a separate Si APD for each pump wavelength, as described above, accommodates the 1.25-ns FW1%M of each individual pump channel but supports an overall transmission rate of 1.6 GHz with low ISI. Fig. 11 (a) show the response histogram of each APD in the optical-sampling up-conversion system for a repetitive signal pattern "11111111". For each APD, the detection window is larger than FW1%M of APD response, so the ISI is greatly diminished. To illustrate both the temporal demultiplexing and the ISI in this system, Fig. 11 (b) shows the response histogram of each of the two APDs for a repetitive signal pattern "10010110". The

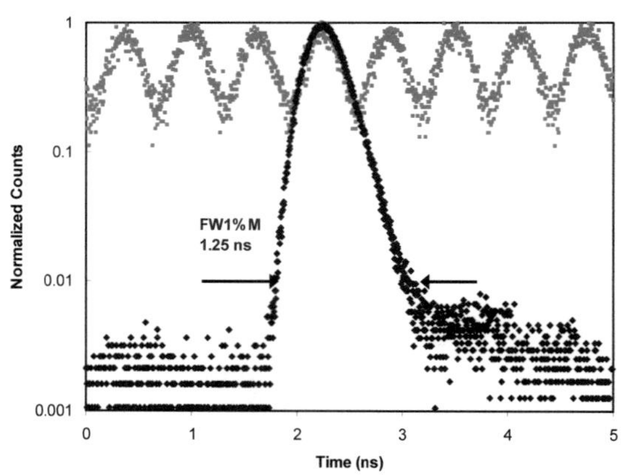

Fig. 10. Response histogram of the up-conversion detector with a single pump wavelength. The response histogram of single pulse (black) shows the FW1%M is 1.25 ns and its temporal resolution is insufficient to resolve, with low ISI, the repetitive data pattern "11111111" at 1.6 GHz (grey).

APD 1 receives the signal at odd time bins, resulting in the pattern "1001" and APD 2 receives the signal at even time bins resulting in the pattern "0110", and the original signal can be reconstructed from the data recorded by the two APDs. To measure the ISI in the optical sampling up-conversion system under conditions found in a typical QKD system we also drove the signal with a 1.6 Gb/s pseudo-random data pattern. After comparing the received data to the original data, the error rate was found to be approximately 1.2 %. Subtracting the error rate caused by the imperfect extinction ratio of the modulator and the intrinsic dark counts of APDs, the error rate caused by ISI is less than 1%.

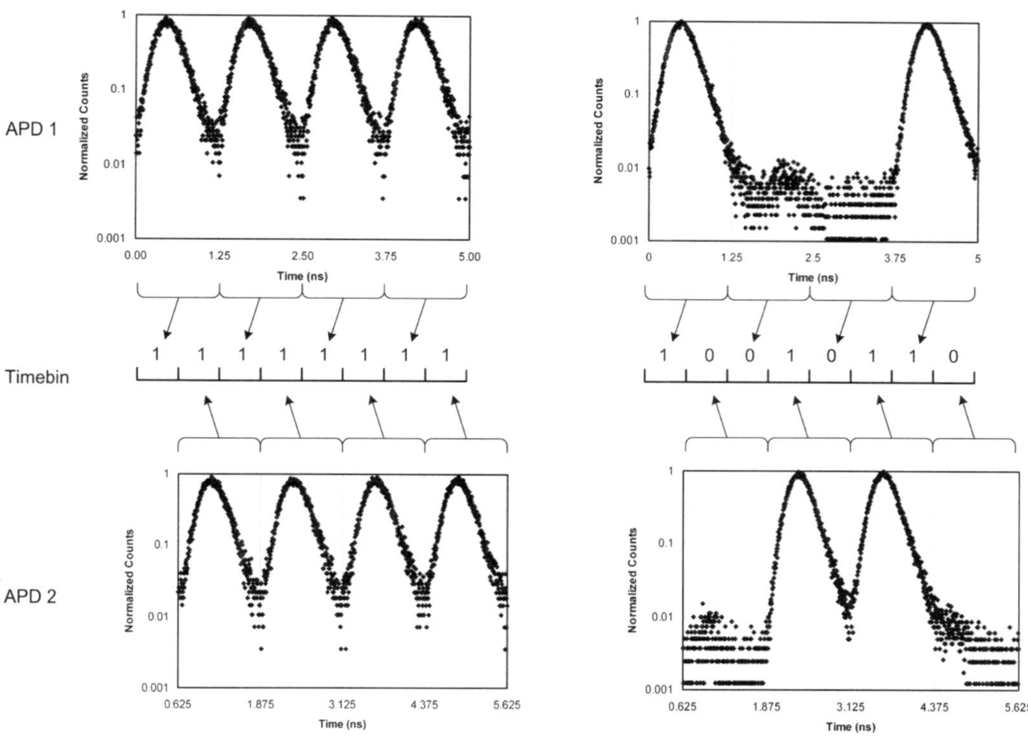

(a) (b)

Fig. 11. Response histogram of the up-conversion detector with two spectrally and temporally distinct pump pulses (a) response histogram of APD 1 and APD 2, for a repetitive signal pattern "11111111" at 1.6 GHz. (b) response histogram of APD 1 and APD 2, for a repetitive data pattern 10010110 at 1.6 GHz.

The above experimental results demonstrate that an up-conversion single-photon detector with two spectrally and temporally distinct pump pulses can operate at transmission rates that are twice as fast as can be supported by its constituent APDs. Further sub-division of the APD's minimum resolvable period (e.g. the FW1%M) is possible with more pump wavelengths and a corresponding number of Si APDs, allowing further increases in the maximum supported transmission rate of the single-photon system. However, the ability to increase the temporal resolution is ultimately limited by the phase-matching bandwidth of the nonlinear waveguide and available pump power.

Fourier analysis shows that shorter pulse duration corresponds to a broader frequency bandwidth. Considering only transform limited Gaussian pulses, the relationship between the pulse duration and spectral bandwidth for such "minimum uncertainty" pulses is given by [Donnelly and Grossman, 1998]:

$$t_{FWHM} \cdot \omega_{FWHM} = 4\ln(2)\,, \tag{6}$$

where t_{FWHM} and ω_{FWHM} are the FWHM of temporal width and frequency bandwidth, respectively. For the pump wavelengths in our experiment (~1550 nm), pulse widths shorter than 3 ps correspond to frequency bandwidths larger than 1.2 nm, which covers most of the 3-dB quasi-phase matching bandwidth of our 1-cm PPLN waveguide and thus precludes any other up-conversion pump wavelengths. A 100-ps pump pulse corresponds to a transform-limited bandwidth of 0.035 nm, in which case the waveguide used in our experiment could support more than 10 pump channels with greater than 50% quasi-phase matching efficiency. In this case, its temporal resolution can be increased by one order-of-magnitude compared to an up-conversion detector with just one pump wavelength. To provide uniform detection efficiency across all temporal regions, the pump power can be reduced in the well-phase-matched regions to match the conversion efficiency in the outlying spectral regions.

As the pump wavelengths become closer together, or if a shorter nonlinear waveguide is used to increase the quasi-phase matching bandwidth, technical issues associated with obtaining high optical powers in each pump, and efficient spectral separation of the up-converted photons become significant. We note that novel nonlinear crystal structures, such as chirped gratings or adiabatic gratings [Suchowski et al. (2010)] can provide broad bandwidth and relatively high conversion efficiency. With these new technologies, we believe it is reasonable to consider an up-conversion single-photon detector using spectrally and temporally distinct pump pulses with temporal resolution better than 10 ps. It should be noted that this scheme is not only suitable for up-conversion detectors using Si APDs; other single-photon detectors with better temporal resolution, such as SSPDs, can also be integrated into the scheme for further improvement of their temporal resolution.

4. Conclusion

Frequency up-conversion single photon detector technology is an efficient detection approach for quantum communication systems at NIR range. Traditionally, an up-conversion single photon detector uses CW pumping at a single wavelength. In CW pump mode, the pump power is usually set at a level where the conversion efficiency is the highest. In that case, the noise counts caused by the SRS in the waveguide might induce high error rates in a quantum communication system. An up-conversion single photon detector with a pulsed pump can reduce the noise count rate while maintaining the conversion efficiency. Furthermore, in a CW pump mode, the temporal resolution is determined by the timing jitter of the Si APD used in the detection system. A multiple wavelength pumping technique adds a new wavelength domain into the upconversion process. The data detected within a period of the Si APD's time jitter can be projected into the wavelength domain so that the spectrally and temporally distinct pulse pumping increases both the temporal resolution and the system data transmission rate.

5. Acknowledgement

The authors would like to thank for the support from NIST Quantum Information Initiative. The authors also thank Dr. Alan Mink, Dr. Joshua C. Bienfang and Barry Hershman for their supports and discussions.

6. References

Bennett, C. H. (1992). Quantum cryptography using any two nonorthogonal states. *Phys. Rev. Lett.*, Vol. 68, pp 3121-3124

Diamanti, E.; Takesue, H.; Honjo, T.; Inoue, K. & Yamamoto, Y. (2005). Performance of various quantum-key-distribution systems using 1.55-μm up-conversion single-photon detectors. *Phys. Rev. A,* Vol. 72, 052311

Donnelly, T. D. and Grossman, C. (1998) Ultrafast phenomena: A laboratory experiment for undergraduates. *Am. J. Phys.* Vol. 66, pp 677-685

Fejer, M.; Magel, G.; Jundt, D. & Byer, R. (1992). Quasi-phase-matched second harmonic generation: tuning and tolerances. IEEE J. Quantum Electron. Vol.28, pp 2631-2654

Gol'tsman, G. N.; Okunev, O.; Chulkova G.; Lipatov, A.; Semenov, A.; Smirnov, K.; Voronov, B. & Dzardanov, A. (2001). Picosecond superconducting single-photon optical detector. *Appl. Phys. Lett.* Vol. 79, pp 705-707

Hadfield, R. (2009). Single-photon detectors for optical quantum information applications, *Nat. Photonics*, Vol. 3, pp 696-705

Hamamatsu. (2005). Near infrared photomultiplier tube R5509-73 data sheet.

Langrock, C.; Diamanti, E.; Roussev, R. V.; Yamamoto, Y.; Fejer, M. M. & Takesue, H. (2005). Highly efficient single-photon detection at communication wavelengths by use of upconversion in reverse-proton-exchanged periodically poled LiNbO3 waveguides. *Opt. Lett. Vol.* 30, pp. 1725-1727

Ma, L., Slattery, O. and Tang, X. (2009) Experimental study of high sensitivity infrared spectrometer with waveguide-based up-conversion detector. *Opt. Express* Vol 17, pp 14395–14404.

Martin, J. & Hink P. (2003) Single-Photon Detection with MicroChannel Plate Based Photo Multiplier Tubes. *Workshop on Single-Photon: Detectors, Applications and Measurement Methods, NIST.*

Mink, A.; Tang, X.; Ma, L.; Nakassis, T.; Hershman, B.; Bienfang, J. C.; Su, D.; Boisvert, R.; Clark, C. W. & Williams, C. J. (2006). High speed quantum key distribution system supports one-time pad encryption of real-time video. *Proc. of SPIE,* Vol. 6244, 62440M,

Mink, A., Bienfang, J., Carpenter, R., Ma, L., Hershman, B., Restelli, A. and Tang, X. (2009) Programmable Instrumentation & GHz signaling for quantum communication systems. *N. J. Physics,* Vol. 11: 054016,

Pelc, J. S., Langrock, C., Zhang, Q. and Fejer, M. M. (2010) Influence of domain disorder on parametric noise in quasi-phase-matched quantum frequency converters. *Opt. Lett.,* Vol. 35, pp 2804-2806

Restelli, A., Bienfang, J. C., Mink, A. and Clark, C. (2009) Quantum key distribution at GHz transmission rates. *Proc. of SPIE* Vol. 7236, 72360L,

Smith, R. G. (1972) Optical power handling capacity of low loss optical fibers as determined by stimulated Raman and Brillouin scattering. *Appl. Opt.* Vol. 11, pp. 2489-2494

Suchowski, H., Bruner,B. D., Arie, A. and Silberberg, Y. (2010) Broadband nonlinear frequency conversion. *OPN* Vol. 21, pp 36-41

Tanzilli, S.; Tittel, W.; Halder, M.; Alibart, O.; Baldi, P.; Gisin, N. & Zbinden, H. (2005). A photonic quantum information interface. *Nature,* Vol 437, pp 116-120

Thew, R. T.; Tanzilli, S.;, Krainer, L.; Zeller, S. C.; Rochas, A.; Rech, I.; Cova, S.; Zbinden, H. & Gisin, N. (2006). Low jitter up-conversion detectors for telecom wavelength GHz QKD. *New J. Phys.* Vol. 8, pp 32.

Vandevender, A. P. & Kwiat, P. G. (2004). High efficiency single photon detection via frequency up-conversion. *J. Mod. Opt.,* Vol. 51, 1433-1445

Wiza, J. (1979). Microchannel plate detectors. *Nuclear Instruments and Methods* Vol. 162: pp 587-601

Xu, H.; Ma, L.; Mink, A.; Hershman, B. & Tang, X. (2007). 1310-nm quantum key distribution system with up-conversion pump wavelength at 1550 nm. *Optics Express,* Vol 15, No.12, pp 7247- 7260

Part 2

Photodiode for High-Speed Measurement Application

5

Low Scattering Photodiode-Modulated Probe for Microwave Near-Field Imaging

Hamidreza Memarzadeh-Tehran, Jean-Jacques Laurin and Raman Kashyap
École Polytechnique de Montréal
Department of Electrical Engineering
Montreal, Canada

1. Introduction

The space surrounding a radiating or scattering object is often divided into three regions, namely reactive near-field, near-field (NF) or Fresnel region and far-field (FF) or Fraunhoffer zones. In addition, the term "very-near-field" region is sometimes defined as very close to the antenna (e.g., antenna aperture). There are no abrupt boundaries between these three zones, however there are some commonly used definitions. For antennas with a size comparable to the wavelength (λ), the NF to FF boundary is calculated as $r \approx 2D^2/\lambda$, where D is the maximum dimension of the radiating device and r is the distance between the device and observation point.
The most widespread use of near-field measurement is in antenna diagnostics. In this case, fields are sampled near the antenna, typically in the Fresnel region, and a NF-to-FF transformation is used to obtain the radiation patterns (Petre & Sarkar, 1992). Rather than extrapolating away from the antenna, another possible application consists of reconstructing the field and current on the radiating device. This may require sampling within the reactive near-field region, i.e., with $r < \lambda$. Such *in-situ* near-field diagnostics have been made on antennas (Laurin et al., 2001), microwave circuits (Bokhari et al., 1995) and device emissions (Dubois et al., 2008). They can also be used to measure the wave penetration into materials and their radio-frequency (RF) characterization purposes (Munoz et al., 2008). Dielectric properties reconstruction (Omrane et al., 2006) is another use of NF measurement. Measuring the coupling between components of microwave circuits (Baudry et al., 2007), calculating FF radiation pattern of large antennas (Yan et al., 1997), and testing for electromagnetic compatibility EMC (Baudry et al., 2007) and EMI (Quilez et al., 2008) are among the other uses of NF measurement.

1.1 Statement of the problem— Obtaining accurate NF distribution

In applications such as the source or dielectric properties reconstruction, an ill-posed inverse problem has to be solved. The solution process is highly sensitive to noise and systematic measurement error. Accurate and sensitive NF measurement systems therefore need to be designed and implemented. Typically, NF imagers suffer from three important issues: limited accuracy and sensitivity, long measurement durations and reduced dynamic ranges, all of which depend on the measuring instruments and components used.

1.2 Modulated Scatterer Technique (MST)—An accurate approach for NF imaging

The distribution of near fields can be acquired using a *direct* (Smith, 1984) or an *indirect* (Bassen & Smith, 1983) technique. In the *direct* methods a measuring probe connected to a transmission line (e.g., coaxial cable) scans over the region of interest. The transmission line carries the signals picked-up by the probe to the measurement instruments. The major drawback associated with such technique is the fact that the fields to be measured are short-circuited on the metallic constituents of the transmission line. Multiple reflections may also occur between the device under test (DUT) and the line (Bolomey & Gardiol, 2001) resulting in perturbed field measurement. Moreover, flexible transmission lines such as a coaxial cables, which are widely used in microwave systems, do not always give accurate and stable magnitude and phase measurements (Hygate, 1990). This phenomenon in turn leads to inaccurate measurement, particularly where the measuring probe has to scan a large area. In contrast, *indirect* methods (Justice & Rumsey, 1955) are based on scattering phenomenon and require no transmission lines. Instead, a scatterer locally perturbs the fields at its position and the scattered fields are detected by an antenna located away from the region of interest, so as to minimize perturbation of the fields. This antenna could be the DUT itself (i.e., monostatic mode, in which case the signal of interest appears as a reflection at the DUT's input port) or an auxiliary antenna held remotely (i.e., bistatic mode). The variations of the received signals induced by the scatterer are related to the local fields at the scatterer's positions and are interpreted as the field measurement (magnitude and phase) by means of a detector. The *indirect* method employs a scatterer which is reasonably small, does not perturb the radiating device under test but is sufficiently large so that it is able to perturb the field up to the system's measurement threshold. Thus, a trade-off has to be made between accuracy and sensitivity. The *indirect* method suffers from limited dynamic range and sensitivity (King, 1978).

To overcome the drawbacks mentioned above, a technique known as the modulated scatterer technique (MST) was proposed and developed. MST was addressed and generalized by Richmond (Richmond, 1955) to remedy the drawbacks of both the *direct* and *indirect* methods. Basically, it consists of marking the field at each spatial point using a modulated scatterer, which is called the MST probe (Bolomey & Gardiol, 2001). This technique brings some outstanding advantages in the context of NF imaging such as eliminating the need to attach a transmission line to the measuring probe and improving the sensitivity and dynamic range of the measurement. From the point of view of probe implementation, tagging the field (modulation) can be done either electrically (Richmond, 1955), optically (Hygate, 1990), and sometimes, mechanically (King, 1978). Unlike optical modulation, the other modulation techniques somehow show the same disadvantages as the *direct* method. In an electrically modulated scatterer a pair of twisted metallic or resistive wires carry modulation signals to the probe. The presence of these wires may perturb the field distribution near the DUT, resulting in inaccurate measurements, whereas in an optically modulated scatterer (OMS) the modulating signal is transferred with an optical fiber that is invisible to the electromagnetic radio-frequency signal (Hygate, 1990). Thus, it can be assumed that it will only weakly influence the DUT's field distribution to be measured.

In this chapter, the design and implementation of a NF imager equipped with an array of optically modulated scatterer (OMS) probes that is able to overcome the drawbacks associated with the conventional *direct* and *indirect* methods are addressed. Additionally, a method to improve the dynamic range of the NF imager using a carrier cancellation technique is discussed.

2. Photodiode-loaded MST probe— Optically modulated scatterer

An OMS probe includes a small size antenna loaded with a light modulated component. The modulation signal is carried by an optical fiber coupled to the photoactivated component. It is switched ON and OFF at an audio frequency causing modulation on the antenna load impedance, which results in a corresponding modulation of the fields scattered by the probe. In the bistatic configuration the scattered field is received by an auxiliary antenna, as illustrated in Fig. 1. In the monostatic case the antenna under test is used to receive the modulated signal. In the following, the design and implementation of an optically modulated scatterer (OMS) is explained and discussed. Criteria for antenna type and modulator selection, tuning network design and implementation, and an OMS probe assembly will be also covered. Finally, the probe is characterized in terms of sensitivity, accuracy, and dynamic range.

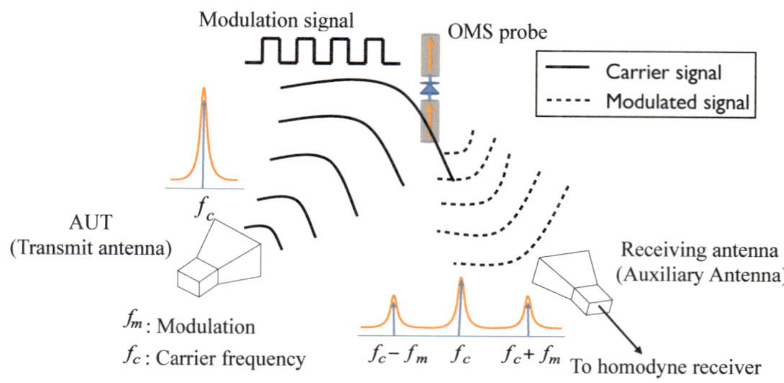

Fig. 1. Schematic of an MST-based NF imager in bistatic mode.

2.1 Antenna type

In practice, there is a limited number of antenna types that can perform as MST probes. Dipoles, loops, horns and microstrip antennas have been reported. The leading criterion to select the type of antenna is to keep the influence of the probe on the field to be measured as small as possible. The concept of a "minimum scattering antenna" (MSA) provides us with an appropriate guideline for selecting the scattering antenna. Conceptually, an MSA is invisible to electromagnetic fields when it is left open-circuited (Rogers, 1986) or connected to an appropriate reactive load (Iigusa et al., 2006). The horn and microstrip antennas do not fulfill MSA requirements due to their bulky physical structures and large ground plane, respectively, which cause significant structural-mode scattering regardless of antenna termination. The short-dipole (length$< \lambda/10$) and small-loop approach the desired MSA characteristics. A dipole probe might be a better choice because of its simpler structure. Moreover, a loop probe may measure a combination of electric and magnetic fields if it is not properly designed (King, 1978).

2.2 Modulator selection criteria

From the concept of AM modulation, we can introduce modulation index m as the ratio of the crests (1+μ) and troughs (1-μ) of the modulated signal envelope, where μ is the level of AM-modulation (King, 1978). Therefore, m can be defined as:

$$m = \frac{crest - trough}{crest + trough} \quad (1)$$

Assuming two states of the modulator with load impedance Z_{ON} and Z_{OFF}, and a probe impedance Z_p=Z_{dipole}+Z_{tn}, where Z_{tn} stands for the tuning network impedance[1], the modulation index of the signal scattered by the probe is given by (King, 1978):

$$m = \frac{|Z_p + Z_{ON}| - |Z_p + Z_{OFF}|}{|Z_p + Z_{ON}| + |Z_p + Z_{OFF}|} \quad (2)$$

whereas the ratio of the currents flowing in the probe terminals in both states is given by:

$$CR \equiv \frac{|I_{ON}|}{|I_{OFF}|} = \frac{|Z_p + Z_{OFF}|}{|Z_p + Z_{ON}|} \quad (3)$$

We can thus write:

$$m = \frac{1 - CR}{1 + CR} \quad (4)$$

If a small resonant probe is used, the real and imaginary parts of Z_p can be made very small, and possibly negligible compared to Z_{ON} and Z_{OFF}, such that:

$$m \approx \frac{|Z_{ON}| - |Z_{OFF}|}{|Z_{ON}| + |Z_{OFF}|} \qquad CR \approx \frac{|Z_{OFF}|}{|Z_{ON}|} \quad (5)$$

The maximum possible magnitude of the modulation index occurs when $CR = 0$ ($m = 1$) or $CR \rightarrow \infty$ ($m = -1$). Ideally, it is desired to maximize $|m|$ in order to have the strongest possible sideband response for a given level of a measured field. The selected modulated load should have either $|Z_{ON}| \gg |Z_{OFF}|$ or $|Z_{OFF}| \gg |Z_{ON}|$. In other words, input impedance of the device in the ON and OFF states should differ significantly. The results that will be presented in the next sections were obtained with probes based on a photodiode manufactured by Enablence (PDCS30T). This device was selected due to its high impedance variation as a function of input light level at a target test frequency of 2.45 GHz. The input impedance of the photodiode was measured on a wafer probing station using a calibrated Agilent 8510C vector network analyzer for different optical power levels (no light, and with a sweep from -10 dBm to 13 dBm) in the 2-3 GHz frequency range. The optical power in this measurement, was applied to the photodiode via an optical fiber, which was held above its active area by an accurate x-y positioning device.

Fig. 2a shows the impedance magnitude, revealing saturation for light power greater than +6 dBm (NB. In this figure we use the following definition $dB\Omega \equiv 20log_{10}\Omega$). The impedance of the diode in the "no-light" or OFF state and +6 dBm or ON state is shown in Fig. 2b. The diode can be modelled approximately by a series RC circuit, with $R_{OFF} = 38.8\Omega$ and $C_{OFF} = 0.31pF$. In the ON state, a similar model with $R_{ON} = 15.8\Omega$ and $C_{ON} = 13.66pF$ can be assumed. These models are approximately valid in a narrow frequency band centered at 2.45GHz. According to Equation 3, at 2.45 GHz these measured data lead to CR=13.38 (22.5 dB) and $m = -0.86$.

[1] It is assumed that this network consists of a series reactance in this example but other topologies are of course possible.

It is worth mentioning that the model used for this photodiode (i.e., series RC connection), it is only valid for small-signal operation. The photodiode switch-ON and breakdown voltages are 1.5 V and 25 V respectively. In addition, the maximum optical power should not exceed 10 dBm to prevent nonlinear operation.

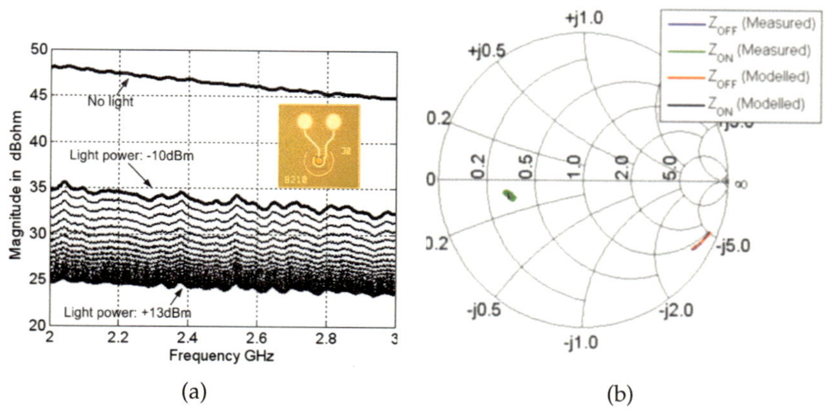

Fig. 2. (a) Input impedance magnitude of the photodiode (PDCD30T manufactured by Enablence), and (b) Input impedance (normalized to 50Ω) of the photodiode chip in the 2-3 GHz range with and without illumination. The measurement results and those obtained with a model of the photodiode are compared.

2.3 Selection of OMS probe length

Usually, the scatterer (i.e., OMS probe) should have minimum interaction with the source of the fields to be measured. The dynamic range of the measurement system depends on the minimum and maximum field levels the probe is able to scatter, and the detection threshold and saturation level of the receiver. Achieving a high dynamic range necessitates using a larger scatterer at the expense of oscillations in field measurements and deviation from the true field. In general for electrically small probes, the smaller the dimension of the scatterer the smaller the expected disturbance, but at the cost of lower sensitivity. Smaller probes also lead to better image resolution. Thus, a trade-off has to be made between the dynamic range on one side and the resolution and sensitivity of the probe on the other side. The first MST dipole probe reported by Richmond (Richmond, 1955) had a length of 0.31λ. Liang et al. used a length ranging between 0.05λ-0.3λ in order to make fine and disturbance-free field maps (Liang et al., 1997). Measured electromagnetic fields were also reported in (Budka et al., 1996) for operation in the 2-18 GHz band using MST probes that are 150 μm, 250 μm, and 350 μm long. A length of 8.3 mm was used by Hygate (Hygate, 1990) for signals below 10 GHz. Nye also used 3 mm and 8 mm MST probes at f=10 GHz to obtain NF maps of antennas or any passive scatterers (Nye, 2003). The probe presented here has a length of $\lambda/12$ at a design frequency of 2.45 GHz. The impedance of the printed short dipole at this frequency, as obtained by method of moment, is $Z_p = 1.22 - j412\Omega$.

In order to ensure that a $\lambda/12$ dipole probe not only meets the requirements of MSA but also has a negligible influence on the field to be measured, let us consider the measurement mechanism by MST probe using a network approach, as demonstrated in Fig. 3. The AUT in this figure acts as a radiating source and also a collecting antenna (i.e., port #1), and the scatter represents a measuring probe which is loaded with Z_L at port #2 (e.g., input impedance of the modulator) (King, 1978). Using of the impedance matrix of the passive network we can write:

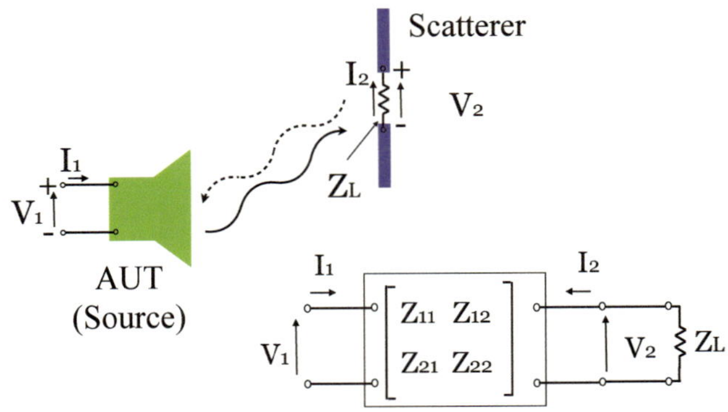

Fig. 3. Modelling of measurement mechanism using network approach, monostatic implementation.

$$V_1 = Z_{11}I_1 + Z_{12}I_2 \tag{6}$$
$$V_2 = Z_{21}I_1 + Z_{22}I_2 \tag{7}$$

The current induced in the probe (i.e., I_2) yields a voltage $V_2 = -I_2 Z_L$ on port 2. One can obtain Equation 8 by solving Equation 7 for V_1:

$$V_1 = \left(Z_{11} - \frac{Z_{12}Z_{21}}{Z_{22} + Z_L} \right) I_1 \tag{8}$$

It is also assumed that the voltage on port 1 in the absence of the scatterer is given by $V_1^0 = Z_{11}^0 I_1$, where Z_{11}^0 is the input impedance of the AUT. Then, by subtracting it from Equation 8, it yields,

$$V_1 - V_1^0 = \Delta V_1 = \left[(Z_{11} - Z_{11}^0) - \frac{Z_{12}Z_{21}}{Z_{22} + Z_L} \right] I_1 \tag{9}$$

It has been assumed that current I_1 fed to the AUT is unchanged in the two cases. Based on Equation 9, it can be shown that the measuring probe has two separate effects at the receiver's voltage, namely, the effect due to its physical structure (i.e., structural mode) and its loading (i.e., antenna mode). On the right hand side, the first term is present even when the probe is left open-circuited (i.e., when $Z_L \to \infty$), that results from the probe's structural mode. The second term appears when the probe loading (i.e., Z_L) is finite or zero, allowing current to flow in port 2. This contribution is therefore called the antenna mode. Only the latter term is modulated in MST-based probes. The first term is present and varies when the probe is moved from one measurement point to another but those variations are slow compared to the rate of modulation. It can thus be assumed that they will not affect the measurement at the modulation frequency. By considering an open-circuited scatterer (i.e., $Z_L \to \infty$), ΔV_1

gives $(Z_{11} - Z_{11}^0)I_1$; this represents the variation of the induced voltage across the AUT's terminal compared to the case in absence of the scatterer. Ideally, it is expected that ΔV_1 will vanish for MSA antennas, i.e., structural mode radiation is vanishingly small. Now, in order to

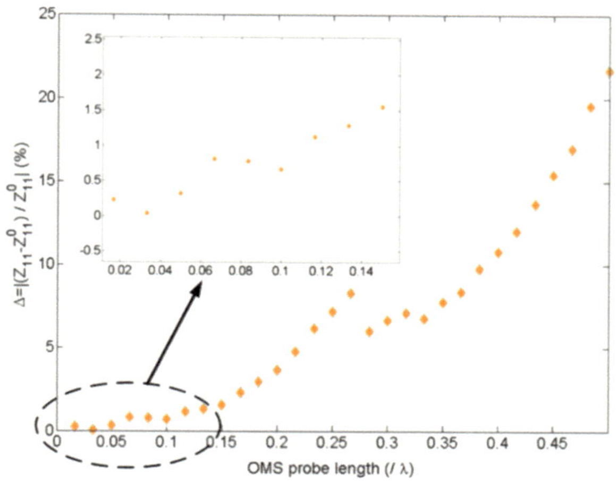

Fig. 4. AUT impedance variation due to the probe structural modes, as a function of the probe length.

investigate whether the chosen length (i.e., $\lambda/12$) for the OMS probe fulfills the requirements of the MSA antenna, we performed a simulation in Ansoft HFSS, a 3D full wave finite element solver, wherein, a planar dipole with a length of $L = 10mm$, width of $w = 1mm$ and a center gap of $g = 100\mu m$ was considered. The dipole was positioned in front of the aperture of a horn antenna operating at a test frequency of 2.45 GHz. Then, the value of $\Delta = \frac{Z_{11} - Z_{11}^0}{Z_{11}^0}$ versus the length for probe was calculated. The results plotted in Fig. 4 show that Δ varies by less than 1.5% for probes shorter than 0.15λ. Therefore, an OMS probe consisting of a short dipole with length of $\lambda/12$ can be considered as a good MSA when it is used to characterize this horn antenna.

2.4 Tuning network design

As shown in (King, 1978), scattering by the probe can be increased by adding an inductive reactance in series with the capacitive short-dipole (i.e., $Z_p = Z_{dipole} + j\omega L$) so that a resonance occurs in one of the two states. The inductance value should be chosen such that the numerator or the denominator in Equation 3 is minimized, leading to an increased modulation index. This effect, however, is frequency selective.

The value of the inductance should make the loaded short dipole resonant when the light is ON (denominator of Equation 3 minimized) and increase its impedance when the light is OFF (or vice versa). To find the optimum inductance value, one may try to maximize CR. Fig. 5 represents CR versus inductance. The inductance of 25 nH associated with the peak in the curve is referred to as the optimal point of the tuning network and it can be seen that the maximum CR is close to the estimated value 22.5 dB calculated in Section 2.2. The minimum of CR near $L = 42nH$ also leads to a local maximum of $|m|$ but it is not as high.

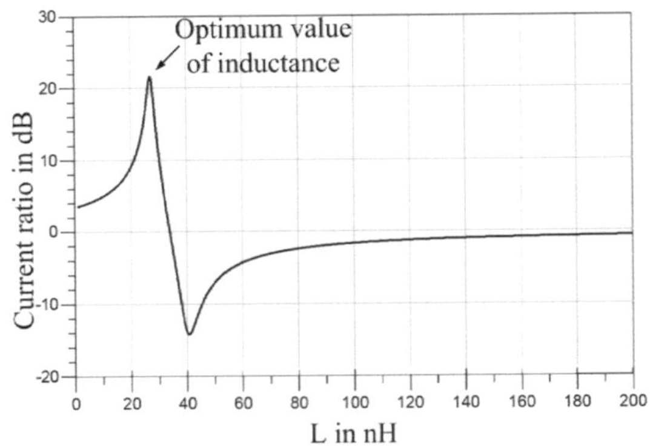

Fig. 5. Current ratio versus the inductance value used for tuning.

3. Matching network impact on the OMS probe performance

The impact of the tuning network on the probe performance is presented here. The difference between the scattered field when the dipole is in ON and OFF states (i.e. $Z_{OFF} = 38.8 - j206.2\Omega$ and $Z_{ON} = 15.9 - j4.8\Omega$) at 2.45 GHz was calculated versus frequency for two cases: with and without considering a tuning network in an OMS probe structure. To do this, a method of moment code was developed to calculate the ON and OFF states scattered field in the 1-4 GHz frequency range.

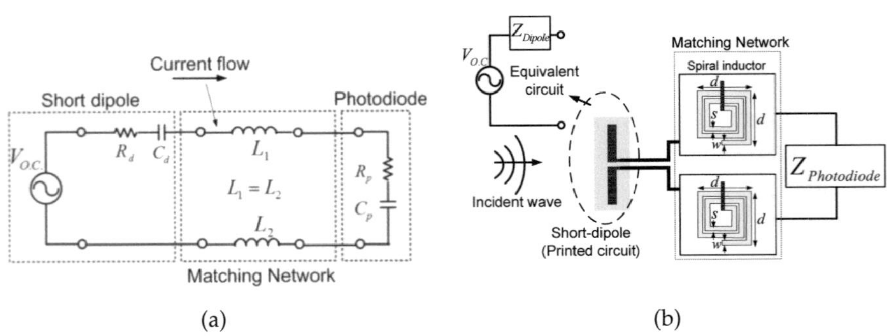

Fig. 6. (a) Schematic depicting the equivalent circuit of the OMS probe, wherein $R_d = 1.22\ \Omega$, $C_d = 0.15\ pF$, $R_p(ON) = 15.85\ \Omega$, $C_p(ON) = 13.65\ pF$, $R_p(OFF) = 38.78\ \Omega$, $C_p(OFF) = 0.31\ pF$ and $L1 = L2 = 12.7\ nH$, and (b) Matching network for the proposed OMS probe (d=0.99 mm, s=63.5 μm and w=50.8 μm). Dipole length: 1 cm. Drawing is not to scale.

In this model (see Fig. 6a), the scattered field was calculated 1 cm away from the dipole when a uniform plane wave illumination is considered. The results shown in Fig. 7 exhibits a significant improvement of about 23 dB in scattered field when the tuning network is added. As a consequence, the sensitivity of the OMS probe is significantly improved. The two peaks on the solid curve correspond to resonances that occur in the ON and OFF states of the OMS probe.

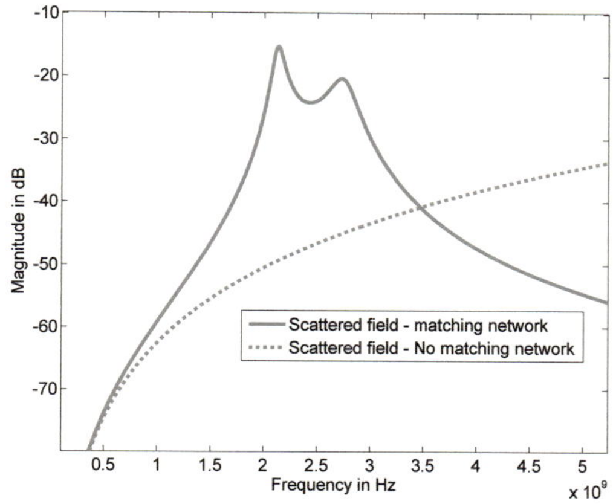

Fig. 7. Frequency response of an OMS probe: Solid line probe with tuning network and dashed line probe without tuning network.

4. OMS probe fabrication

The OMS probe was fabricated on a thin ceramic substrate (alumina) with a thickness of 250 μm, a relative permittivity of 10.2 and $tan\delta = 0.004$. An optical fiber is coupled to the active surface of the photodiode using a precision positioning system by monitoring photo-induced DC current while the fiber is moved to find the optimal position. Finally, the fiber is permanently fixed by pouring epoxy glue when in the position corresponding to the current peak. In addition, in order to prevent any damage to the coupling by mishandling the probe, a strain relief structure made of a low permittivity material ($\epsilon_r \approx 2.7$) is added. Fig. 8 shows the photograph of the completed probe assembly. The dimensions of the ceramic substrate are 7 mm and 15 mm. The tuning element is implemented with two spiral inductors (see Fig. 6b). Each inductor occupies an area of 1mm×1mm. The photodiode area is $0.2mm^2$. Wire-bonding provides the electrical contacts between the photodiode and the inductor terminals on the substrate.

(a) 3D view (b) Top view

Fig. 8. Photograph of the implemented OMS probe.

5. Validating the fabrication process

Once the OMS probe is fabricated, including fiber coupling, it is necessary to verify whether it operates at the frequency at which it was designed. As the photodiode saturates at an input power of +6 dBm (see Fig. 2), no further modulation index change is anticipated beyond this point.
The OMS probe was tested by exposing it to a constant power electric field (e.g., near a horn antenna or microstrip transmission line) at 2.45 GHz. An optical signal (waveguide of 1.3 μm) modulated at $\sim 100KHz$ with a power between -10 dBm to 13 dBm was applied to the OMS probe. The sidebands were recorded during this measurement at the input port of the horn using a spectrum analyzer. Fig. 9 illustrates the results obtained by this experiment. It can be seen that the level of the sidebands (normalized to its maximum) increases linearly with the optical power when it is smaller than +6 dBm. As expected, beyond this limit the probe is not able to scatter more fields. This test not only confirms that the probe is operating at a desired working point but it also shows the quality of the fiber/photodiode coupling.

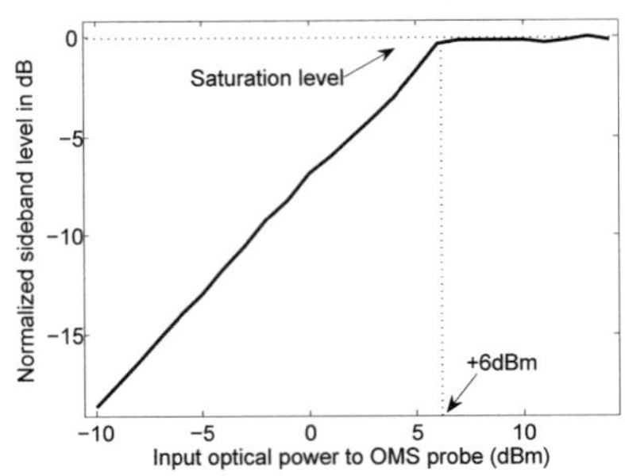

Fig. 9. Variation of sideband power level (dB) versus input optical power (dBm) to the OMS probe.

6. Omnidirectional and cross-polarization characterization

6.1 Omnidirectional response

A desirable feature for a near-field probe is to be able to measure a specific component of the E or H field. In the case of a short dipole it is the component of the E field parallel to the dipole axis, independently from the direction of arrival of the incoming wave(s). For a thin-wire dipole, rotational symmetry of the response about the dipole axis is expected. In practice the presence of a substrate, the flat strip geometry of the dipole and the presence of the dielectric support structure break the symmetry. A detailed model of the probe including these elements was simulated with Ansoft-HFSS as shown in Fig. 10a. In these simulations, the probe is on the z-axis and centered at the origin. A near-field plot of E_z (Co-pol.) and E_ϕ (Cross-pol.) on a 36 mm circle and in plane $z = 0$ are shown in Fig. 10b. The probe operates as a transmit antenna but the response in the receive mode is the same due to reciprocity. The results show a fluctuation of less than 0.45 dB in the desired E_z component, and very low level of cross-polarization.
Rotational symmetry of the response was also studied experimentally with the setup shown in Fig. 11a. In this case, the probe operates in the receiving mode and it is located near the

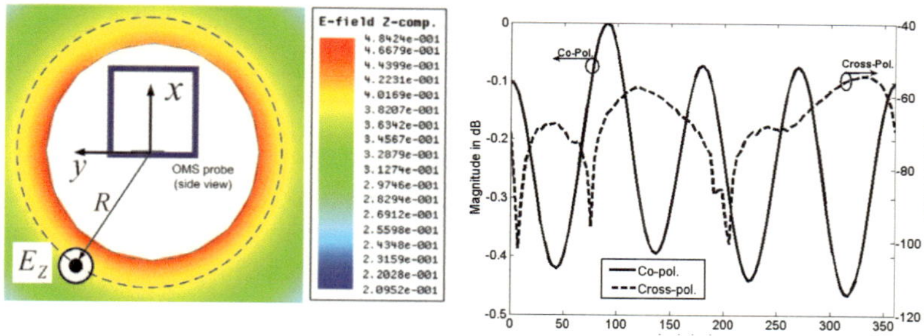

(a) Magnitude (b) Phase

Fig. 10. Schematic of the OMS probe when investigated for omnidirectivity characteristic. Co-polarized (E_z solid line) and Cross-polarization (E_ϕ dashed line) radiation of the OMS probe in the H-plane at a distance of 36 mm from the probe axis, as predicted by HFSS (the data is normalized with respect to the maximum value of E_z).

aperture of a transmitting horn antenna. The experiment was done by rotating the OMS probe about its axis while recording the power levels of the sidebands on a spectrum analyzer. The measured pattern at a distance of 12.2 cm (one free-space wavelength) shown in Fig. 11b exhibits a fluctuation of about 0.6 dB. The figure also shows simulation results obtained with HFSS. In this case, the magnitude of the difference between the horn's S_{11} parameter, in the absence and the presence of the rotated probe, is plotted. The experimental and simulated curves were normalized to make the comparison easier. In the simulation results, the effect of the dielectric substrate and support structure is barely perceptible. On the contrary, the experimental curve does not exhibit such a good rotational symmetry, as a difference of 0.6 dB can be observed between the maximum and minimum values. It is believed that this fluctuation may be due to mutual interactions between the probe rotation fixture and the horn antenna, which were not taken into account in the simulations.

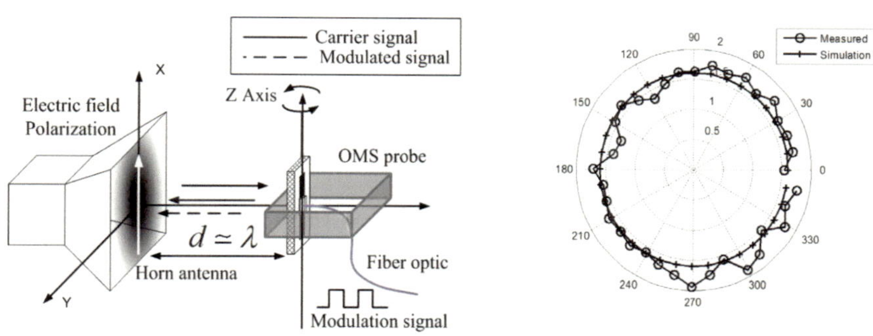

(a) (b)

Fig. 11. The setup for testing the omnidirectional performance of an OMS probe (a). Measured radiation pattern in the probe H-plane at a distance of one wavelength from the illuminating waveguide (magnitude in dB) (b).

6.2 Cross polarization

According to Fig. 10a, the cross-polarization of the OMS probe is give by Equation 10.

$$E_{\phi} = E_{cross-pol.} = -E_x \sin(\phi) + E_y \cos(\phi) \tag{10}$$

HFSS simulations predicts a cross-polarization rejection of more than 55 dB for the OMS probe. To verify this result experimentally, the coupling between two identical open-ended WR-284 rectangular waveguides that face each other (Fig. 12) was measured. Although rectangular waveguides already have very good on-axis cross-polarization rejection, it was further improved by inserting a grid of parallel metal-strips (3 strips per cm) printed on a thin polyimide substrate (thickness of 5 mil and a relative permittivity of 3.2). These polarizers were mounted on the apertures of the transmit and receive waveguides. The strips, illustrated on the Tx waveguide in Fig. 12, are oriented perpendicular to the radiated field. The Tx waveguide did not show significant change of the return-loss after adding the polarizers. In the experiment, the apertures were aligned and set one wavelength apart from each other. Then, the OMS probe was mounted on a fixture made of foam transparent to microwaves ($\epsilon_r \approx 1$) and was inserted between the aperture of the waveguide as illustrated in Fig. 12. The setup operated in a bistatic mode, i.e. the sidebands generated by the OMS probe were measured at the output port of the receive waveguide. Measurements were made with the receive waveguide rotated about its axis by 0 and 90 degrees; the level of the sidebands introduced by the probe changed by 60.55 dB. This should be considered as a lower bound on the probe-induced cross-polarization, as the cross-polarization rejection of the polarizers is not infinite in practice.

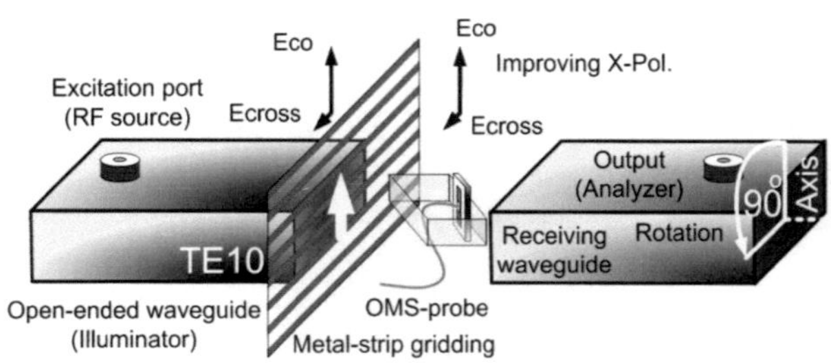

Fig. 12. Setup to measure co-to-cross polarization (E_{ϕ}) rejection of the OMS probe (only one of the polarizer sheets is shown for clarity).

7. OMS probe frequency response

The frequency response of the OMS probe was assessed by using it in a monostatic scheme. The probe was inserted in a rectangular WR-284 rectangular waveguide and aligned with the main component of the E-field. With the photodiode in the OFF state, the waveguide was connected to a calibrated vector network analyzer through a 3-stub tuner that was adjusted to give the minimum possible reflection coefficient (less than -65 dB) over the tested frequency band. Then, an optical power level of +6 dBm was applied to drive the photodiode in the ON state. The difference between the complex reflection coefficient at the tuner's input port in both states was then normalized to have the maximum at 0 dB. The results displayed in Fig. 13 show two peaks. It is believed that they are due to the different resonance frequencies of $Z_p +$

Z_{ON} and $Z_p + Z_{OFF}$ (see Fig.7). In fact, if a simple capacitor model is assumed for the short dipole in free-space, resonance frequencies of 2.53 GHz and 3.09 GHz can be calculated for the ON and OFF states respectively. Results from a simulation done with the thin-wire method of moment are also shown in the figure. In the simulation, the probe is illuminated with a uniform plane wave in free space. This shows the normalized difference between the squared scattered field taken 1 cm away from the dipole in the two states. The results also exhibit a double peak response. In the measurement, the resonance observed in the waveguide are shifted to lower frequencies. This shift is thought to be due to imperfections in the construction and uncertainty in the substrate's constitutive material parameters. Furthermore, the value of Z_p in free-space is not the same as in the waveguide where the dipole is interacting with the metallic walls. Finally, as these reflection differences are obtained by subtracting very similar measured values, the results are susceptible to measurement and simulation inaccuracies. Both curves exhibits a maximum sensitivity near the design frequency of 2.45 GHz. Finally, the waveguide measurement process described above was simulated in HFSS. The reflection coefficient difference shown in Fig. 13 exhibits peaks near 2.7 GHz and 3 GHz. It should be noted that this curve is derived from differences between S_{11} values with a variation smaller than 5×10^{-5} in magnitude. Therefore, the frequency shift compared to the other two curves may be partly due to simulation inaccuracies.

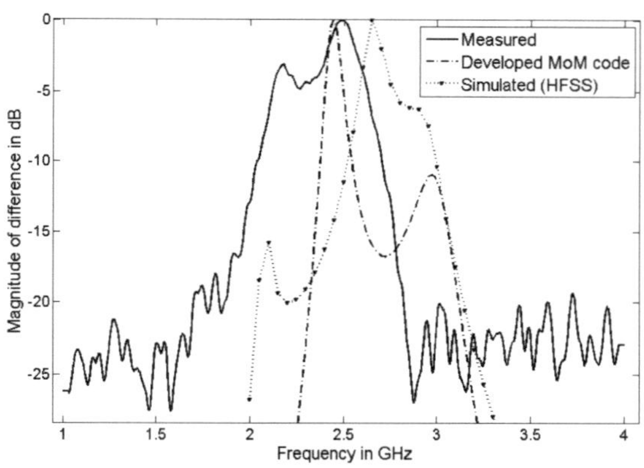

Fig. 13. Difference of frequency response for the OMS probe in ON and OFF states: solid line is the measured reflection coefficient; dashed line is the simulated scattered field; dotted line is simulated reflection coefficient.

8. OMS probe performance validation

In order to verify the performance of the developed OMS probe, it was set to measure the electric field distribution near a 50-Ω microstrip transmission line. The test was made in a monostatic setup, where the measured signal is proportional to the square of the complex electric field ($v \propto E^2$). The transmission line was fabricated on a Rogers substrate (RO3035) with a relative permittivity of 3.8 and a thickness of 60 mils (Fig. 14).

The rapidly varying fields near the line are highly suitable to assess the resolution and the dynamic range of the measurement system. In this measurement, the probe is scanned across the microstrip line at a height of 3 mm above it, and measures the transverse electric field distribution along x (i.e., E_x) (Fig. 14). The transmission line was terminated with a matched load.

To validate the measurement results, we also included the field distribution of the transmission line predicted by HFSS (Fig. 15).The results obtained from simulation need to be post-processed to take into account the effect the finite length of the measuring probe. This topic will be discussed in Section 9.

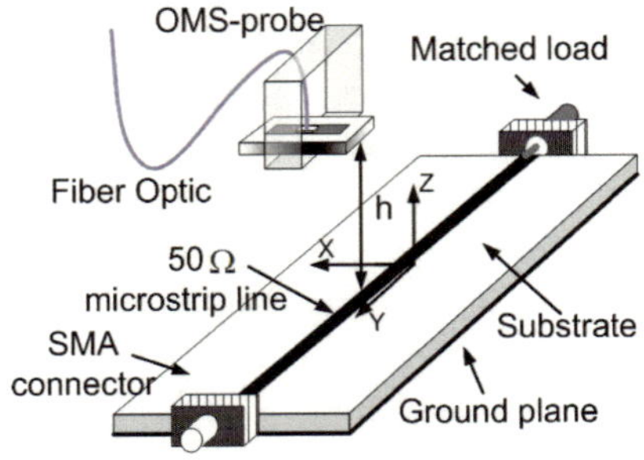

Fig. 14. Schematic of the probe and microstrip transmission line under test.

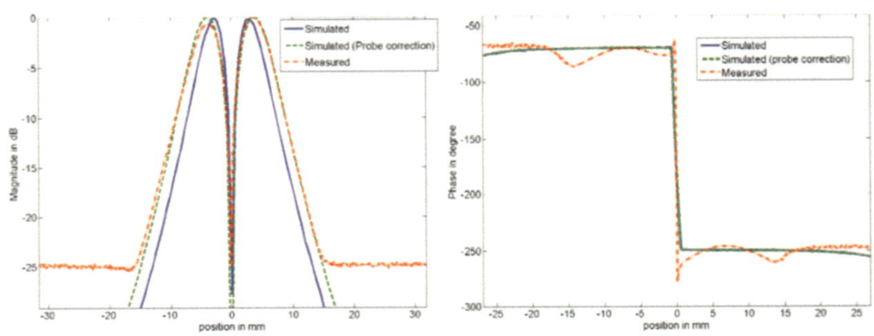

(a) Magnitude (b) Phase

Fig. 15. Measurement results and simulations (magnitude and phase) of electric field (E_x) at h =3mm.

8.1 Taking the square root: sign ambiguity removal

When the NF imager operates in monostatic mode, the measured fields are obtained by taking the square root of the measured data. The square root of a complex signal $v = X_I + jX_Q$ has two solutions and it is necessary to select the proper one. The procedure might be straightforward when the measured field takes nonzero values. In this case it is possible to ensure continuity of the phase distribution in the whole data set. In contrast, sign retrieval is not an easy task if nulls occur (i.e., $E = 0$) at some locations. In these cases, no clear method has been addressed to choose the sign of the square root correctly. However, a technique was reported in (Hygate & Nye, 1990) for some particular cases.

In the case of the microstrip line considered here it is well known that transverse electric field (E_x in Fig. 14) has a null on the strip's symmetry plane and a different sign on both sides. Thus, even if choosing the sign of the electric field on either side is impossible without *a priori* knowledge. It is assumed that when a contour with zero E field is crossed, the phase changes by π.

9. Probe correction

The short dipole implementing the probe has a finite length. Therefore, the measured data is not representative of the fields at a point but rather of the integral of the weighted field along the probe. To take this effect into account, we used the induced *e.m.f* method for calculating the induced voltage across the probe's terminal generated by an incident E-field (see Fig. 16a). In this method, we need to know the current distribution (J) on the probe when it is radiating, i.e., acting as a transmitting antenna. Since the length of the probe is shorter than 0.1λ, one can assume that J can be approximated by triangular current distribution, as shown in Fig. 16b.

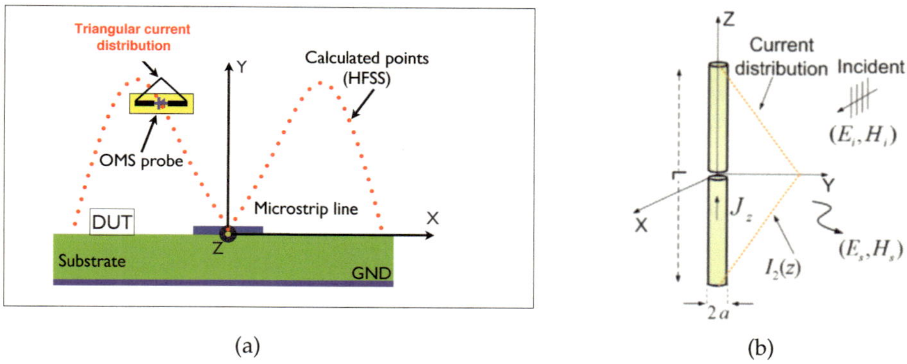

Fig. 16. (a) Schematic showing the effect of a probe length on the field to be measured, and (b) Geometry for calculating the induced current on the OMS probe.

$$J_{probe} = J_{probe}(0)\left[1 - \frac{2|z|}{L}\right]u_z \tag{11}$$

The measured field is given by the field-current convolution for every point using Equation 12 (see Fig. 16b).

$$V_{probe} = -\frac{1}{J_{probe}(0)}\int_L \bar{E}_i . u_z J(z) dl \tag{12}$$

This equation was used to process the field calculated by HFSS in Fig. 15. The simulations, after applying convolution, probe correction, are in very good agreement with the measurements, both in magnitude and in phase plots, which proves the excellent performance of the probe (see Fig. 15). Within the ± 15 mm interval, the average difference between the simulated (with probe correction) and measured fields was 6.4% in magnitude and 3.2 degrees in phase. It is worth mentioning that the probe correction does not alter phase information in this example due to uniformity of the phase on both sides of the x=0 plane.

10. Sensitivity

The sensitivity of the measurement system is not only dependent on the modulation index of the loaded probe but also on the sensitivity and noise floor of the receiving equipment measuring the sideband signal. In the monostatic configuration, the magnitude of this signal is proportional to $\Delta\rho$, the difference in the AUT reflection coefficient in the photodiode's ON and OFF states. It can be proven that for a monostatic test configuration this difference is proportional to S_{21}^2 (Equation 13), where S_{21} represents coupling between the AUT and probe ports (Fig. 17) (Bolomey & Gardiol, 2001).

$$\Delta\rho \propto S_{21}^2 \rightarrow \Delta_\rho = K_1 S_{21}^2 \tag{13}$$

Furthermore, the field incident on the probe is proportional to S_{21}, i.e.,

$$E \propto S_{21} \rightarrow E = K_2 S_{21} \tag{14}$$

Using Equations 13 and 14, we can obtain:

$$E = K\frac{\Delta\rho}{S_{21}} \tag{15}$$

The sensitivity of the system to electric field can be given in terms of the minimum possible reflection coefficient that can be accurately measured, namely $\Delta\rho_{min}$. Consequently, the sensitivity of the system is simply given by

$$E_{min} = |K\frac{\Delta\rho_{min}}{S_{21}}| \tag{16}$$

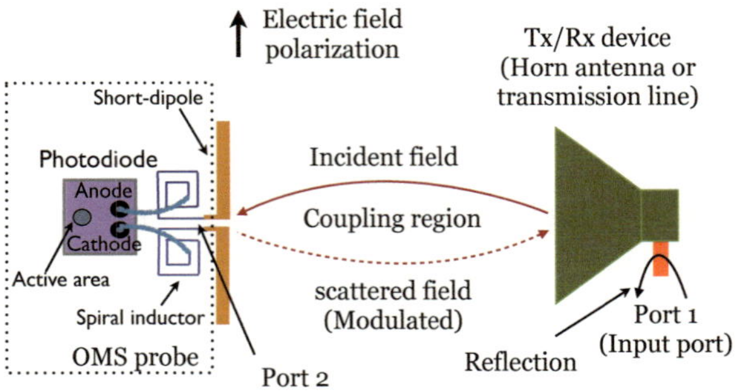

Fig. 17. Drawing of the setup used to measure sensitivity of the OMS probe.

where $K = K_2/K_1$ is a constant. The field sensitivity will therefore depend on the AUT. For a radiating structure, we expect a higher value of S_{21} and therefore a better sensitivity, than for a guiding structure such as a microstrip line. To illustrate this, we have estimated the sensitivity for two structures: a horn antenna operating at 2.45 GHz and the microstrip line terminated with a matched load. By simulation, we obtained the field incident on the probe for an incident input power of 1 watt at the DUT's input port. The same configuration was then repeated experimentally, that is to say with the probe located at the same point as in the simulations. With the probe in this fixed position to keep S_{21} constant, the incident power was reduced with an attenuator until the receiver's noise floor was reached. The field sensitivity was then calculated by scaling the E-field value obtained in simulations by the square root of the threshold power level (in watts) measured experimentally. In the case where the probe was in the aperture of the horn (large S_{21}), the sensitivity was 0.037 V/m. When it was at a height of 3 mm above the microstrip line (small S_{21}), the sensitivity degraded to reach 54.3 V/m. This large difference illustrates a weakness of the monostatic configuration for characterizing non-radiating structures.

11. Optically modulated scatterer (OMS) probes array: Improving measurement speed in a NF imager

A linear array of seven OMS probes was developed in order to improve the measurement speed of the NF imager. In the array, the probes are laid in parallel along a line perpendicular to the probes' axes (see Fig. 18). The probes were mounted on a piece of planar foam ($\varepsilon_r \approx 1$) with a spacing of 3 cm between the probes. The foam has a thickness of 1.2 cm and is very rigid. It also prevents the array from vibrating when a very fast measurement is made. The array is moved mechanically along one direction, while being moved electronically (as well as mechanically if finer measurement resolution is required) in the orthogonal direction so as to scan a $2D$ grid. Thus, this arrangement divides by seven the number of mechanical movements in only one direction. It is shown in (Cown & Ryan, 1989) that not only the probe translations by the positioning system but also the switching time between the probes remarkably slow down scanning of the NF imager. Thus, to achieve faster measurements it is necessary to pay attention to both aspects simultaneously.

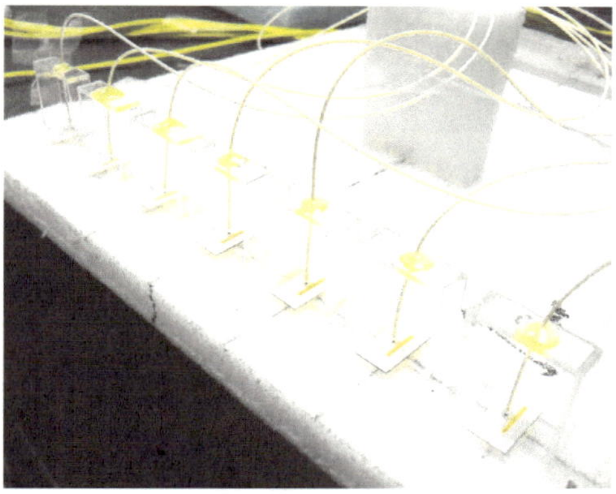

Fig. 18. Photography of the developed array of seven OMS probes.

11.1 Laser diodes array: custom-designed optical switch

In practice, it is necessary to use an optical switch in order to send a modulation signal to the designated probe in the array.

To this end, an array of controlled laser diodes (see Fig. 19) was designed and developed. Each laser diode is individually connected to a probe. A digital controller was also implemented to provide proper signaling to the probe. The controller produces a reference signal used by the lock-in amplifier (LIA). The stability of this reference prevents phase jitter in the measured data.

The electronically switched feature of the array not only increases the measurement speed but also eliminates cross-talk between the outputs, which was observed with a mechanical optical switch. As a result, we obtained a 14-times improvement in the measurement time compared to the setup reported in (Tehran et al., 2009).

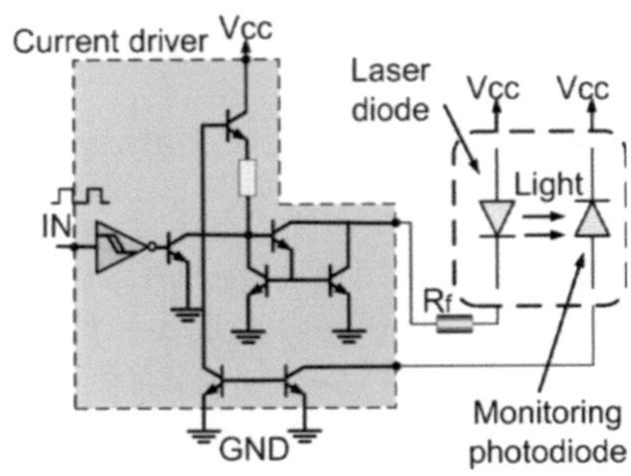

Fig. 19. Schematic showing a laser diode and its driver.

11.2 The developed NF imager equipped with array of OMS probes

Fig. 20 demonstrates the NF imager incorporating all of its essential parts namely, microwave electronic, and optical circuitries necessary to transmit/receive and process the scattered fields by an OMS probe in the NF imager. The microwave part consists of an RF source, an active circuit equivalent to a conventional I-Q demodulator and a carrier canceller circuit. Base-band analog and digital parts include a lock-in amplifier (LIA), model SR830 manufactured by Stanford Research Systems, which provides signal vector measurement (magnitude and phase), a current driver exciting and controlling a laser diode, and a digital controller that generates the reference signal required by an LIA and also that addresses the RF SPDT switch. This controller also sends commands to the laser diodes modulating the OMS probes. The whole setup is controlled by a computer software developed using LabView.

12. Validating the NF imager

12.1 Array calibration

It is practically impossible to make a set of identical OMS probes. Differences in the responses of the probes can be caused by differences in the photodiode characteristics, materials used, optical fiber/photodiode coupling quality and many other factors (Mostafavi et al., 2005). In order to quantify these differences in the probes, we performed a simple monostatic field probing experiment in which the seven probes are set to measure the E field at the same fixed point near a DUT. The obtained results are then used to compute a complex correction factor (CF) corresponding to each probe using Equation 17.

$$CF = \frac{E_{ref}}{E_{Probe\#i}}; i = 1,7 \tag{17}$$

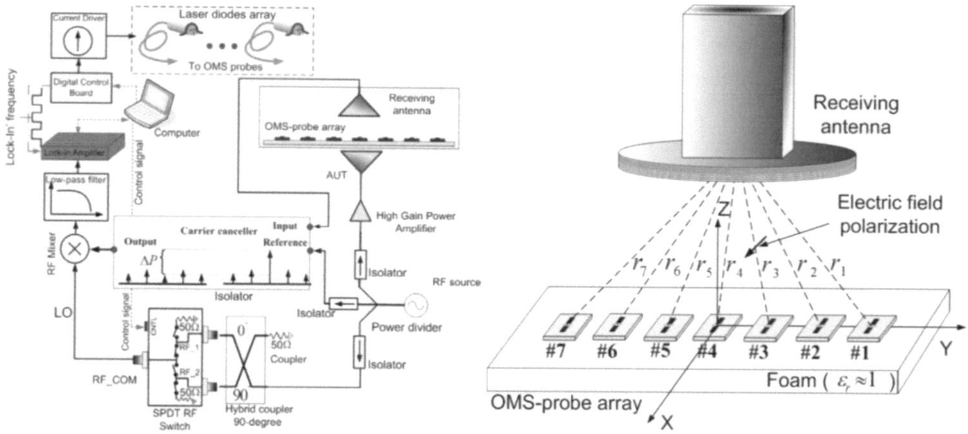

(a) Near-field imager circuitry (b) The imager receiving part

Fig. 20. (a) Near-field imager microwave circuitry configured for bistatic operation, and (b) receiving part of the imager incorporating the OMS probe array and a WR-284 rectangular waveguide.

Probe #	CF	$\lvert CF \rvert$	$\angle CF(deg)$
1	0.8704+j0.0218	0.8706	1.4347
2	0.9645-j0.0806	0.9678	4.7724
3	0.959-j0.0511	0.9603	-3.050
4	1+j0	1	0
5	1.0007-j0.0091	1.0007	-0.5210
6	1.0252+j0.0258	1.0255	1.4415
7	1.0432+j0.0406	1.0439	2.2287

Table 1. The measurement results of a known field using individual probes (all measurements have been normalized to the reading of probe #4).

In this experiment, an antenna with a highly concentrated near-field distribution was used as a DUT. This antenna incorporates a cylindrical waveguide loaded with a dielectric material having a dielectric constant about 15. This dielectric part concentrates the fields over a small area where the probe under test is located, while weakly illuminating the other probes (which are switched OFF). The probes are positioned within the illuminated region near the antenna and the fields in the E-plane of the illuminating antenna are scanned. Ideally, it is expected that the probes will measure the same field distribution. However, due to the factors mentioned earlier they do not. Therefore, as an effective compensation technique, a probe in the array is used as a reference (e.g., probe#4, central) to which the rest of the probes are weighted by a complex number (e.g., correction factor). The correction factors can be obtained for several points and averaged to get a better agreement between the responses of the probes. The computed correction factors based on the method explained here, are listed in Table 1. The effect of applying correction factors on the measurement results will be discussed later.

12.2 Receiving antenna compensation

In the bistatic test setup, the receiving part of the NF imager incorporates an auxiliary antenna (AA) to pick up the scattered fields and send them to the coherent detector, as illustrated in Fig. 20b (see also Fig. 1). During the scan, the AA is moved together with the array and its phase centre has a minimum distance from the central probe (i.e., probe#4). In this configuration, the rest of the probes are placed symmetrically on both sides of probe #4. As can be observed in Fig. 20b, the scattered fields propagate along different paths to reach the AA (i.e., r_i; i=1, 7). Each probe is also seen by the AA with a different view. Then, the picked-up signals will not be identical even if all the probes are exposed to the same fields. So, we need to compensate the measured data (raw data) for the NF radiation pattern of the AA. In principle, the simple compensation method described in the previous subsection should suffice. In practice however it has been observed that the coupling between each probe and the AA slightly varies when the probes are moved near the AUT, even if the AA is maintained at a fixed position with respect to the probe array. This variation comes from mutual interaction of the AA and probe array with the AUT, which is not constant during the scan. A method to compensate for this effect is introduced in the next paragraph. We first set the AA in Fig. 20b

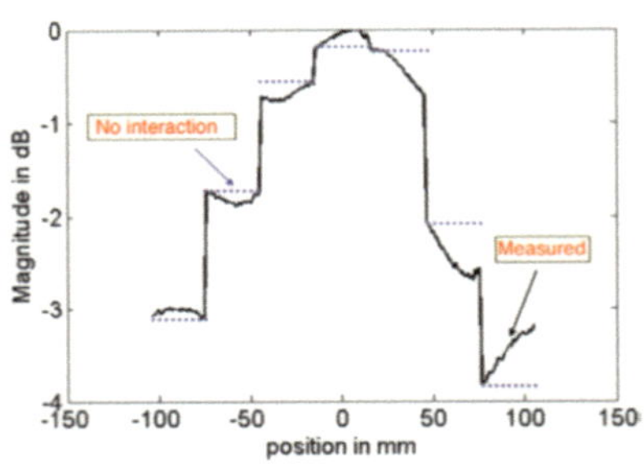

(a) Magnitude

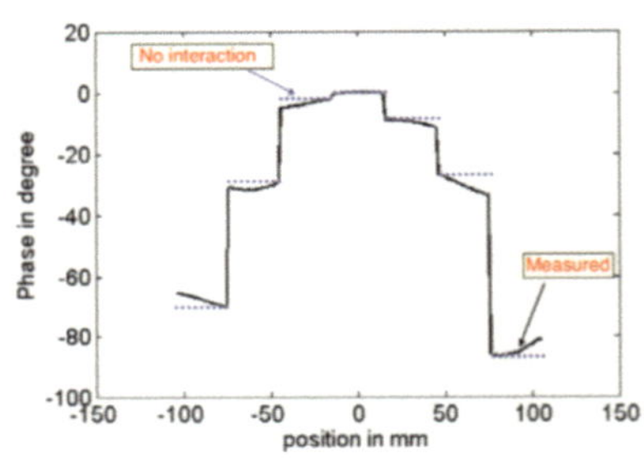

(b) Phase

Fig. 21. The measurement result obtained in the test to compensate for the radiation pattern of the receiving antenna; (a) magnitude and (b) phase of the normalized measured E field by the AA in the monostatic setup.

to operate as an illuminator in a monostatic mode (TX/RX device). During this test the AUT is passive and terminated with a matched load. In this experiment, the probes are addressed successively and then moved to a new position until the array scans the region of interest above the AUT. Ideally, a flat response is expected over the region scanned by each probe, but given the interaction of the array with the surrounding objects, including the passive AUT, and the interaction between probes, the measured results are not constant, as illustrated in Fig. 21. In this test the AUT was a horn antenna and the array was scanned at a height of 30 mm (i.e., $\lambda/4$) above the aperture. The ideal results, i.e., with no interaction between the probes with the AUT and the AA are shown by broken line in Fig. 21. The asymmetry of the curves occurs because of discrepancies in the probes of the array, displacement of probes and misalignment. Even though each probe is at a fixed distance and angle from the receiving antenna, significant variation can be observed when a 30 mm interval is scanned. The results also demonstrate the importance of the compensation before any comparison is made to validate the imager's results. After this test, the E-field measurements of the AA at each position of the array and for each probe (i.e., E_{AA}) are used to correct the NF measurements obtained for the AUT in the bistatic setup, i.e.,:

$$E_{AUT} = \frac{E_{AUT,Bistatic}}{E_{AA}} \quad (18)$$

13. OMS probes array: validation results

The electric field distribution of a planar inverted-F antenna (PIFA, Fig. 22) radiating at 2.45 GHz was measured in bistatic and monostatic modes on a plane located at $\lambda/4$ above the antenna's ground plane (Fankem & Melde, 2008). Such an antenna is commonly used in portable devices (e.g., cellphone) and communication systems.

Fig. 23 shows 2D measurement of the AUT E-field distribution after compensation for probes' differences, receiving antenna radiation pattern and variations of interactions with the AUT. The magnitude plot shows a dynamic range of ~25 dB over a scan area of 240 mm by 210 mm. Fig. 24 shows E- and H-planes NF cuts of the PIFA, including the measured magnitude and phase. For validating the results obtained by the imager, all measurements are compared with simulations and also to the field distribution obtained by the imager operating in the monostatic mode. All curves (i.e., magnitude of E-field) are in good agreement with each other except that of the monostatic measurement, which deviates from the true field starting from -20 mm toward negative x values.

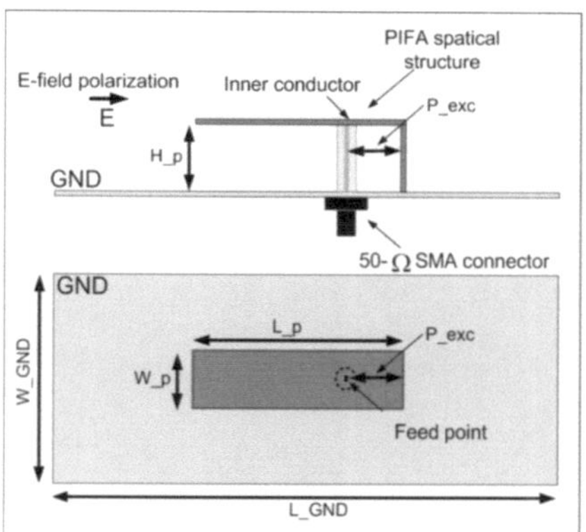

Fig. 22. Antenna under test (AUT). PIFA antenna operating at 2.45 GHz with measured return-loss of about 12 dB; the physical dimensions of the PIFA are as follows: L_p=27 mm, W_p=13 mm, H_p=7 mm, P_{exc}=7 mm, W_{GND}=70 mm and L_{GND}=137 mm.

In all cases the measured phase information in the E-plane of the PIFA are in good agreement over the whole x interval. In order to quantify the difference between the measurement results and the simulated distribution of the PIFA, the mean square error of the data was calculated. The error associated with E-plane and H-plane cuts are 0.12% and 0.06%, respectively, with respect to simulations. The benefit of probe correction in the bistatic case is clearly visible in the H-plane results.

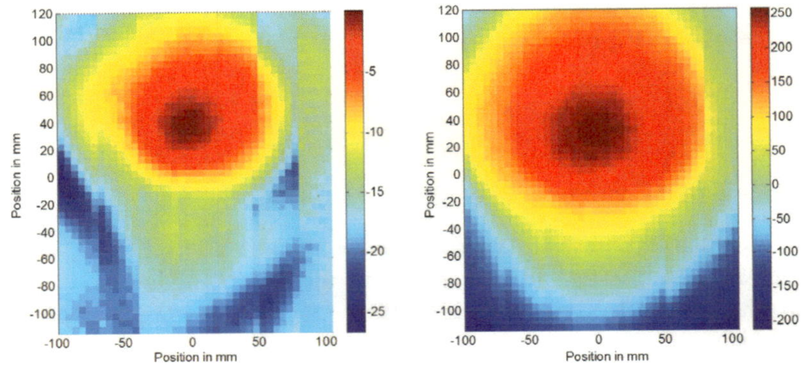

(a) Magnitude (b) Phase

Fig. 23. 2-D map of electric field distribution measured (compensated results) at a distance of $\lambda/4$ above AUT; (a) magnitude (dB) and (b) phase (deg.).

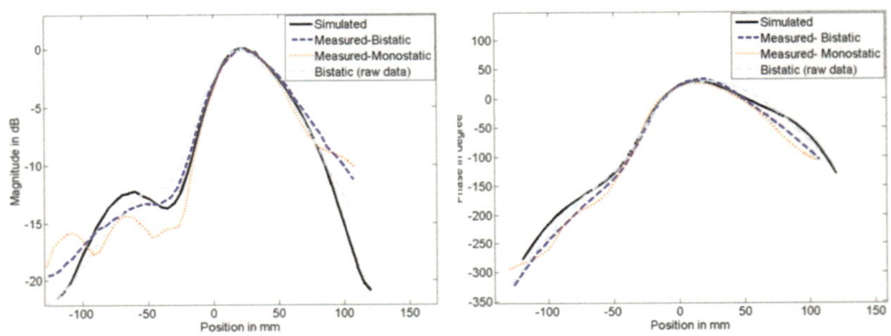

(a) E-plane (magnitude) (b) E-plane (phase)

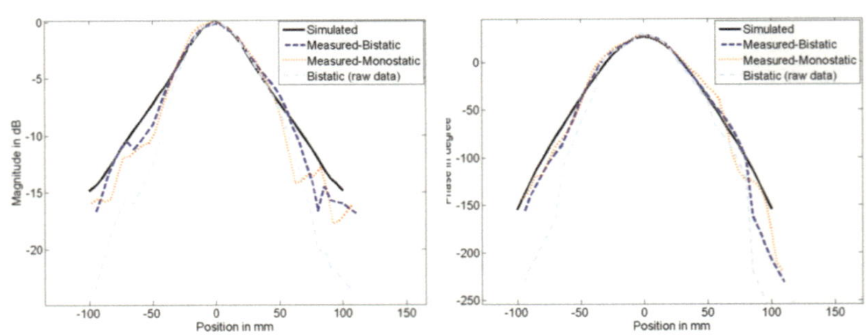

(c) H-plane (magnitude) (d) H-plane (phase)

Fig. 24. E- and H-plane cuts of the measured E-field at distance of $\lambda/4$ from the PIFA antenna's ground plane; (a) and (c) magnitude (dB), and (b) and (d) phase (deg.).

14. Carrier cancellation: NF imager dynamic range and linearity improvement

In an MST-based NF imager the received signals (modulated) consist of a carrier and sidebands. Although the probe reflects the field at the carrier frequency, this does not affect the measurements, which are performed at the sideband frequencies. The carrier is generally stronger (e.g., $\sim 50dB$) than the sidebands in the modulated signal. Nonlinear behaviour such as saturation and compression in the receiver occurs at higher powers of the carrier, particularly beyond -3 dBm.

To overcome these problems, the NF imager can be equipped with an automated circuit, so-called the carrier canceller, in order to eliminate the carrier from the received signals and leave the sidebands intact. This allows us to amplify the signal applied to AUTs, leading to a higher sideband level. As a consequence, the signal-to-noise (SNR) ratio of the modulated signals at the coherent detector's front-end improves considerably, resulting in larger dynamic range. The carrier at the receiving port is minimized by combining the received signal with a 180-degree out-of-phase tone at the carrier frequency whose magnitude and phase are adjusted adaptively.

14.1 High-dynamic range NF imager: Example of application

In order to demonstrate the high dynamic range of the NF imager when it is equipped with the carrier cancellation circuit, the E-field distribution above a bandpass filter operating at 2.45 GHz was measured. The measurement was performed 3 mm above the traces at 2.45 GHz. Figure 25a shows photograph of the filter under test. The results shown in Figure 25b-c include both magnitude and phase of E_x. It can be seen, that the power propagate through the filter to reach the matched load. The achieved dynamic range of the near-field map is approximately 80 dB.

15. Conclusions

This chapter addressed and discussed the design and implementation of a NF imager based on the modulated scatterer technique (MST) in detail. The imager consists of several optically modulated scatterer (OMS) probes that are very accurate, highly sensitive and also frequency selective. Each OMS probe was optimized to operate at 2.45 GHz. This probe, guarantees almost perturbation-free measurements. It can be implemented with low-cost commercial off-the-shelf photodiodes. The OMS probes were also studied and verified for omnidirectionality and cross-polarization rejection performance. The probes showed an absolute deviation of about ± 0.3 dB with respect to an omnidirectional response. The co-to-cross polarization ratio was measured and found to be better than 60 dB. The frequency response of the probe was studied theoretically and experimentally in order to qualify the performance of the matching network and to assess its impact on the frequency response of the OMS probes. The performance of the probes was validated by measuring the NF distribution of a 50 Ω microstrip transmission line. The measurements were compared with results of simulations using HFSS. The results also showed that the sensitivity of the measurement system can be as low as 0.037 V/m. Error associated with magnitude and phase measurement of respectively 6.4% and 3.2 degrees were observed. By developing a linear array of OMS probes, the measurement speed for an E-field measurement was increased more than 14 times compared to that obtained with a commercially available opto-mechanically switched systems. To improve the accuracy of measurements using the array, the raw measurement data were corrected using the proposed calibration technique, to compensate for uncertainties in the probes' responses. The E-field measurements made with the developed imager were in good agreement with the simulations and were very rapid. Benefiting from carrier cancellation, the isolation between the input and output ports of the imager was improved by about 60 dB. This enabled us to increase the signal fed to the NF imager and reach an overall dynamic range of 85 dB, i.e., an increase of 18 dB compared to the case when the NF imager is not equipped with the carrier cancellation circuit.

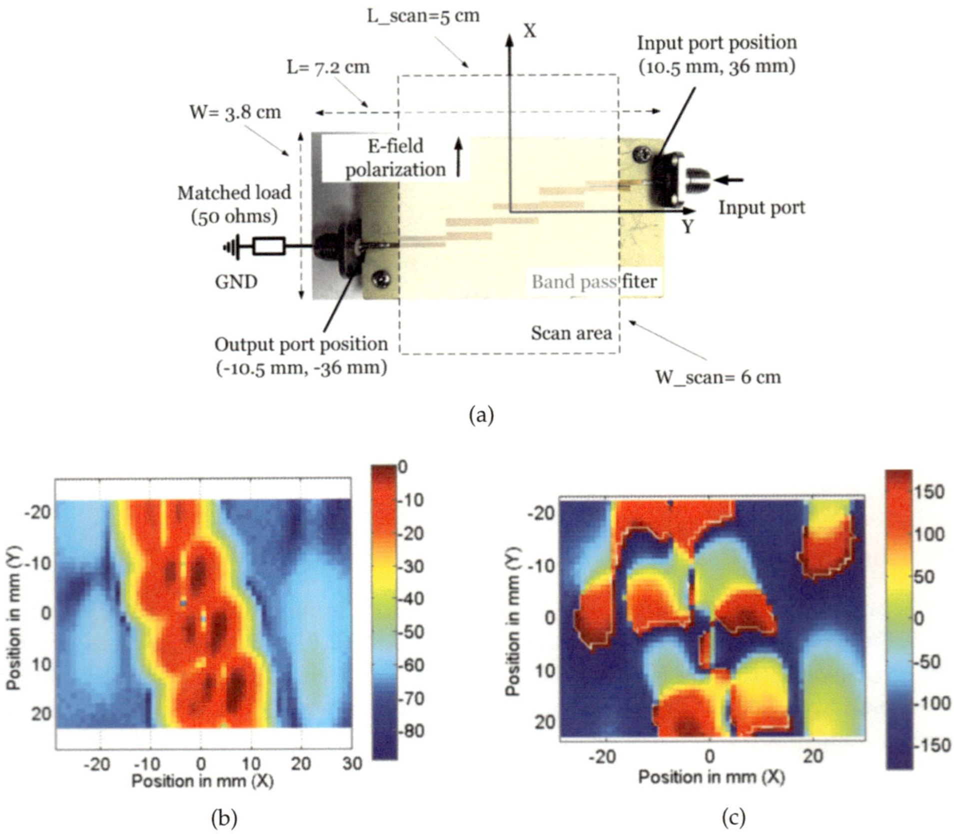

Fig. 25. The measurement results of the transverse E-field above a bandpass filter at 2.45 GHz; (a) magnitude and (b) phase.

16. References

Bassen, H. & Smith, G. (1983). Electric field probes–a review, *Antennas and Propagation, IEEE Transactions on* 31(5): 710–718.

Baudry, D., Arcambal, C., Louis, A., Mazari, B. & Eudeline, P. (2007). Applications of the near-field techniques in emc investigations, *Electromagnetic Compatibility, IEEE Transactions on* 49(3): 485–493.

Bokhari, S., Zurcher, J.-F., Mosig, J. & Gardiol, F. (1995). Near fields of microstrip antennas, *Antennas and Propagation, IEEE Transactions on* 43(2): 188–197.

Bolomey, J. & Gardiol, F. (2001). *Engineering applications of the modulated scatterer technique*, Artech House.

Budka, T., Waclawik, S. & Rebeiz, G. (1996). A coaxial 0.5-18 ghz near electric field measurement system for planar microwave circuits using integrated probes, *Microwave Theory and Techniques, IEEE Transactions on* 44(12): 2174–2184.

Cown, B. & Ryan, C.E., J. (1989). Near-field scattering measurements for determining complex target rcs, *Antennas and Propagation, IEEE Transactions on* **37**(5): 576–585.

Dubois, T., Jarrix, S., Penarier, A., Nouvel, P., Gasquet, D., Chusseau, L. & Azais, B. (2008). Near-field electromagnetic characterization and perturbation of logic circuits, *Instrumentation and Measurement, IEEE Transactions on* **57**(11): 2398–2404.

Fankem, B. & Melde, K. (2008). Nested pifas for dual mode of operation: Gps and global communications, *Antennas and Wireless Propagation Letters, IEEE* **7**: 701–705.

Hygate, G. (1990). Measuring microwave fields with a minimum of perturbation, *Precision Electromagnetic Measurements, 1990. CPEM '90 Digest., Conference on*, pp. 198–199.

Hygate, G. & Nye, J. F. (1990). Measuring microwave fields directly with an optically modulated scatterer, *Measurement Science and Technology* 1(8): 703–709.

Iigusa, K., Sawaya, T., Taromaru, M., Ohira, T. & Komiyama, B. (2006). Experimental proof of electrically invisible state of inductively loaded dipole and proposal of electrically invisible meander-lines, *Antennas and Propagation, IEEE Transactions on* 54(11): 3374–3382.

Justice, R. & Rumsey, V. (1955). Measurement of electric field distributions, *Antennas and Propagation, IRE Transactions on* 3(4): 177–180.

King, R. J. (1978). *Microwave Homodyne System*, Wiley New York.

Laurin, J.-J., Zurcher, J.-F. & Gardiol, F. (2001). Near-field diagnostics of small printed antennas using the equivalent magnetic current approach, *Antennas and Propagation, IEEE Transactions on* 49(5): 814–828.

Liang, W., Hygate, G., Nye, J., Gentle, D. & Cook, R. (1997). A probe for making near-field measurements with minimal disturbance: the optically modulated scatterer, *Antennas and Propagation, IEEE Transactions on* 45(5): 772–780.

Mostafavi, M., Bolomey, J.-C. & Picard, D. (2005). Experimental study on compensation of array element pattern of collinear dipole array sensor, *EICE Trans Commun, Inst Electron Inf Commun Eng.* E88-B(8): 3314–3316.

Munoz, K., Perrey, A. & Zoughi, R. (2008). Potential application of the modulated scatterer technique to multilayered material evaluation and health monitoring, *Instrumentation and Measurement Technology Conference Proceedings, 2008. IMTC 2008. IEEE*, pp. 1643–1647.

Nye, J. (2003). A simple method of spherical near-field scanning to measure the far fields of antennas or passive scatterers, *Antennas and Propagation, IEEE Transactions on* 51(8): 2091–2098.

Omrane, B., Laurin, J.-J. & Goussard, Y. (2006). Subwavelength-resolution microwave tomography using wire grid models and enhanced regularization techniques, *Microwave Theory and Techniques, IEEE Transactions on* 54(4): 1438–1450.

Petre, P. & Sarkar, T. (1992). Planar near-field to far-field transformation using an equivalent magnetic current approach, *Antennas and Propagation, IEEE Transactions on* 40(11): 1348 –1356.

Quilez, M., Aragon, M., Atienza, A., Fernandez-Chimeno, M., Riu, P. & Silva, F. (2008). A near-field probe for in situ emi measurements of industrial installations, *Electromagnetic Compatibility, IEEE Transactions on* **50**(4): 1007–1010.

Richmond, J. (1955). A modulated scattering technique for measurement of field distributions, *Microwave Theory and Techniques, IRE Transactions on* 3(4): 13–15.

Rogers, P. (1986). Application of the minimum scattering antenna theory to mismatched antennas, *Antennas and Propagation, IEEE Transactions on* **34**(10): 1223–1228.

Smith, G. (1984). Limitations on the size of miniature electric-field probes, *Microwave Theory and Techniques, IEEE Transactions on* 32(6): 594–600.

Tehran, H., Laurin, J.-J. & Kashyap, R. (2009). A low-perturbation near-field imager equipped with optical mst probes, *Antennas and Propagation, 2009. EuCAP 2009. 3rd European Conference on*, pp. 3649–3653.

Yan, B., Saoudy, S. & Sinha, B. (1997). A low cost planar near-field/far-field antenna measurement system, *Antennas and Propagation Society International Symposium, 1997. IEEE., 1997 Digest*, Vol. 1, pp. 152–155 vol.1.

6

Single Shot Diagnostics of Quasi-Continuously Pumped Picosecond Lasers Using Fast Photodiode and Digital Oscilloscope

Michal Jelínek, Václav Kubeček and Miroslav Čech
Czech Technical University in Prague
Czech Republic

1. Introduction

Ultrashort laser pulses with duration in the range between 10 and 100 ps play important role in many applications, such as microsurgery, micromachining, laser ranging, nonlinear optics etc. Methods enabling precise measurement of such pulses duration are therefore essential. Common silicon photodiodes used in combination with ordinary oscilloscopes cannot be used for such a precise measurement because of low temporal resolution of the whole system. Therefore, several sophisticated methods based on optical or electro-optical effects were developed in the past decades and are still widely used (Diels, 1995; Keller, 2007; Rulliere, 2003). The first method based on nonlinear optical effect is measurement of second order autocorrelation function of the measured light pulse. This function is always symmetric and does allow to obtain detailed information on the exact pulse shape and pulse duration is calculated assuming the certain pulse temporal profile. This method, which can be used either for single or repetitive pulses, is in principle very precise but has also several disadvantages. The whole measuring system has to be aligned precisely and the measured beam has to enter the system in accurately aligned angle. Precision depends also on control of the delay line used for scanning autocorrelation measurement. Because the resulting pulse length is in this case calculated from many points, it is clear that it represents average value of the real pulses lengths and is difficult to evaluate pulse duration stability or study some other effects, i.e. pulse shortening under specific conditions. The single shot autocorrelators on the other hand can measure the autocorrelation function from only single laser shot but exact interpretation is not also unique. The only direct picosecond pulses measurement method based on electro optical effect uses a streak camera. This method allows to measure single pulse duration and shape but several consecutive pulses in the pulse train cannot be measured with sufficient temporal resolution. It is also necessary to carefully align whole measuring system and perform its rigorous calibration including readout.

Repetitive signals as mode locked pulse trains from continuously pumped lasers can be measured using sampling oscilloscopes in combination with fast detectors enabling resolution in units of picoseconds. For signals with repetition rates below several kHz sampling oscilloscopes cannot be used and only real time oscilloscopes with lower resolution are available. In the past few years fast real time oscilloscopes with analog bandwidth up to

40 GHz primarily designed for digital signal processing became available. In combination with sufficiently fast photodiode (commercially available with bandwidth up to $\sim 60\,\mathrm{GHz}$) a powerful tool for single ultrashort laser pulse characterization appeared. The photodiode - oscilloscope system for measurement of such laser pulses has several advantages. This procedure is simple to perform because the laser beam can be easily directed on the photodiode and very precise alignment of the measuring system is not necessary. The measurement is also repeatable, quick, and moreover duration of several pulses in the train can be recorded simultaneously from the single laser shot which enables investigation of special effects, such as pulse shortening in the laser output pulse train (Kubecek, 2009). Furthermore, oscilloscopes offer sophisticated functions and e.g. pulse duration statistics from thousands of pulses can be studied.

The subject of this chapter is theoretical and experimental study of the diagnostics of picosecond laser single pulses and pulse trains with repetition rates below 100 Hz using measuring system consisting of a fast real time oscilloscope and InGaAs PIN photodiode. In the first section physics of detection and general properties of photodiodes are described. In the second section theoretical analysis of pulse duration measurement method using oscilloscope - photodiode system is performed. Measuring system impulse response based on calculation from the corresponding datasheet parameters is discussed and subsequently resulting value is calculated for the given system. The third section deals with experimental verification of the described measurement method and calculated impulse response. Three laboratory laser systems based on neodymium (Nd) doped active materials and operated in quasi-continuously pumped regime were studied. The lasers were passively mode-locked by the semiconductor saturable absorbers and generated output trains consisted of 5 to 100 pulses. For some measurements single pulse was extracted. The pulses had duration in the range between 10 and 200 ps depending on the laser system. The pulse duration was measured simultaneously by our oscilloscope - photodiode system and at least one of precise optical measuring system (autocorrelator or streak camera) to investigate how short pulses can be measured with sufficient accuracy. Subsequently, it was possible to use the measuring system not only for instantaneous estimation of duration of the pulse shorter than corresponding impulse response, but moreover to investigate pulse duration stability using oscilloscope sophisticated statistics functions. In addition, two tested laser systems were operated under special regime resulting in pulse duration shortening along the output pulse train, which was possible to study precisely at each individual laser shot.

2. Physics of detection and photodiodes

Photodiodes represent one of fundamental light detection devices and play almost un-substitutable role in many applications, where the time and amplitude characteristic of the incoming light pulses has to be investigated or further exploited. Among the main advantages of photodiodes belong ease of use, fast time response, sensitivity at sufficiently wide spectral region, reasonably low thermal and electrical noise, and small dimensions enabling integration in electro-optics devices. Photodiodes are mostly used for detection of laser light pulses in many applications, such as telecommunications, sensing, security, and laser systems monitoring.

Light detection by a semiconductor material is based on the well-known phenomena of photon absorption (Saleh, 2007). If the incident photon energy exceeds the band gap energy of the semiconductor, an electron-hole pair is released. These charge carriers move by diffusion to places with lower concentration and may contribute to electrical current, if the semiconductor is connected into some electrical circuit. Since the diffusion rate is slow, these photo-conductors cannot be used as sufficiently fast photodetectors.

The situation changes when a photodiode formed by PN junction is used. Between the P and N regions, there is a depletion layer consisted of positively and negatively charged fixed ions. These ions form electrical field in the direction from N to P. If the photon is absorbed in this depletion layer, the electrons and holes are accelerated by the described electrical field. In this layer the carriers move by drift process which depends on the electrical field and is much quicker than diffusion. The depletion layer can be further extended applying reverse bias voltage on the photodiode which significantly decreases the carrier transit time. Reverse bias voltage (below the breakdown threshold) increases the noise component formed by dark current. In relation to the detector speed, capacitance formed by the depletion layer has to be treated and kept as low as possible, because it significantly contributes to the overall capacitance of the detector therefore influences its RC constant. Depletion layer capacitance is proportional to its area and inversely proportional to its thickness. Because the area relates to the photodiode responsivity, there has to be a trade-off between these values and fast photodiodes have lower sensitivity.

In order to extend the depletion layer, the intrinsic-doped layer can be inserted between P and N regions forming PIN diode. This leads to reduction of the necessary reverse bias voltage to several volts and simultaneously to rise the detector speed because most of photons are absorbed in this layer and generated carriers drift through this layer accelerated by the bias voltage. Furthermore, thicker depletion layer decreases the junction capacitance.

In the following section photodiode parameters with emphasis on the spectral region and response time will be discussed from the point of material and construction parameters. Requested spectral region ranging approximately from 500 to 1500 nm and covers the most of wavelengths used in solid state lasers enabling ultrashort pulses generation. Spectral region of the photodiode is mainly limited by its material parameter - the absorption layer band gap. As the photon absorption does not occur when the incident photon energy is less than absorption layer band gap E_g, the long-wavelength component λ_{cutoff} is limited by $\lambda_{cutoff} = hc/E_g \Rightarrow \lambda_{cutoff}$ [μm] $= 1.24/E_g$ [eV]. Therefore, common silicon (Si) photodiodes can be used for wavelengths only up to $\sim 1.1\,\mu$m. There is also second well-known semiconductor-compound material - gallium arsenide (GaAs). In the pure form, its long-wavelength region is limited to about 850 nm. By adding other component, the wavelength region can be significantly extended. Usage of many compounds has been published but in the commercially available photodiodes mainly Indium (In) component is used in the discussed spectral region. Pure InAs has λ_{cutoff} of about 3.4 μm and depending on its concentration in $In_xGa_{1-x}As$ compound, the wavelength range can be tuned. Mainly used compound has $x \sim 0.5$ determining λ_{cutoff} of $\sim 1.7\,\mu$m (Bitter, 2000).

Usage of GaAs-based material for the PIN photodiode construction has also other advantage in the charge carrier mobility. The electron mobility in GaAs is about five times higher than in Si while the hole mobility is comparable (Gibbons, 1987). As for construction parameters, it has already been said that in order to obtain fast response time it is necessary to keep the depletion layer capacitance as low as possible. Because of the mentioned reasons commercially available photodiodes based on GaAs / InGaAs have higher frequency bandwidth in comparison with the silicon photodiodes. The fastest photodiodes for telecommunication have to be based on InGaAs because of its desirable spectral response around 1.5 μm.

Besides these standard-type PIN photodiodes there is an extensive effort on the new photodiodes concepts development and utilization mainly in the telecomunication technique. Novel high-speed and high-power photodiodes with bandwith higher than 100 GHz were demonstrated. These new configurations are aimed to overcome the main disadvantages of the classical PIN photodiodes.

The bandwidth-efficiency and saturation current of the photodiode can be improved using the Dual-Depletion Region (DDR) detector (Effenberger, 1996). The depletion region of this structure consists of a InGaAs absorption layer and a InP drift (buffer) layer. This leads to the reduction of the junction capacitance due to increased depletion layer thickness. Moreover, the electrons must travel across both InGaAs and InP layers, whereas much slower holes must travel only across the InGaAs layer resulting in equalization of the transit times for electrons and holes.

Improved approach is the Uni-Travelling Carrier (UTC) photodiode (Ishibashi, 2000; Nagatsuma, 2011). This device utilizes only electrons as the active carriers. The active region consists of a p-InGaAs absorption layer and a wideband i-type carrier-collection layer. Photo-generated minority electrons diffuse from the absorption layer to the depleted collection layer, where they are accelerated and transported to the contact. A diffusion barrier made in the p-InGaAs layer prevents the electrons from diffusing into the p-contact while the generated holes can diffuse to the contact material. The photo-response is therefore determined by the electron diffusion time in the absorption layer which can be very thin leading to high bandwidth. Furthermore, the output peak current increases linearly with increasing input energy, and the waveform does not significantly change until it reaches the saturation point. In comparison with the waveform of the PIN photodiode, which consists from two current components (initial fast component attributed to electron transport and the slow tail caused by hole transport).

In order to increase the quantum efficiency of the standard (vertically-illuminated) PIN photodiode, which depends on the absorption layer thickness, a waveguide detection scheme was used, where the light absorption is perpendicular to current collection (Malyshev, 2004). Using this scheme it is possible to achieve high absorption efficiency and high speed simultaneously due to the decoupling of the efficiency from the absorption layer thickness. Efficient operation of these photodiodes based on the thin film germanium on silicon was successfully demonstrated (Wang, 2008).

Another approach leading to high-speed photodetectors is based on the metal-semiconductor Schottky junction, mainly developed in the MSM (Metal-Semiconductor-Metal) photodiode (Berger, 1996; Kache, 2005). Its structure is comprised of back-to-back Schottky diodes that use an interdigitated electrode configuration on the top of the active absorbing layer. This construction leads to low capacitance in comparison with the standard PIN photodiodes and therefore the MSM photodiode response speed is mostly limited by the transit time of the photo-generated carriers. Different materials may serve as the active layer and besides IR and visible spectral range also UV detectors were demonstrated successfully (Liu, 2010).

3. Picosecond pulse measurement using oscilloscope - photodiode system

3.1 Measuring system response

This chapter is aimed at measurement of ultrashort light pulses with duration between 10 and 200 ps generated by mode-locked lasers. The measurement is performed using oscilloscope - photodiode system and therefore overall response time of these both components has to be treated. Response time of each electrical component is limited by its electrical frequency bandwidth which is usually defined at the frequency where a sinusoidal output signal amplitude is attenuated to about 70 % of its original value (or in other terms the signal power is attenuated to 50 %), also known as -3 dB point. Let's consider the pulse width (duration) as full width at half maxima (FWHM) of light intensity. For the pulse width measurement the laser pulse shape can be in good approximation considered as Gaussian and therefore sum of square calculation of the real pulse duration $FWHM_{REAL}$ can be used:

$$FWHM_{REAL} = \sqrt{FWHM_{MEAS}^2 - FWHM_{SYSTEM}^2} \quad (1)$$

where $FWHM_{MEAS}$ is measured pulse width and $FWHM_{SYSTEM}$ is the minimal pulse width (impulse response or response to a Dirac delta function) of the measuring system. This instrumental constant can be calculated as

$$FWHM_{SYSTEM} = \sqrt{FWHM_{OSC}^2 + FWHM_{PD}^2} \quad (2)$$

where $FWHM_{OSC}$ is the oscilloscope minimal pulse width (given by the oscilloscope impulse response) and $FWHM_{PD}$ is the photodiode minimal pulse width. Into this calculation also influence of cables and connectors can be included when their frequency bandwidth cannot be neglected in comparison with other measuring components bandwidth. Similar theorem can be used for the calculation of rise time (RT) as the system response on the step input signal (Johnson, 1993; Keller, 2007).

3.2 Minimal pulse width and rise time of the measuring system components

Minimal pulse width FWHM and rise time RT of the measuring components can be calculated for the given frequency bandwidth f_{3dB} according to the formulas

$$FWHM = \frac{K_{FWHM}}{f_{3dB}}, \; RT = \frac{K_{RT}}{f_{3dB}} \quad (3)$$

Constant K varies in range from 0.3 to 0.5 according to the step or impulse response and also according to calculation performed for the oscilloscope or photodiode.

3.2.1 Photodiode response

The photodiode rise time can be calculated for the given electrical bandwidth using $K_{RT} = 0.35$. This value is based on the RC element step response and derived for example in (Johnson, 1993; Keller, 2002). It is also accepted by most of photodiode manufacturers who state rise time value in datasheets according to this calculation. The photodiode impulse response can be calculated using $K_{FWHM} = 0.312$ which is also derived in (Keller, 2002) and assumes the Gaussian response.

3.2.2 Oscilloscope response

Calculation of the oscilloscope response is little bit more complicated because the constant K relates to the oscilloscope type. Generally, two types of oscilloscopes are distinguished according to their shape of frequency response. Analog oscilloscopes and digital oscilloscopes with 3 dB bandwidth less than 1 GHz usually have Gaussian frequency response. This type of frequency response slowly decreases towards -3 dB attenuation and further. For this type of oscilloscopes is possible to use similar constants as for the photodiode. However, fast digital oscilloscopes with higher bandwidth have maximally flat or so-called brick wall response. In this type the frequency response is much flatter below defined -3 dB frequency but then drops rapidly. These oscilloscopes have several advantages. Signals with frequencies below f_{3dB} can be measured more precisely and higher frequency signals do not produce sampling alias errors. However, for this type of oscilloscope the assumptions and formulas for the Gaussian pulse are not valid precisely. Also the K_{RT} value is higher and ranging from 0.4 to 0.5 (Agilent, 2011; Tektronix, 2009). The K_{FWHM} is not precisely defined and according to uncertainty of K_{RT} its value can be just estimated.

3.2.3 Calculation of the measuring system impulse response

According to the previous analysis, the step and impulse response of the measuring system used in our experiments can be calculated. The measuring system consisted of the PIN photodiode EOT ET-3500 (EOT, 2011) connected by high frequency SMA cable to the real time oscilloscope LeCroy SDA-9000 (LeCroy, 2009) [1]. Datasheet parameters of the photodiode are as following: cutoff frequency >15 GHz, rise and fall time <25 ps, spectral range 1000 - 1650 nm, responsivity 0.88 A/W @ 1550 nm, active area diameter 32 μm, junction capacitance 0.12 pF. Important datasheet parameters as well as rise time and minimal pulse width calculation are summarized in Table 1.

Components & System	Datasheet values f_{3dB} [GHz]	Datasheet values RT [ps]	Step response RT [ps]	Impulse response K_{FWHM}	Impulse response FWHM [ps]	Impulse response K_{FWHM}	Impulse response FWHM [ps]
LeCroy	9.0	<49	49	0.44	49	0.40	45
EOT	15.0	<25	25	0.31	21	0.31	21
System	-	-	55	-	53	-	50

Table 1. Datasheet values and calculated system response $FWHM_{SYSTEM}$.

The system rise time can be calculated directly from the given datasheet values. The minimal pulse width is calculated from the given frequency bandwidth. For the oscilloscopes two values of K_{FWHM} are used. The first value of 0.44 is the same as used for the step response calculation, the second value of 0.4 is the estimation based on fact that the impulse response for the Gaussian systems is about 0.9 times shorter than step response (Andrews, 1989). It can be seen that the calculated minimal FWHM given by the impulse response is about 50 ps.

[1] Trade names are used to specify the experimental setup only.

4. Experimental investigation of picosecond laser pulses

4.1 Experimental determination of the measuring system minimal FWHM

Minimal FWHM was determined experimentally using an experimental fiber laser generating mode-locked pulses at 1.5 μm with duration less than 2 ps (measured by the autocorrelator). Pulse of this duration can be assumed as Dirac delta function for our measuring system. In order to avoid nonlinearities in the photodiode and oscilloscope, during all the measurements the oscilloscope vertical resolution was set at 5 mV/div and the signal amplitude was about 20 mV. The oscilloscope bandwidth was set to maximal analogue bandwidth of 9 GHz with the sampling frequency of 40 GS/sec. The oscilloscope enables two regimes of waveform acquisition and display - linear (only measured points are displayed) and *sin(x)/x* (approximation by this function). It was experimentally found that the FWHM measurement difference using these two acquisition regimes is about 1 ps and can be neglected. Therefore, most of further described measurements were performed in the linear acquisition regime.
There are two possibilities how to determine the FWHM of the measured pulse. The first is use of build-in function of the oscilloscope - Width at 50 %. The oscilloscope also enables to show histogram or statistics of these measured values. The second possibility is to save the data and perform a curve fit by Gaussian function. It has been found that using a Gaussian fit is for our pulses adequate and the determined FWHM of a such pulse with duration below 80 ps is about 18 % shorter than the value measured by the oscilloscope. Because of this uncertainty, most of FWHM values presented below were determined by the Gaussian fit of the measured pulse shape. All the presented values represent average value of about 100 pulses.
Recorded pulse from the 1.5 μm fiber laser with duration of 2 ps using *sin(x)/x* waveform approximation is shown in Fig. 1. The width measured by the oscilloscope was 75.5 $\pm$ 1.5 ps.

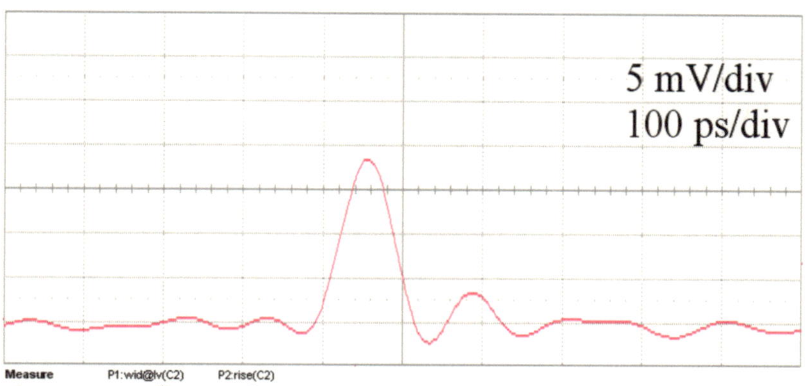

Fig. 1. Oscilloscope trace of the measured 2 ps long pulse using *sin(x)/x* approximation.

In Fig. 2 similar pulse recorded in the linear acquisition regime is shown. The width measured by the oscilloscope was 76 $\pm$ 2 ps. According to the Gaussian fit the pulse width was 63 $\pm$ 2 ps. There is a difference of about 13 ps in comparison with theoretically calculated minimal FWHM of $\sim$ 50 ps given in Table 1 which can be explained by uncertainity of used constants K, datasheet values, and influence of the cable and connectors.
There is also the second possibility to determine the $FWHM_{SYSTEM}$ using the longer pulse with known duration $FWHM_{REAL}$ and from the measured $FWHM_{MEAS}$ to calculate the

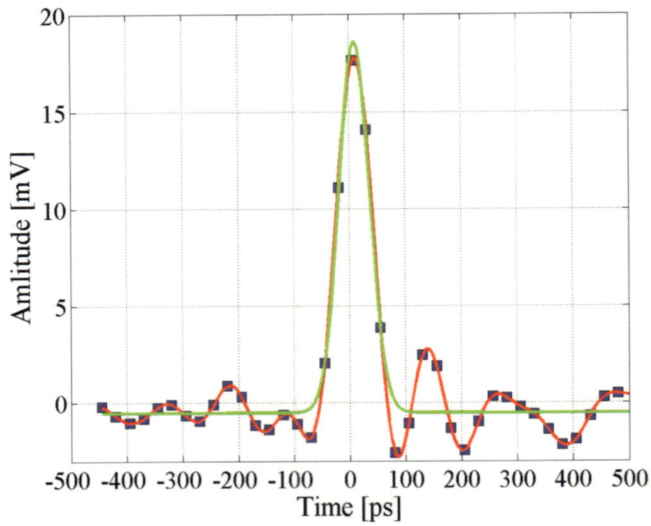

Fig. 2. Measured 2 ps long pulse (dots) and its Gaussian fit (green line) and Spline fit (red curve).

system response. In our experiments we have used a laboratory designed mode-locked Nd:YAG laser providing stable 22 ± 2 ps pulses (measured by the streak camera and autocorrelator) with repetition rate of 10 Hz at the wavelength of 1.06 μm (Jelinek, 2011; Kubecek, 2011). The laser system schematic is shown in Fig. 3. From the Gaussian fit of the measured pulse the width of 64 ± 2 ps was determined and using this value the $FWHM_{SYSTEM}$ of 60 ps was calculated. This value is in good agreement with experimentally determined value of 63 ps obtained using fiber laser.

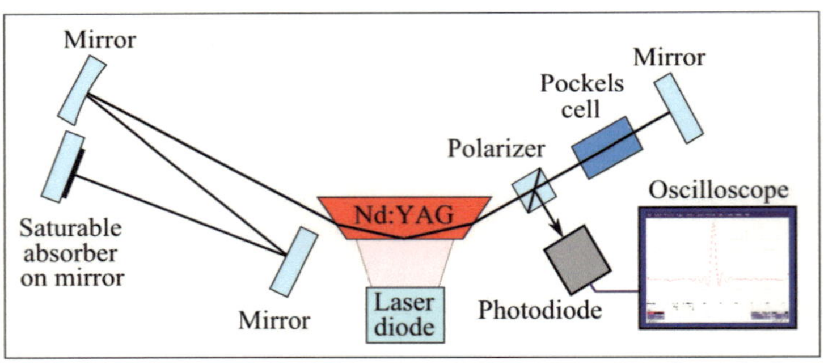

Fig. 3. Schematic of the Nd:YAG laser system generating 22 ± 2 ps pulses.

4.2 Duration estimation of pulses shorter than system impulse response time

In order to determine how short pulses can be reliably measured using our calibrated measuring system, pulses generated by two other passively mode-locked laser sources were measured and the real pulse width was calculated using both constants $FWHM_{SYSTEM}$. The first source was continuously pumped and mode locked Nd:YAG laser generating pulses in range of 17 to 21 ps (measured by the autocorrelator) with repetition rate of 110 MHz. The second source was quasi-continuously pumped and mode-locked $Nd{:}GdVO_4$ laser generating after cavity dumping from the Q-switched trains single pulses with duration of 56 ps (measured by the autocorrelator and streak camera) at the repetition rate of 30 Hz (Kubecek, 2010). Calculated pulse widths are shown in Table 2 and also in Fig. 4 together with calibration curves for both $FWHM_{SYSTEM}$ constants.

Laser	Pulse width FWHM [ps]				
	Measured real (autocorrelator or streak)	LeCroy value	Gaussian approximation (our value)	Calculated value for $FWHM_{SYSTEM}$: 60 ps	63 ps
Er fiber CW ML[2]	2	76 ± 2	63 ± 2	19 ± 7	-
Nd:YAG SP ML[2]	22 ± 2	79 ± 2	64 ± 2	22 ± 5	11 ± 8
Nd:YAG CW ML[2]	17 - 21	79 ± 2	66 ± 3	27 ± 7	20 ± 9
$Nd{:}GdVO_4$ SP ML[2]	56 ± 8	90 ± 6	82 ± 12	56 ± 16	52 ± 18

Table 2. Measured and calculated pulse widths for all studied laser sources.

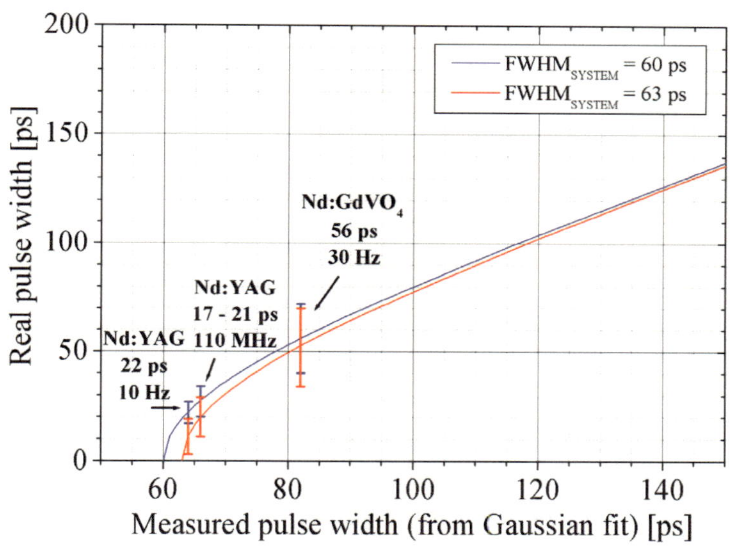

Fig. 4. Calibration curves for our measuring photodiode-oscilloscope system (for $FWHM_{SYSTEM}$ of 60 and 63 ps) and calculated real pulse durations of three measured laser sources

It can be seen that the real pulse width calculation from the measured $\sim$20 ps pulses is possible, but error up to 50 % may be introduced according to $FWHM_{SYSTEM}$ constant choice and the uncertainty of the measurement and the Gaussian fit. The real pulse width calculation for $\sim$50 ps pulses is more realistic and for both calibration curves (for different $FWHM_{SYSTEM}$ constants) does not introduce significant error. The uncertainty originates mainly from the laser stability itself.

[2] ML: mode-locking, CW: continuous wave, SP: single pulse

4.3 Single pulse duration stability investigation

The oscilloscope - photodiode system can be used for the single pulse duration stability investigation. An example of such measurement is shown in Fig. 5. Duration of the single pulses from the mode-locked Nd:$GdVO_4$ laser was measured using oscilloscope's build-in function and histogram from ~ 2000 successive pulses was shown. In spite of the fact that using the oscilloscope - photodiode system there may be some uncertainty in the absolute pulse width calculation, the width stability from many pulses can studied.

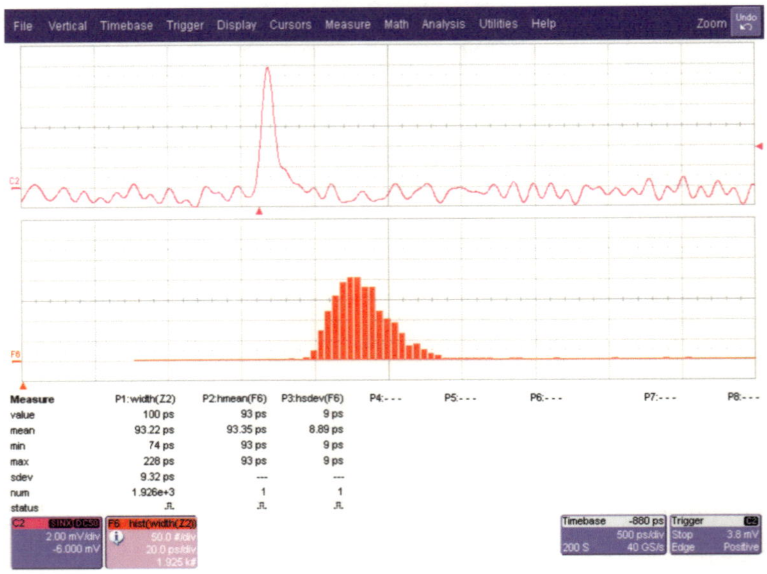

Fig. 5. Single pulse stability investigation using the oscilloscope statistical functions. Upper trace: measured pulse, lower trace: pulse width histogram from ~ 2000 successive pulses.

4.4 Investigation of the pulse shortening along the Q-switched mode-locked train

Using the oscilloscope - photodiode system it is possible to measure not only the temporal and energetic stability of the single pulses, but moreover to study some special effects, such as pulse width shortening along the laser output train containing tens to hundreds of pulses. Investigation of such effect in single output train cannot be performed by available optical measuring methods. As it was mentioned in the previous chapter, in spite of the fact that using the oscilloscope - photodiode system there may be uncertainty in the absolute pulse width, the pulse shortening effect studied in two pulsed laser systems can be clearly observed. The first laser system was based on Nd:$GdVO_4$ active material and passively mode locked by the semiconductor saturable absorber. The active medium was quasi-continuously pumped by the laser diode at repetition rate of 30 Hz. The 30 μJ laser output pulse train consisted of 12 pulses and its oscillogram is shown in Fig. 6. Lower traces show details of the highest pulse - pulse no. 3 in the train and later pulse no. 9. Fig. 7 shows plotted dependence of pulse duration evolution along the train measured from single laser shot and recalculated. It can be seen that the pulse duration decreased from the initial 160 to 55 ps at the end of the train (Kubecek, 2010).

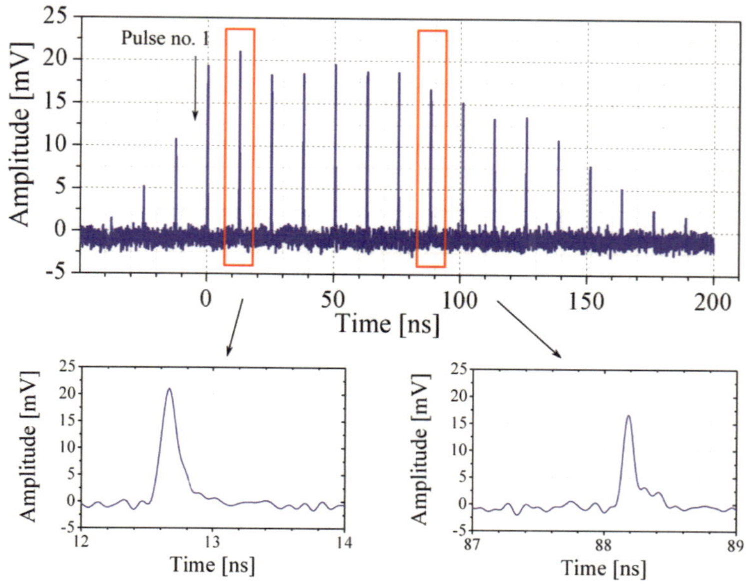

Fig. 6. $Nd:GdVO_4$ laser system output pulse train oscillogram (upper trace) and zoomed pulses no. 3 and 9 (lower traces).

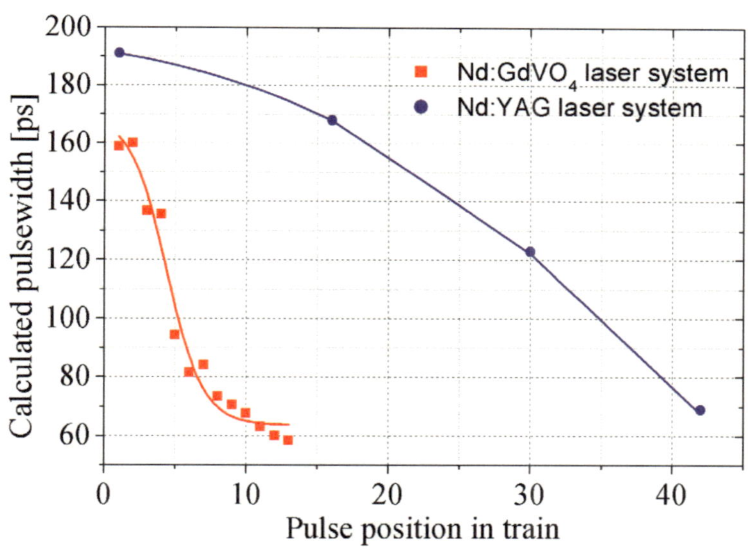

Fig. 7. Calculated pulse duration evolution along the trains generated by the $Nd:GdVO_4$ and Nd:YAG laser systems.

Similar pulse shortening effect was also observed in the output train of the passively mode-locked Nd:YAG laser with passive negative feedback when output trains containing hundred of pulses can be generated. Stretched 100 ns long pulse train shown in Fig. 8 has total energy of 60 μJ and contains $\sim$40 pulses. The pulse duration evolution along this train is shown in Fig. 7. The pulse shortening effect from original 190 ps in the beginning of the train to the final 70 ps was observed (Kubecek, 2009) resulting from the combined effect of the saturable absorber nonlinear transmission and passive negative feedback due to the beam defocusing via two-photon absorption in GaAs substrate of the semiconductor saturable absorber (Agnesi, 1992).

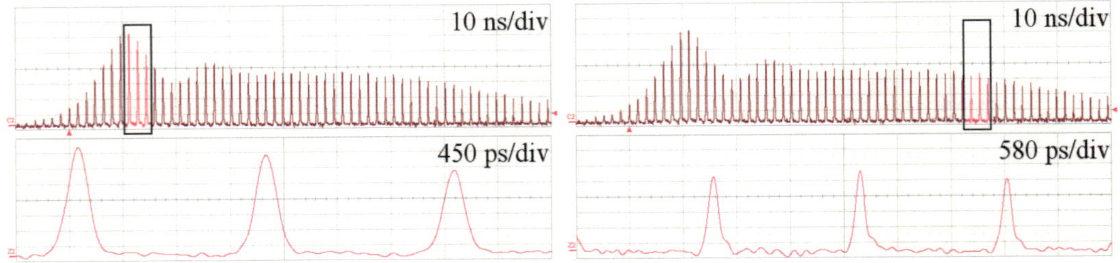

Fig. 8. Nd:YAG laser system output pulse train oscillogram (upper traces) and details of the pulse shapes from the beginning and end of the train (lower traces).

5. Conclusion

The aim of this chapter was the investigation of capabilities of the photodiode - oscilloscope measuring system for the single shot diagnostics of quasi-continuously pumped picosecond lasers. After the introduction, physics of light detection and photodiodes with emphasis on the response time of the PIN photodiodes was shortly discussed. In the third section, the oscilloscope - photodiode measuring system response and minimal pulse width was theoretically analyzed. According to this analysis, calculations based on datasheet values were performed for the used system consisting of the real time digital oscilloscope LeCroy SDA-9000 and PIN photodiode EOT ET-3500. The minimal pulse width (FWHM of the impulse response) of 50 ps was estimated. In the next section, this minimal pulse width was measured experimentally. Dependence of the width on different oscilloscope settings and waveform fitting was discussed. Measured minimal pulse width resulted in values between 60 and 63 ps and according to these results two calibration curves were obtained. In order to determine how short pulses can be reliably measured using the calibrated measuring system, pulses generated by two other laser sources were measured and their real widths were calculated and compared. It has been shown that even for pulses shorter than the minimal pulse width the useful real pulse width estimation can be obtained. Measurement and subsequent width calculation of the pulses with the duration comparable to the minimal pulse width can be performed with sufficient precision. The advantages of the calibrated measuring system were demostrated on the study of the laser pulse width stability and also on the investigation of the special effect - pulse shortening along the laser output pulse train.

6. Acknowledgements

The authors gratefully acknowledge the assistance of Pavel Honzatko, PhD and the consultations with David Vyhlidal.
This research has been supported by the Czech Science Foundation under grant No. 102/09/1741, the research projects of the Czech Ministry of Education MSM 6840770022

"Laser Systems, radiation and modern optical applications" and ME 10131 "Picosecond solid state lasers and parametric oscillators for sensors of rotation and other physical quantities."

7. References

Agilent. Evaluating oscilloscope bandwidths for your application. *Agilent Application note 1588*, URL: cp.literature.agilent.com/litweb/pdf/5989-5733EN.pdf.

Agnesi, A., et al. (1992). Generation of Extended Pulse Trains of Minimum Duration by Passive Negative Feedback Applied to Solid-state Q-Switched Lasers. *IEEE Journal of Quantum Electronics*, Vol. 28, No. 3, 710–719.

Andrews, J.R. (1989). TDR, step response and S parameter measurements in the time domain *Picosecond pulse labs - Application note AN-4*, URL: www.picosecond.com/objects/AN-04.pdf.

Berger, P.R. (1996). MSM Photodiodes *IEEE Potentials*, 25–29.

Bitter, M. (2000). *InP/InGaAs pin-Photodiode Arrays for Parallel Optical Interconnects*, Dissertation to the Swiss Federal Institute of Technology Zuerich, Switzerland.

Diels, J.-C. & Rudolph, W. (1995). *Ultrashort laser pulse phenomena*, Academic Press, ISBN 0-12-215492-4, USA.

Effenberger, F.J. & Joshi, A.M. (1996) Ultrafast, Dual-Depletion Region InGaAdInP p-i-n Detector. *Journal of lightwave technology*, Vol. 14, No. 8, 1859–1864.

Electro-Optics Technology, Inc. 10GHz GaAs and InGaAs Photodetectors datasheet, Jan 2011.

Gibbons, G. (1987). Gallium arsenide electronics. *Phys. Technol.*, Vol. 18, 5–16.

Ishibashi, T., et al. (2000). InP/InGaAs uni-traveling-carrier photodiodes. *IEICE Trans. Electron.*, Vol. E83-C, 938Ű-949.

Jelínek, M., et al. (2011). 0.8 mJ quasi-continuously pumped sub-nanosecond highly doped Nd:YAG oscillator-amplifier laser system in bounce geometry. *Laser Physics Letters*, Vol. 8, No. 3, 205–208.

Johnson, H.W. & Graham, M. (1993). *High Speed Digital Design: A Handbook of Black Magic*, Prentice Hall, ISBN 0133957241, USA.

Kache, S. (2005). *Optimization of charge collection efficiency in MSM photodetector.* Master thesis to University of Missouri-Columbia, USA.

Keller, U. (2002). *Ultrakurzzeit-Laserphysik*, Eidgenossische Technische Hochschule (ETH), Zurich, Switzerland.

Keller, U. (2007) Ultrafast solid-state lasers *Laser Physics and Applications, Subvolume B: Laser Systems.* Springer, ISBN 978-3-540-26033-2, Berlin, Germany.

Kubeček, V., et al. (2009). Pulse shortening by passive negative feedback in mode-locked train from highly-doped Nd:YAG in a bounce geometry. *Proc. SPIE*, Vol. 7354, 73540R.

Kubeček, V., et al. (2010). 0.4 mJ quasi-continuously pumped picosecond Nd:GdVO$_4$ laser with selectable pulse duration. *Laser Physics Letters*, Vol. 7, No. 2, 130–134.

Kubeček, V., et al. (2011). Cavity dumping of single 19 ps pulses from passively mode-locked quasi-continuously pumped highly doped Nd:YAG laser. *Proceedings of CLEO Europe*, Munich, Germany, May 2011.

LeCroy. Serial Data Analyzers (9 GHz - 18 GHz) datasheet, 2009.

Liu, K., et al. (2010). ZnO-Based Ultraviolet Photodetectors. *Sensors*, Vol. 10, 8604–8634, ISSN 1424-8220.

Malyshev, S. & Chizh, A. (2004). State of the art high-speed photodetectors for microwave photonics application. *Proc 15th Int. Conf. Microwaves, Radar and Wireless Communications*, Vol. 3, 765–775.

Nagatsuma, T. & Ito, H. (2011). High-Power RF Uni-Traveling-Carrier Photodiodes (UTC-PDs) and Their Applications. *Advances in photodiodes.* InTech, ISBN 978-953-307-163-3, Rijeka, Croatia.

Rulliere, C. (2003). *Femtosecond laser pulses*, Springer, ISBN 0-387-01769-0, USA.

Saleh, B.E.A. & Teich, M.C. (2007). *Fundamentals of photonics*, Wiley, ISBN 978-0-471-35832-9, USA.

Tektronix. XYZs of Oscilloscopes. *Tektronix Application note*, URL: www.tek.com.

Wang, J., at al. (2008). Evanescent-coupled Ge p-i-n photodetectors on Si-waveguide with SEG-Ge and comparative study of lateral and vertical p-i-n configurations. *IEEE Electron Device Letters*, Vol. 29, No. 5, 445–448.

7

A Photodiode-Based, Low-Cost Telemetric- Lidar for the Continuous Monitoring of Urban Particulate Matter

Massimo Del Guasta, Massimo Baldi and Francesco Castagnoli
Istituto Fisica Applicata "Nello Carrara" (IFAC) CNR
Italy

1. Introduction

Photodiodes are widely used in LIDARs (Light Detection And Ranging) (Measures, 1988). In ordinary LIDARs, a pulsed laser source is used to probe the atmosphere, while a fast photomultiplier or Avalanche photodiode (APD) is used to receive the high-frequency return from the atmosphere. APDs are used mainly in the near infrared, where photomultipliers are blind. APDs were used in both analog (Porter et al., 2002) and photon counting mode (Tatsumi & Tadashi, 1999) for the fast detection of IR LIDAR signals. In our simple telemetric LIDAR, a "vintage" technique used in the 1930s for pioneer studies on atmospheric aerosols (Duclaux,1936) and since then seldom reassessed (Meki et al.,1996), has been re-examined for the remote measurement of urban aerosols. Indeed, it represents a simplified and less expensive version of the elastic-backscatter LIDAR for short-range applications in which a continuous monitoring of particulate matter (PM) is required. It meets the requisite of being a simple instrument for the unattended, real time monitoring of PM to be used in urban pollution monitoring networks. For short-distance applications in which aerosols are to be measured within one hundred meters, a telemetric LIDAR can replace an ordinary LIDAR with a cost that is approximately 40 times lower than that of any ordinary LIDAR. The technique consists of illuminating the atmosphere with a laser beam modulated at low frequency, and then collecting the light scattered by aerosols by means of a photodiode array placed at the output of a telescope located at a certain distance from the laser. The observation angle defines the distance of the probed air volume through triangulation; the received intensity is related to PM10 in non-condensing conditions. The instrument is inexpensive, rugged, and suitable for outdoor operation, 24 h/day; it provides, moreover, all-weather measurement of PM with a time resolution of a few minutes. In the prototype, a green laser is modulated (on/off) at 620 Hz and emitted into the atmosphere. The choice of a visible wavelength simplifies both the alignment of the system and the calibration of the system in terms of volume backscatter (ch.2). The light backscattered by clean air and suspended matter is observed by means of a simple refractive telescope placed at a distance of 50 cm. The light received, which is filtered by means of an interference filter, is focused on a photodiode array placed on the telescope-focus surface. Each photodiode receives light scattered from different distances due to the telemetric geometry. A single photodiode may be selected for continuous measurements at a fixed distance, while a cyclic scan of different photodiodes is possible in order to measure it at different distances. A lock-in filter centred at the modulation frequency extracts and amplifies the weak signal produced by molecular air and aerosols. The DC signal produced by the lock-in is easily acquired by the digital electronics, which is based on a Microchip PIC18F6720 microcontroller. The telemetric-LIDAR data are acquired together with meteo and house-keeping data. The same board controls the laser, the meteorological sensors, and a GPS-GSRM module for the remote transmission of data. Remote PM measurements at distances of between a few meters and a maximum of 100 m can be obtained using this instrument. The signal obtained is almost proportional to the mass concentration of urban aerosols, as will be shown in this chapter through comparisons with standard PM10 instruments.

2. Theory of operation

Urban atmospheric aerosol is composed of particles of varying sizes. The size distribution N(r) for LIDAR applications can be modelled as the sum of two lognormal modes (John et al.,1990): the "accumulation mode", composed of mature aerosol particles, and the "coarse mode", composed of dust that has a short life in the atmosphere:

$$\frac{dN(r)}{dr}=\sum_{i=1}^{2}\frac{N_i}{r\,\ln(s_i)\sqrt{2\pi}}\exp\left[-\frac{\ln^2(\frac{r}{r_{mi}})}{2\ln^2 s_i}\right] \tag{1}$$

where r is the particle radius, r_{mi} is the median radius, Ni the total concentration, while s_i is the geometric width for the i-th mode. The elastic-backscatter LIDAR technique (including the telemetric LIDAR described here) measures the light backscattered at almost 180° by gases (Rayleigh scattering) and aerosols. The interpretation of LIDAR measurements in terms of aerosol quantities is based on a simulation of the scattering of the light by means of particles of known composition, shape and size. The scattering by a generic, spherical particle is described by the EM field transformation matrix:

$$\begin{bmatrix}E_s(\theta)\\E_p(\theta)\end{bmatrix}=\begin{bmatrix}S_1(\theta) & 0\\0 & S_2(\theta)\end{bmatrix}*\begin{bmatrix}E_{so}\\E_{p0}\end{bmatrix}\frac{e^{i(\omega t-\vec{k}\cdot\vec{R})}}{ikR} \tag{2}$$

where θ is the scattering angle, R is the distance vector, E_p, E_s are the EM field components with polarization parallel and perpendicular, respectively, to the incidence plane, and $S_i(\theta)$ are elements defined by the geometry and composition of the particle. The scattering by homogeneous, spherical particles is formally solved (Mie scattering), and simple series expansions provide good numerical approximations (Van de Hulst, 1957).
The differential scattering cross section, defined by:

$$\beta'(\theta)=\frac{|S_1(\theta)|^2+|S_2(\theta)|^2}{2k^2} \tag{3}$$

is simplified in the case of LIDARs into the backscatter (θ=180°) differential cross section:

$$\beta'(180°) = \frac{|S_1(180°)|^2}{k^2} \tag{4}$$

This quantity represents the power received by the LIDAR per unit of solid angle divided by the power of the incident plane wave for a single scatterer. In the presence of a volume concentration of N [part/m^3] of identical spheres, the volume backscatter coefficient is defined by $\beta = N^*\beta'(180°)$ [m^{-1} sr^{-1}]. In general, both aerosols and gases contribute to the backscatter. After conversion of the backscattered light into electrical signals, the LIDAR signal *V(z)* received from distance *z* from the elastic-backscatter LIDAR can be expressed as (Collis & Russel, 1976) :

$$V(z) = k\ E_0 \frac{1}{z^2}\left[\beta_a(z) + \beta_m(z)\right]\ e^{-2\int_0^{z'}[\sigma_a(z')+\sigma_m(z')]dz'} \tag{5}$$

where Eo is the emitted laser power, *k* is an instrumental constant, and *a* and *m* refer, respectively, to the aerosol and molecular components. The exponential term accounts for the dispersion of the laser beam in the atmosphere due to the extinction by aerosols (σ_a) and gases (σ_m). In the instrument described here, the extinction terms can be disregarded due to the short working distances, and thus the LIDAR signal can be simplified to:

$$V\ (z) \approx kE_0 \frac{1}{z^2}\left[\beta_a(z) + \beta_m(z)\right] \tag{6}$$

By assuming a discrete number of measurement distances, for each j-th measurement distance, and by assuming a constant laser power Eo, the telemetric-LIDAR output is simply proportional to the volume backscattering cross section:

$$V_j = K_j\left[\beta_{aj} + \beta_m\right] \tag{7}$$

The molecular component of the backscatter β_m (Penndorf, 1955; Bucholz, 1995) is practically constant in the first 100 meters of atmosphere, and is numerically well-known (β_m=1.59E-6 [m^{-1} sr^{-1}] at 532 nm and NPT conditions). This quantity is proportional to λ^{-4} In conditions of clean air ($\beta_a << \beta_m$) the constants K_j can thus be obtained from the signal V_j:

$$K_j = \frac{Vj_{clean}}{\beta_m} \tag{8}$$

(being $V_{j\ clear}$ proportional to λ^{-4}, the use of visible, green light makes this calibration easier than using of IR light). After this measurement, the instrument is calibrated in terms of volume backscatter units [m^{-1} sr^{-1}], and the aerosol volume backscatter can be derived from the calibrated measurement as: $\beta_a = \beta - \beta_m$. The aerosol volume backscatter is related, in a complex and nonlinear way, to the concentration, size distribution, shape and composition of the aerosol particles . Scattering simulations performed at 532 and 1064 nm (Kent, 1978; Del Guasta & Marini, 2000) on spherical particles of different nature showed that βa is, roughly, directly proportional to the aerosol wet-mass concentration [μg/m^3] in most of the practical cases encountered in the urban environment. The proportionality factor as obtained by simulation (Del Guasta & Marini, 2000) is found to be $\approx 2\ {*}10^7 \pm 5{*}10^6$

[(ug/m^3)/(1/m sr)] at 532 nm. The aerosol mass concentration was relatively independent from particle size distribution and composition in simulated urban conditions. The simple proportionality between wet-mass aerosol concentration and β_a could be used in the presence of relative humidity in the 70%<RH<98% range and in the absence of strong coarse aerosol (dust) loadings. In all the other cases the uncertainty in the conversion could increase significantly. The main indication resulting from these simulations is that the output of a simple elastic backscatter LIDAR (like the telemetric instrument described here) is roughly proportional to the wet-mass concentration of urban aerosols. The instrument output could thus simply be calibrated in terms of urban aerosol mass concentration in many conditions. This technique has been utilized experimentally with the use of an ordinary backscatter LIDAR (Del Guasta, 2002). The LIDAR-derived PM concentration is affected by much larger uncertainty than standard techniques are. Nevertheless, the capability of this optical technique to provide unattended remote measurements with a time resolution of minutes makes it a precious complementary tool for urban pollution studies. We can also point out that the wet mass concentration measured by LIDAR refers to aerosol in equilibrium with environmental humidity, and it must be corrected by means of a proper aerosol growth factor (McMurry & Stolzenburg, 1989) used together with the measured relative humidity when the dry aerosol mass concentration (assimilable to the well known PM10) is the final, required quantity. In cases of RH<80% the wet and dry aerosol mass concentrations coincide, at least for monitoring purposes.

3. The instrument

3.1 Overview

The system (Fig.1) is basically composed of a modulated laser source and a receiving system. The backscattered light coming from different ranges is focused on different pixels of a photodiode array. The signal from the photodiodes is amplified, filtered and demodulated by means of lock-in electronics. The final DC voltage is acquired and stored in a microcontroller board. The microcontroller communicates with the external PC, and also manages meteorological and ancillary sensors. A description of the different components follows.

Telemetric LIDAR specifications

Laser	532 nm, 100 mW, 620 Hz modulation
Telescope	Ø 80 mm, f=300 mm
Laser-telescope Distance	0.5 m
Interference Filter BW	10 nm
Photodiode (pixel) size	0.9 x 1.45 mm
Photodiode Preamp Gain	1E7 V/A
Lock-in	0.12 Hz BW, 1E6 Gain

3.2 Laser

The laser is a commercial, low-cost solid-state CW Nd-YAG laser, operating in the green, and equipped with on/off 620 Hz modulation. The laser modulation is obtained by means of a 620Hz quartz oscillator provided with two outputs: the first output drives the laser, while the second is the reference signal to the Lock-in. The delay between the two outputs is adjusted in order to nullify the lag between the TTL modulation of the laser and the effective modulation of the laser emission.

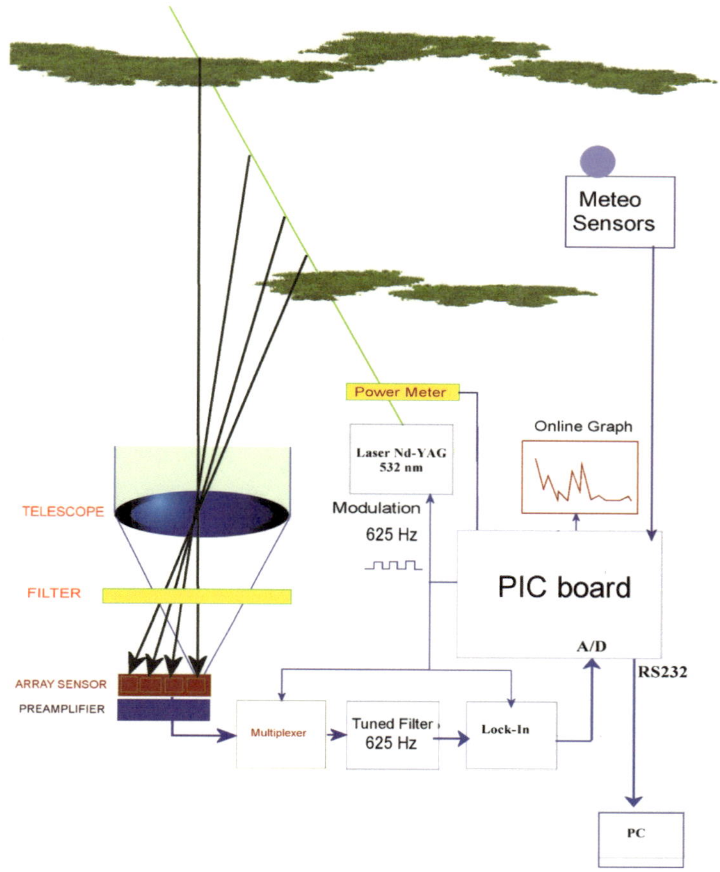

Fig. 1. Schematic of the telemetric-LIDAR

3.3 Receiver optics

The receiver is a refractive telescope equipped with a lens of 300 mm focal length, 80 mm diameter (NA= 3.75). The light cone produced by the lens has a semi-aperture of 8°. The cone crosses an interference filter in order to reduce the background light received by the photodiodes . The broad bandwidth (10 nm) of the interference filter permits the passage of the whole cone of light from the lens without consistent losses, as shown in Fig.2.
The instrument focuses on the photodiode array the light backscattered by air volumes located at distances of between 10 and 100 m, illuminated by a laser beam which is displaced with respect to the optical axis of the telescope. The focusing of the telescope at different distances thus moves along both the telescope axis and off-axis. In order to have all the pixels of the photodiode array on the focus of the telescope for all distances, the array was tilted by 67° with respect to the telescope axis. Fig.3 shows the simulated displacements of the focus at different measurement distances with respect to the focus at infinity f_∞ =(0,0).
A field diaphragm was obtained by gluing a 0.5 mm linear slit directly onto the photodiode array window. In this way, the FOV of the telescope was reduced to ~2 mrad in the direction perpendicular to the telemeter plane. The photodiode physical width (0.9 mm) defines the FOV (~4 mrad) in the direction parallel to the telemetric plane.

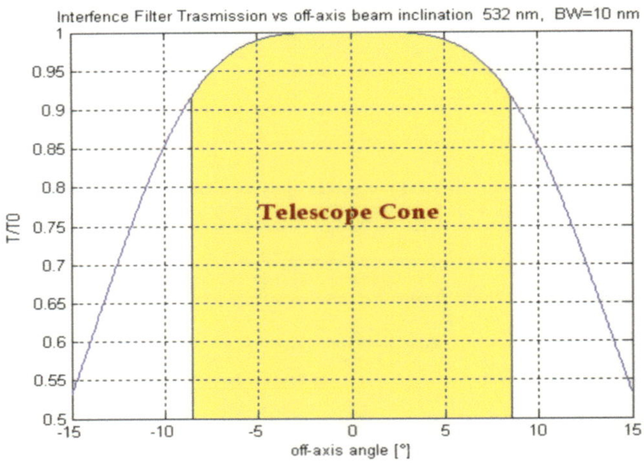

Fig. 2. Interference filter relative transmission for off-axis rays. The aperture of the light cone of the telescope is highlighted

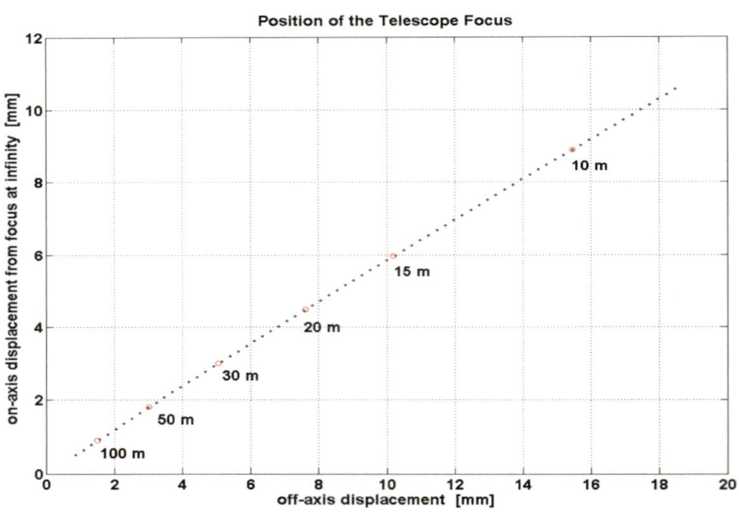

Fig. 3. On-axis and off-axis displacement of the focus for different observation distances

3.4 Measurement range and range resolution

The measurement distance is determined by the angle between the observation direction and the laser beam direction. In this way, by selecting a particular pixel of the array it is possible to receive the light backscattered from distances comprised between 5 and 100 meters. In the configuration described, the field depth (2 x dz) of the instrument is a function of the nominal distance of the measuring volume, as in Fig.4.

As shown in the simulations of Fig. 5, the range resolution of this instrument is similar to or better than that of an ordinary LIDAR for distances below 50 meters. However, it becomes practically useless at distances greater than 80-100 m.

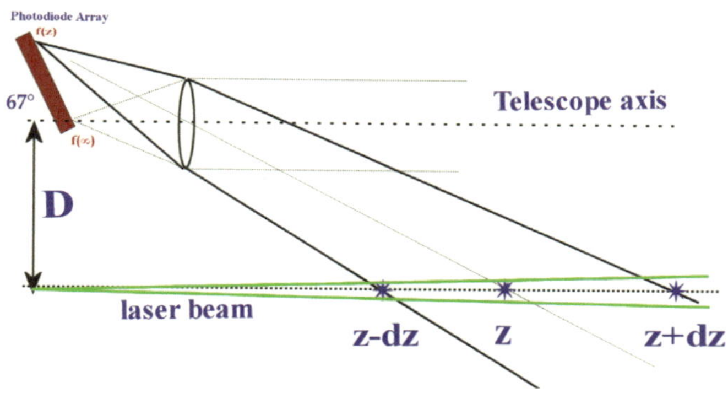

Fig. 4. Field depth calculation

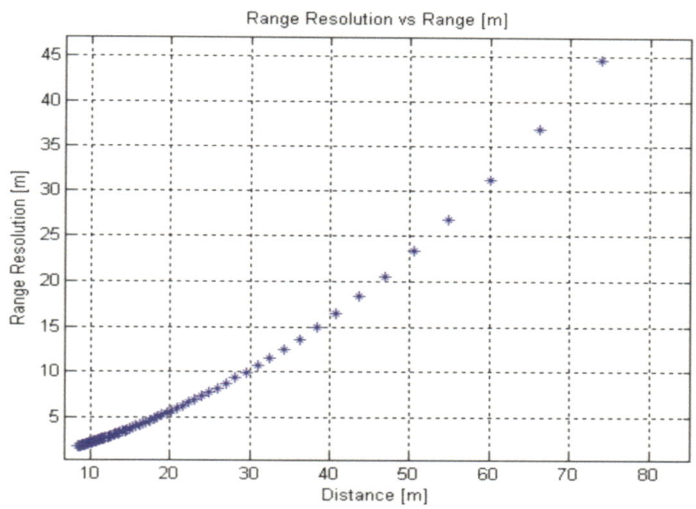

Fig. 5. Field-depth at different measurement distances

3.5 Photodiode and preamplifier

The detector chosen was a 16-element array of silicon photodiodes (Hamamatsu) S4114-16Q. Each pixel is used in photovoltaic mode and preamplified by means of a (Analog Devices) AD743 OPAMP. An ordinary photodiode was chosen for this application for the following reasons:

1. The application is a low frequency one: in this case, ordinary photodiodes are superior to APDs because, since their S/N in the dark is limited by thermal noise, this quantity can be simply increased by increasing the feedback resistor of the preamplifier. I.e. the dark noise expected from a Hamamatsu APD S5344+AD743 was found to be 2.5E-6 V/√Hz @625Hz, more than 6 times greater than the noise of the S4114+AD743 (Fig.6), if we assume the overall gain, wavelength and operating frequency to be the same.
2. This rugged application requires a constant gain with temperature. This condition is easily obtained by means of a photodiode in photovoltaic mode, while APDs always require a thermal stabilization.

Photodiode array characteristics (S4114-16)

Sensitivity	0.35 A/W @ 532 nm
Pixel size	1.45 * 0.9 mm
Dark current	5 pA
Shunt resistance	250 GΩ
Terminal capacity	200 pF
Rise time	0.5 μs
Operating mode	photovoltaic

Opamp characteristics (OPA743)

input noise voltage	4.5E-9
input noise current	2.5E-15
Feedback resistor (thin film Ni Barrier)	10E6 MΩ
Noise Index of Feedback resistor	-20 dB
Feedback Capacity	2.7 pF
Operating temperature	25°C
Bandwidth (-3dB)	3kHz

The correct choice of the components made it possible to limit the dark noise of the photodiode-preamplifier system to the thermal noise of the amplifier feedback resistor (Graeme, 1996). In Fig.6, the different contributions to noise are shown as a function of the frequency.

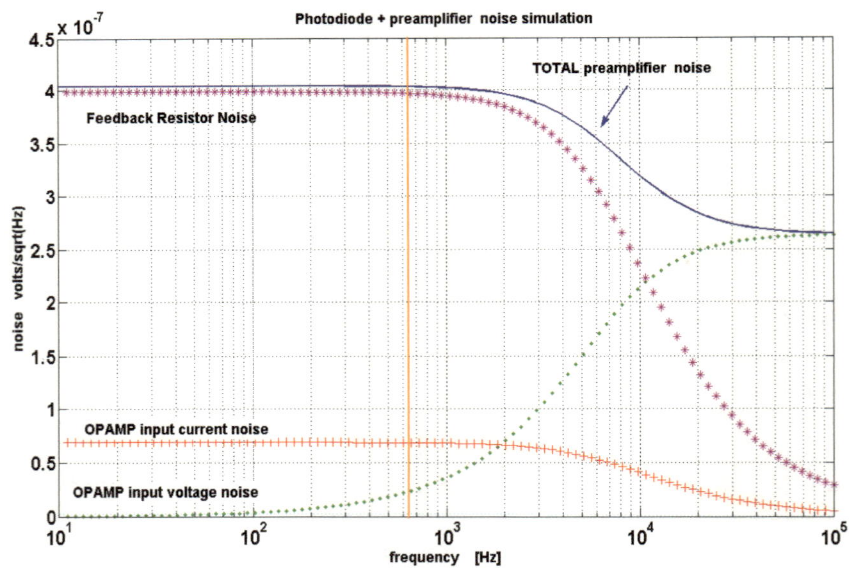

Fig. 6. The noise components of the photodiode-preamplifier as simulated with MATLAB. Yellow line= operating frequency

3.6 Selective filter

The preamplified photodiode pixels are multiplexed into the (unique) lock-in electronics. The choice of the active pixel and thus of the measurement distance is made via software. The multiplexer (MUX) output is the input of a tuned preamplifier that is used to pre-filter the signal for the lock-in board. The selective filter, together with the optical filtering provided by the interference filter, contributes to the reduction of the noise received by the system during daylight operation.

The selective amplifier, which is based on an OPAMP gyrator (Fig.7), provides a narrow-band pass filtering (2.4 Hz (-3 dB) Bandwidth, Gain=5000) of the signal received, that is centered around the 620 Hz modulation frequency. Thanks to an appropriate choice of the passive components, the adopted circuit (Fig.7) was found to be stable with temperature, with a drift of the resonant frequency of less than 0.065 Hz/°C in the temperature range of -25°/+50°C, and a corresponding gain drift of less than 0.05 dB/°C. The phase shift is practically zero in the same temperature range.

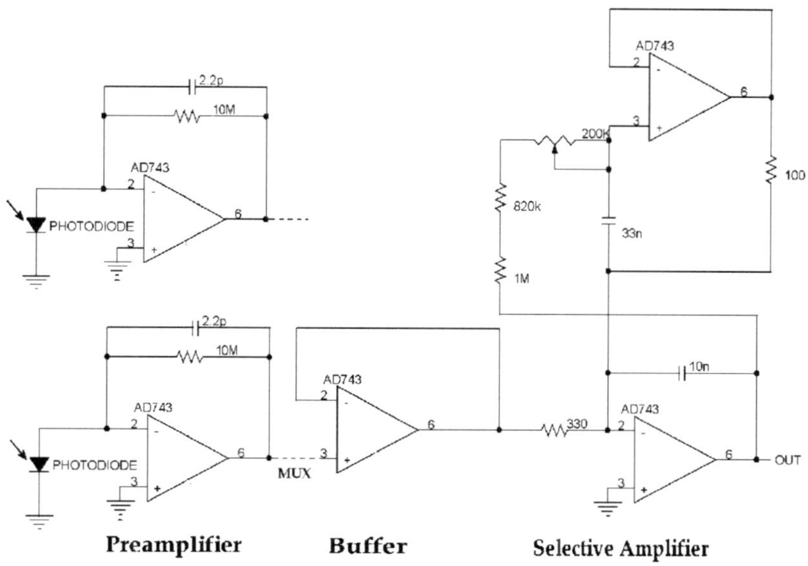

Fig. 7. Schematic of the selective amplifier

3.7 Lock-in

An analog lock-in (Horowitz, 1989), tuned to the modulation frequency (620 Hz), is used to convert the AC signal in phase with the modulating signal into a DC voltage. This lock-in is based on a AD630 modulator-demodulator chip followed by a Sallen-Key, II order active low-pass filter with 0.1 Hz bandwidth and Gain=10. The active filter is followed by a final amplifier with gain=10, and low-pass cutoff of 1 Hz. The final output of the lock-in circuitry is limited to 0-5V in order to match the voltage span of the microcontroller AD inputs.

The step response at the output of the lock-in is shown in Fig.8. It is possible to sample the output 100 msec after the application of the step. The sampling rate of the lock-in output by means of the microcontroller board was thus set at 20 Hz.

3.8 Microcontroller board

The microcontroller board, which controls the whole instrument and acquires the telemetric LIDAR signals, is based on a Microchip PIC18F6720, that is programmed in PicBasicPro. The board manages the whole instrument by converting and averaging the lock-in output, and acquiring meteo and house-keeping data. The board communicates with the PC via RS232, allowing a 20 meters distance between the instrument and the PC. Data are both logged and sent to the remote PC. The same serial line link is used to change the micro controller firmware and to change the measurement settings.

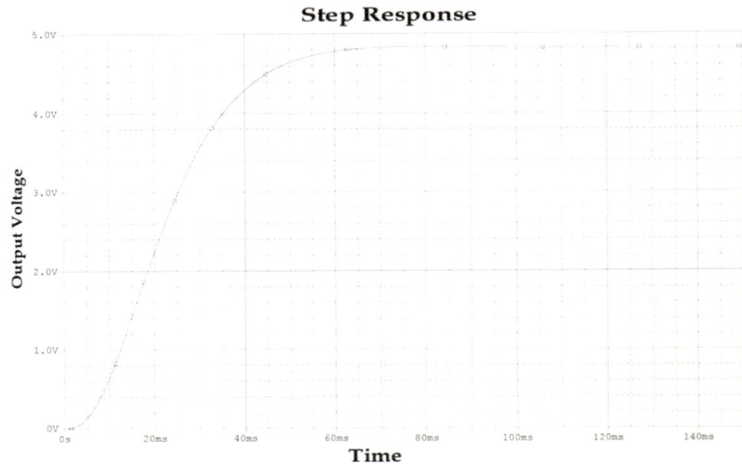

Fig. 8. Step response at the output of the Lock-in

4. Testing the instrument

4.1 Indoor test: signal to noise

MATLAB simulations of noise and signal as expected at the end of the whole chain in conditions of clear atmosphere and in day/night conditions were performed for different operating conditions and distances, as shown in Fig 9. For daylight simulations we assumed a Sun elevation of 30° and a pure molecular atmosphere. In these conditions the sky irradiance observed by the telescope was calculated by assuming pure Rayleigh scattering.

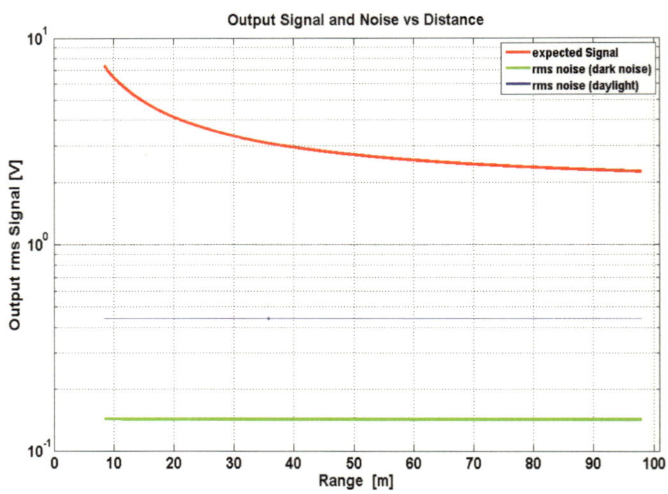

Fig. 9. Clean-air signal and noise as expected from the MATLAB simulations

In the case of nighttime operation, the dark noise of the system was considered. An output average power of 50mW for the laser source, a transmission of 0.7 for the interference filter and an overall optical efficiency of 0.6 for the rest of the optics were assumed.
In the prototype, the measurement of the noise at the output of the complete chain in the dark was 160 mV rms, close to the simulated dark noise in Fig. 9. The different photodiodes showed slight differences in terms of dark noise.
From simulations a S/N>4 was found in daylight conditions for any distance. In practice, when operating the instrument in the atmosphere, a S/N close to 1 was obtained in daylight conditions at the output of the chain. An empirical integration time of 5 minutes was thus chosen for operating the instrument in daytime with a S/N>4.

4.2 Indoor test: inter-calibration of the photodiode outputs

Four pixels of the photodiode array were selected for routine measurements. The pixels corresponded to the following measurement ranges: CH1: 9÷12 m, CH2: 16÷30 m, CH3: 30÷56 m, and CH4: 60÷160 m. The system outputs for these four channels require a cross-calibration because the four measurement volumes are different. The calibration was easily performed by running the instrument horizontally-oriented, and by assuming an homogeneous aerosol loading along the horizontal direction within 100-200 meters from the instrument. The DC offset of each channel is first measured by shutting the laser off, via software. The four channels were thus cyclically measured. After a few days of measurement, the output of the four channels was plotted against a reference channel (CH1). The linear fit of the three scatter plots CH2-CH1, CH3-CH1, and CH4-CH1 provides the inter-calibration constants as shown in Fig.10. By using these constants, the output of all the four channels became proportional to the volume backscatter by the same factor.

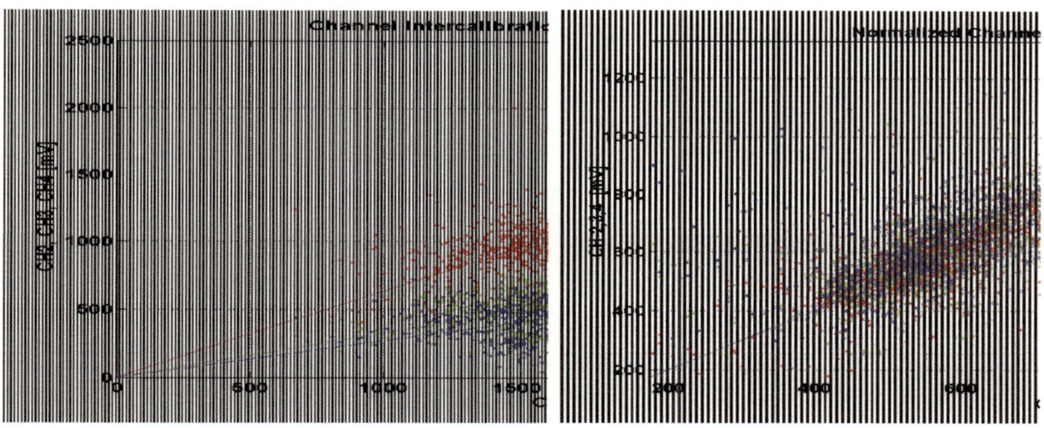

Fig. 10. Scatter plot used for the inter-calibration of the four channels (left) and the same plot after renormalization of the channels.

4.3 Outdoor test in Prato (Italy)

A first prototype of telemetric-LIDAR was first used in a field campaign in Prato (Italy) during the September-October 2005 period. The instrument was installed on the roof of a conventional ARPAT (Agenzia Regionale Ambiente Toscana) monitoring station equipped with a MP101M (Environment S.A.) beta attenuation PM10 instrument. The telemetric-

LIDAR instrument operated 30° above the horizon. Data obtained every 5 minutes for a fixed measurement altitude of 6(±1) meters were compared with the hourly beta-attenuation PM10 data (Fig. 11). The calibration of the telemetric LIDAR output in terms of PM10 was obtained from a linear regression between the hourly-averages of the LIDAR signal and the beta-attenuation data.

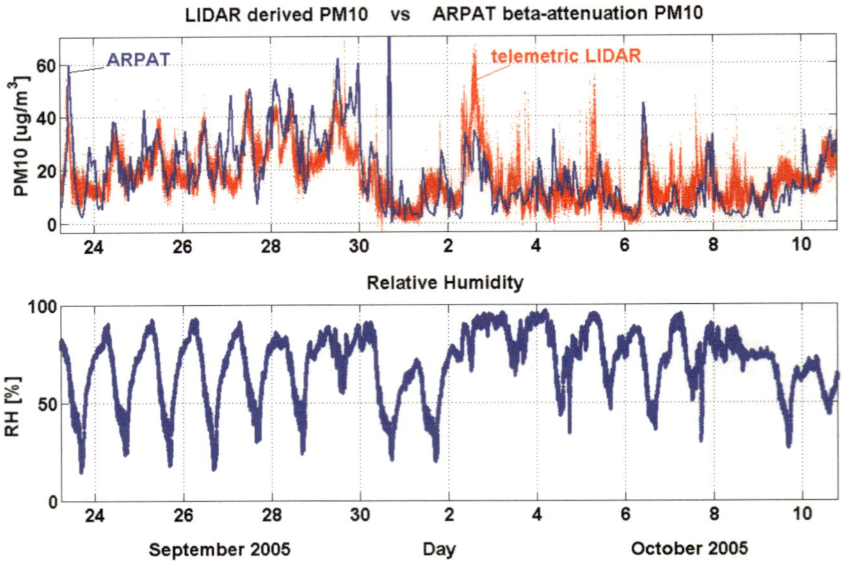

Fig. 11. Comparison between LIDAR-derived and β-attenuation PM10 data (Prato, Italy).

Relative humidity is also plotted
The comparison of the two time series shows a general agreement even if some peaks are uncorrelated. On 2-5 October rain events occurred, a fact which explains the overestimation of PM in the LIDAR data. LIDAR measured the wet PM, which in this case was larger than the dry PM measured by the MP101M instrument. On the other hand, in the case of high RH the sampling head of the MP101M cut out the large water droplets, thus leading to an underestimate the PM concentration. Furthermore, the beta attenuation instrument is not reliable when used with integration times as short as one hour.

4.4 Outdoor measurements in Florence (Italy)

A second, rugged prototype for fully outdoor operation was developed in 2005. In this prototype, the window of the telescope was equipped with two optical sensors that were developed at IFAC CNR: one sensor detected, from inside the box, the dust-cover of the external face of the optical window. This is an important measurement in the urban environment, because sooty/oily particles tend to stick rapidly onto the window, thus increasing the optical losses of the instrument and spoiling the calibration. The other sensor detected dew forming on the optical window in the case of foggy weather. Data from both sensors were managed and stored together with the other data. The Telemetric LIDAR was equipped with a GPS-GSRM-GPS module for the remote transmission of data and alarms. In the case of a dirty or wet window, an alarm message was automatically sent to IFAC via E-mail and SMS. The new prototype was also equipped with a laser power sensor for the continuous monitoring of the laser power. This data was stored together with the other LIDAR data, and was used for the normalization of the LIDAR data (eq.6). In the case of a laser failure, an E-mail message was automatically sent to IFAC. On-board meteorological sensors for wind, relative humidity, and temperature completed the instrumentation. Meteorological data were managed and stored together with the other data. All data were sent via FTP to IFAC at the end of each day.

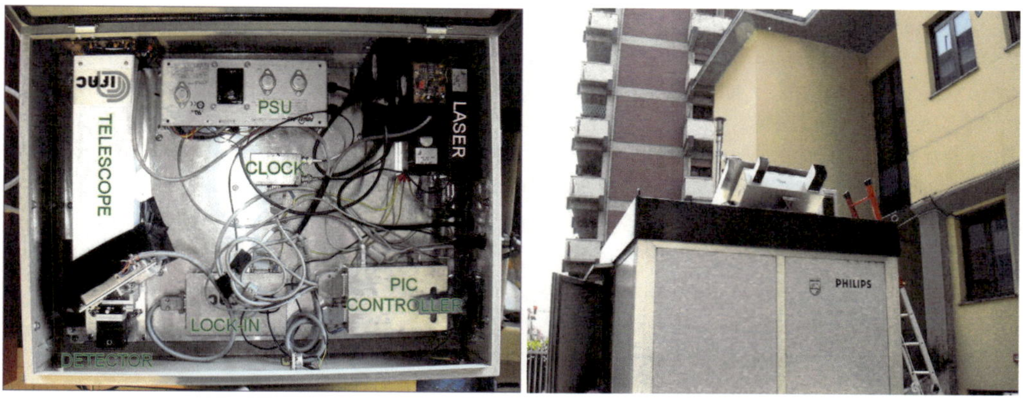

Fig. 12. The instrument opened (left) and installed on the roof of the ARPAT, PM-monitoring station in Florence (I) (right)

This prototype was installed in 2006 on the roof of an ARPAT station (Via Ponte alle Mosse, Florence (Italy), where it was in operation until the end of 2007. The instrument operated on a 45° slant above the horizon, for a fixed measurement distance of 8(±1) meters. ARPAT provided daily gravimetric PM10 data. In Fig.13, a time series of one month of telemetric-LIDAR data is compared with gravimetric PM10 data. The upper plot shows the LIDAR calibrated signal averaged over 10 minutes. The PM10 daily gravimetric data are shown as

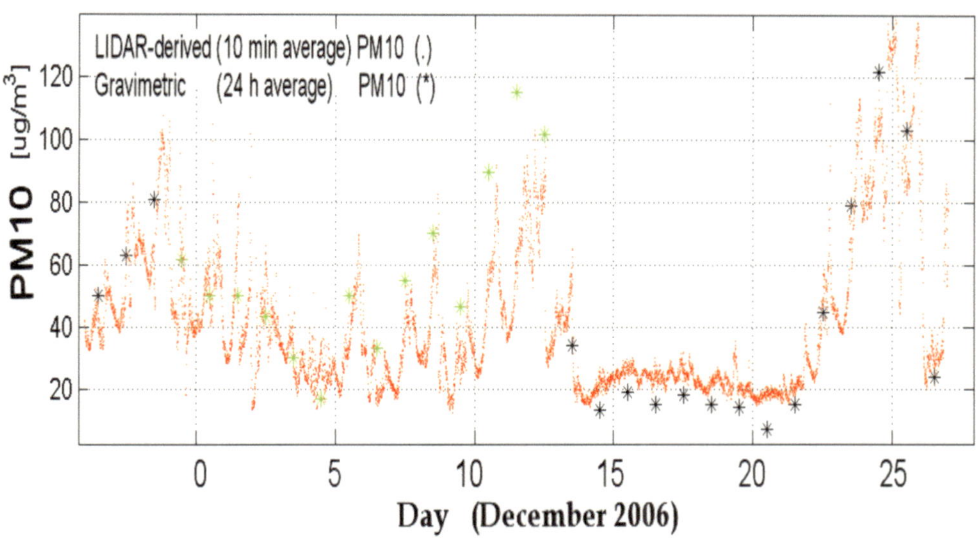

Fig. 13. PM10 as derived from the telemetric-LIDAR and from gravimetric data.

symbols. The ARPAT monitoring station collected PM10 and PM2.5, alternatively, for 15-day periods. Black stars indicate genuine PM10 measurements, while green stars indicate PM10 values calculated from PM2.5 data by applying a constant, empirical factor of 1.4, as suggested by ARPAT.
Until December 14 the diurnal cycle of PM, related to traffic, is evident in the telemetric-LIDAR data. During the 14-21 December period, northern winds cleaned-up the PBL (Planetary Boundary Layer) (Stull, 1988) and prevented the formation of inversion layers, thus reducing the PM10 and cancelling its diurnal cycle. During the 22-26 December period, a strong thermal inversion occurred, which led to high PM10 concentrations. The comparison between the LIDAR-derived and gravimetric data was unfortunately undermined by the different time resolutions of the two types of measurements. The LIDAR information had to be degraded to 24-hour averages in order to compare it with the daily gravimetric data, the only official information available in many Italian towns. The said comparison was used to obtain an empirical system calibration, as shown in Fig.14.

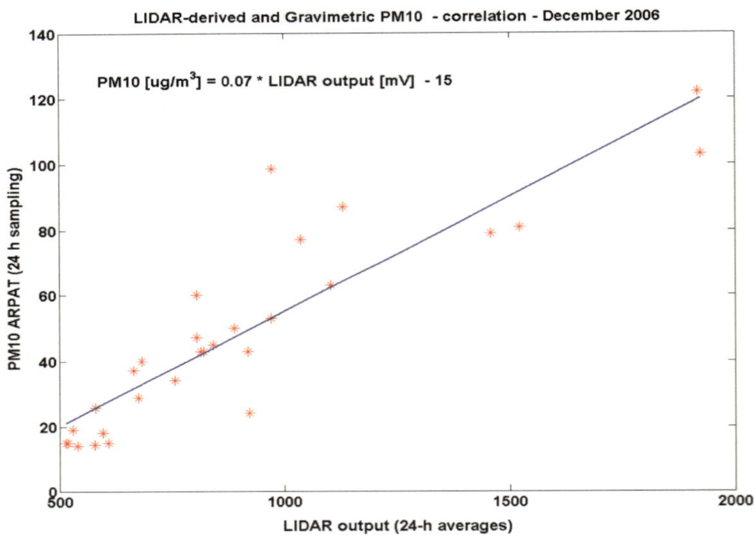

Fig. 14. Correlation between LIDAR-derived and gravimetric PM10 values. 24-hour averages. The Pearson linear correlation coefficient is R= 0.89, p<0.001. The experimental calibration factor for the telemetric-LIDAR is reported.

5. Conclusions

We described an application for a Silicon photodiode array consisting in a low-cost, rugged instrument for the continuous remote monitoring of urban particulate matter (PM). The experimental tests confirmed the optical and electronic simulations, which suggested the possibility of measuring PM in the urban environment 24 hours/day within a range of several tens of meters, with a time resolution of 5-10 minutes. The instrument is a candidate tool for complementing ordinary gravimetric PM10 measurements, with the advantage of offering a high temporal resolution and the absence of pumps or other moving parts. The instrument was found to be suitable for unattended operation and much less expensive than any ordinary PBL LIDAR. When used within a range of a few tens of meters, thanks to its high spatial resolution the instrument could be utilised for the continuous remote monitoring of PM emitted by smokestacks, power plants, and in all those cases in which the relative humidity is non-saturating and the typology of the emitted particles is known.

6. References

Bucholtz A. (1995). Rayleigh-scattering calculations for the terrestrial atmosphere, *Appl. Opt.* 34, pp 2765-2773, *ISSN*: 1559-128X

Collis R.T.H.,Russell P.B. (1976), Lidar measurement of particles and gases by elastic backscattering and differential absorption,In: *Laser Monitoring of the Atmosphere, Topics in Applied Physics* Vol. 14, Hinkley Ed., pp. 71–151, Springer, ISBN 038707743X, Berlin

Duclaux. (1936), J. Phys. Radiat. 7, S. 361. Referenced in: P.S. Argall, Sica R.J., LIDAR In: Hornak J.P.,(2002), *The Encyclopedia of Imaging Science and Technology*, Wiley, ISBN: 978-0-471-33276-3, New York

Measures R.M. (1988). *Laser remote chemical analysis*, John Wiley & Sons Eds., ISBN: 047181640X, New York

Meki K. (1996). Range-resolved bistatic imaging LIDAR for the measurement of the lower Atmosphere, *Opt. Lett* . 21,17, pp.1318-1320,(1996), ISSN 0146-9592

Del Guasta M. (2002). Daily cycles in urban aerosols observed in Florence (Italy) by means of an automatic 532-1064 nm LIDAR. *Atmos. Env.* 26, pp. 2853-2865, ISSN1352-2310

Del Guasta M., Marini S. (2000). On the retrieval of urban aerosol mass concentration by a 532 and 1064 nm LIDAR, *J. of Aerosol Sci.* , 31, 12, pp. 1469-1488, ISSN0021-8502

Graeme J.G. (1996). *Photodiode Amplifiers: op amp solutions*, Chp.5, McGraw-Hill Professional Publ., ISBN 0-07-024247-X

Horowitz, P.&H., Winfield J.(1989). *The Art of Electronics*. pp.1032-1033, Cambridge University Press. ISBN 0-521-37095-7, New York

John W., Wall S.M., Ondo J.L., Winklmayr W. (1990). Modes in the size distributions of atmospheric inorganic aerosols. *Atmos. Environ.* 24, 9, pp.2349-2359, ISSN1352-2310

Kent G.S. (1978). Deduction of aerosol concentrations from 1.06 μm lidar measurements. *Appl. Opt.* 12, 23, pp. 3763-3773, ISSN: 1559-128X

McMurry P.H., Stolzenburg M. (1989). On the sensitivity of particle size to relative humidity for Los Angeles aerosols. *Atmos . Environ.*, 23, pp. 497-507, ISSN: 1352-2310

Penndorf, R. (1957) Tables of the refractive index for standard air and the Rayleigh scattering coefficient for the spectral region between 0.2 and 20.0 um and their application to atmospheric optics, *J. of the Optical Soc. Of America*, Vol.47, N°2, pp.176-182, ISSN 0036-8075

Porter, J. N., Lienert B. R., Sharma S. K., Hubble H. W., (2002). A Small Portable Mie-Rayleigh Lidar System to Measure Aerosol Optical and Spatial Properties. *J. Atmos. Oceanic Technol.*, 19, pp 1873–1877. ISSN: 0739-0572

Stull R.B. (1988) *An introduction to boundary layer meteorology*, Kluwer Academic Publishers. ISBN 90-277-2768-6, Dordrecht (NL)

Tatsumi K. Tadashi I . (1999). Characteristics of Lidar Signal Using Silicon Avalanche Photodiode Single Photon-Counting Module. *Rev. of Laser Engin.*27;3;pp 190-193, ISSN 0387-0200

Van de Hulst H.C., (1998). *Light scattering by small particles,* Wiley & sons Inc., *ISBN* 0471293407, New York

Part 3

Photodiodes for Biomedical Application

8

The Photodiode Array: A Critical Cornerstone in Cardiac Optical Mapping

Herman D. Himel IV[1], Joseph Savarese[2] and Nabil El-Sherif[2,3]
[1]*Duke University, Durham, NC*
[2]*VA New York Harbor Healthcare System, Brooklyn, NY*
[3]*Downstate Medical Center, State University of New York, Brooklyn, NY*
USA

1. Introduction

The human heart pumps oxygenated blood to the organs and extremities in order to maintain normal physiologic function, while simultaneously pumping deoxygenated blood to the lungs for reoxygenation. Coordinated contraction of individual cardiac myocytes provides the mechanical force necessary to produce sufficient pressure and ensure that distant organs and extremities remain oxygenated. Before cardiac myocytes may contract, they must undergo excitation in order to begin the sequence of events which results in an intracellular calcium (Ca_i) rise, which in turn precipitates actin-myosin binding and ultimately results in contraction. The electrical signature of this series of events is reflected in the cardiac action potential (AP), a segment of a transmembrane voltage (V_m) recording which indicates electrical excitation (depolarization) and relaxation (repolarization) of the myocardium.

The duration, amplitude, upstroke velocity (dV_m/dt), and overall morphology of the cardiac AP are important markers of the electrical status of the heart. Studies of the cardiac AP have provided important insights into the mechanisms which drive the transition from a normal, healthy heartbeat toward a deadly cardiac arrhythmia.

Early recordings of the cardiac AP were obtained using microelectrodes (Coraboeuf & Weidmann, 1949a; Coraboeuf & Weidmann, 1949b; Draper & Weidmann, 1951; Sano et al., 1959; Sano et al., 1960; Weidmann, 1951). Although this method was highly effective in tracking temporal changes in the V_m of individual cells, the method could not be easily applied to the problem of tracking excitation over a region of tissue. Extracellular electrode mapping offered a partial solution to this problem and was sufficient to determine activation times in regions of tissue, but with this method the details of repolarization were lost and had to be estimated using indirect indicators. Further, this method required that the electrodes be in direct contact with the tissue. This made defibrillation studies difficult, since large amplitude defibrillation shocks typically obscure the details of activation during electrical recordings. Monophasic action potential (MAP) recordings were capable of elucidating the details of repolarization without damaging tissue, and have even been recorded in the beating human heart using a cardiac catheter (Shabetai et al., 1968). However they too were restricted by having little or no spatial resolution and could not be placed in close contact with each other. As with extracellular electrodes, MAP recordings also require that the electrodes be placed in contact with the tissue.

With the emergence of V_m-sensitive dyes in the 70's, it became possible to interrogate cardiac tissue optically (Salama, 1976), and soon afterward optical methods were developed to interrogate multiple spots simultaneously in a small (~cm^2) region of tissue. Since then the field of cardiac optical mapping (COM) has greatly expanded in scope, from relatively simple early recordings using one or relatively few spots (Morad & Dillon, 1981; Salama, 1976) to highly complex optical systems. These include high spatiotemporal resolutions systems (Choi et al., 2007), panoramic systems (Kay et al., 2004; Rogers et al., 2007), and systems which are capable of interrogating electrophysiological activity beneath the surface (Byars et al., 2003). In addition, several labs have used photodiode-based optical mapping systems to map V_m and Ca_i simultaneously, on both the whole heart (Choi & Salama, 2000; Lakireddy et al., 2006; Laurita & Singal, 2001; Pruvot et al., 2004) and in monolayer cell cultures of cardiac myocytes (Fast, 2005; Fast & Ideker, 2000; Lan et al., 2007).

Cardiac optical mapping systems have greatly increased our understanding in nearly all areas of cardiac electrophysiology, from basic studies of conduction patterns (Cabo et al., 1994; Knisley & Hill, 1995) and effects of fiber geometry (Knisley & Baynham, 1997; Knisley et al., 1994; Knisley et al., 1999; Neunlist & Tung, 1995) to more clinical studies of defibrillation (Al-Khadra et al., 2000; Fast et al., 2002; Federov et al., 2008; Tung & Cysk, 2007) and ablation therapy (Himel et al., 2007; Perez et al., 2006). Although COM has not yet led to a widely accepted method of three-dimensional cardiac tissue interrogation, there have been significant advances in this area as well. Investigators have successfully used optical surface recordings to determine wavefront orientation beneath the surface (Hyatt et al., 2005; Zemlin et al., 2008), and also to interrogate deeper layers of tissue using transillumination methods (Baxter et al., 2001) and deeper-penetrating, near-infrared fluorescing dyes (Matiukas et al., 2006; Matiukas et al., 2007; Salama et al., 2005).

Photodiode sensors were used in some of the earliest optical recordings of cardiac APs (Morad & Salama, 1979; Salama, 1976), and continue to be used today (Cheng, 2006; Sakai, 2008). Photodiodes function by transferring incoming photonic energy to bound electrons in a semi-conductive material in a transistor configuration. These energized electrons may then cross from one side of the transistor to the other, resulting in a voltage difference between the two sides.

If a wire is connected from one side of the photodiode to the other while the photodiode is receiving photonic energy, current will flow in a linear fashion with respect to the input intensity of the collected light (Scherz, 2007). This makes photodiodes an excellent choice as a detector in COM systems, and this fact has been reflected by their widespread use over the past 30 years. Examples of optical APs and activation maps recorded with a photodiode array-based system are shown in figure 1.

Although other technologies such as CMOS and CCD cameras have recently gained popularity due to their higher spatial resolution, photodiode systems remain in use due to their ruggedness, high signal-to-noise ratios, excellent temporal resolution, versatility, and low cost. Recently, for example, photodiodes and photodiode arrays (PDAs) have been used in the construction of optrodes, a novel technique used to record optical signals from deeper intramural regions within the ventricular wall (Byars et al., 2003; Caldwell et al., 2005; Hooks et al., 2001; Kong et al., 2007).

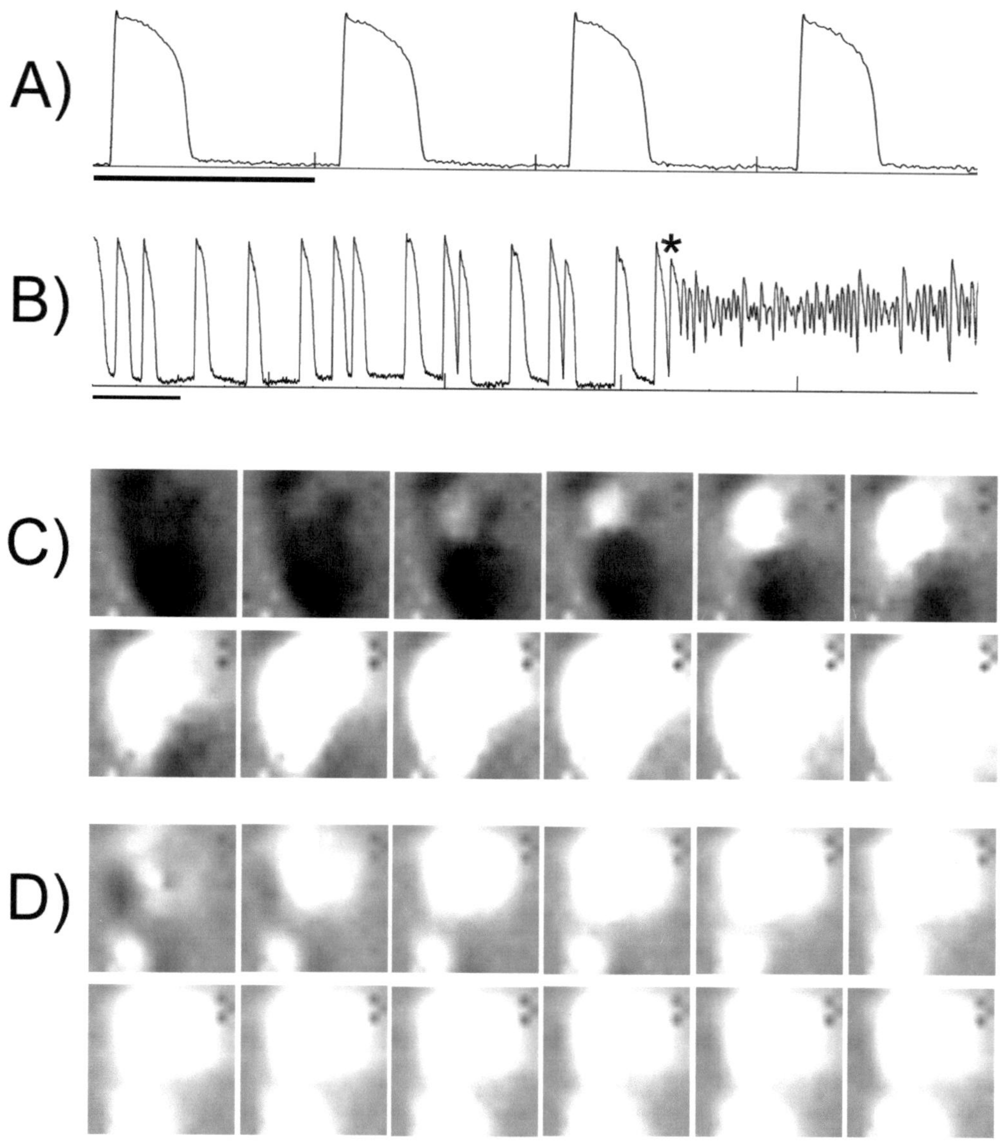

Fig. 1. APs and activation maps for normal and irregular rhythms. For rows A and B, the horizontal bar beneath each recording indicates 1 second. Row A shows APs recorded during basic rhythm. Row B shows APs occurring with irregular diastolic intervals, followed by a long run of a ventricular tachyarrhthmia, triggered by the AP marked with an asterisk. The two rows in section C show a sequence of activation during basic rhythm, while the two rows in section D show a sequence of activation which took place during a premature beat which precipitated a sustained ventricular tachyarrhythmia (note the presence of two distinct activation sites). Frames are read from left to right, and then top to bottom. Each successive frame is 1 ms apart. Lighter areas on the map indicate tissue undergoing activation.

2. Basic principles of cardiac optical mapping

Epi-illumination occurs when the fluorescence emission detector is placed on the same side of the tissue as the excitation source, whereas with trans-illumination the detector and excitation source are placed on opposite sides. For monolayer mapping systems, both epi and trans-illumination are possible since cardiac monolayers are typically only a few tens of micrometers thick. For whole-heart mapping systems which map excitation on the surface of the intact heart preparations, epi-illumination is the preferred method since very little fluorescence is transmitted through the relatively thick myocardial wall.

The tissue being mapped must be illuminated using an excitation source, which excites at least one parameter-sensitive dye in order to elicit a fluorescent signal. Changes in a targeted physiological parameter cause changes in the properties of the dye (e.g., a conformational change in the dye molecules). This results in a change in the emission spectrum of the dye, which is then recorded by a detector (e.g., a PDA), digitized, and stored on a PC for post-experimental analysis.

Changes in fluorescence due to changes in the physiological parameter are often measured as a fraction of the baseline fluorescence. This is an important parameter in optical mapping, and is known as fractional fluorescence ($\Delta F/F$). Fractional fluorescence is useful because it is a way to measure the effectiveness of a particular dye in transducing a physiological change into recordable fluorescent signal. Fractional fluorescence also indicates the general effectiveness of the system, and higher $\Delta F/F$ values are typically accompanied by higher signal-to-noise ratios. Transmembrane voltage is the most commonly studied physiological parameter in optical mapping, but intracellular calcium transients (Ca_iT) have also been studied extensively.

There are several variations of the COM system, however there are basic components that are common to all systems. These basic components include an excitation source, detector, and electronic components used for digitization, filtration, and multiplexing. A schematic for a typical whole-heart mapping system is shown in figure 2.

2.1 Excitation source

The excitation source may be either focused (i.e. laser light) or broadfield illumination (using halogen, tungsten, or more recently, high-power LED sources). Laser and some LED light sources have sufficiently narrow bands so as not to interfere with the fluorescence emission, however broadfield sources should be pre-processed using optical filters in order to decrease the width of their wavelength spectrum before illuminating the target (e.g., the heart). In general, brighter excitation sources lead to higher $\Delta F/F$ values, however the intensity of the excitation source cannot be increased without regard for *photobleaching*, which occurs when the dye emission decreases in intensity due to overexposure to excitation light (Knisley et al., 2000; Kong et al., 2003). A highly stable source is superior to a brighter but noisier source, since the stable source yields greater S/N ratios while allowing longer duration recordings.

2.2 Detector

There are several types of detectors that are currently in use for COM, however the focus of this review is upon those detectors which are photodiode-based. Other detector types will be discussed for the purpose of comparison.

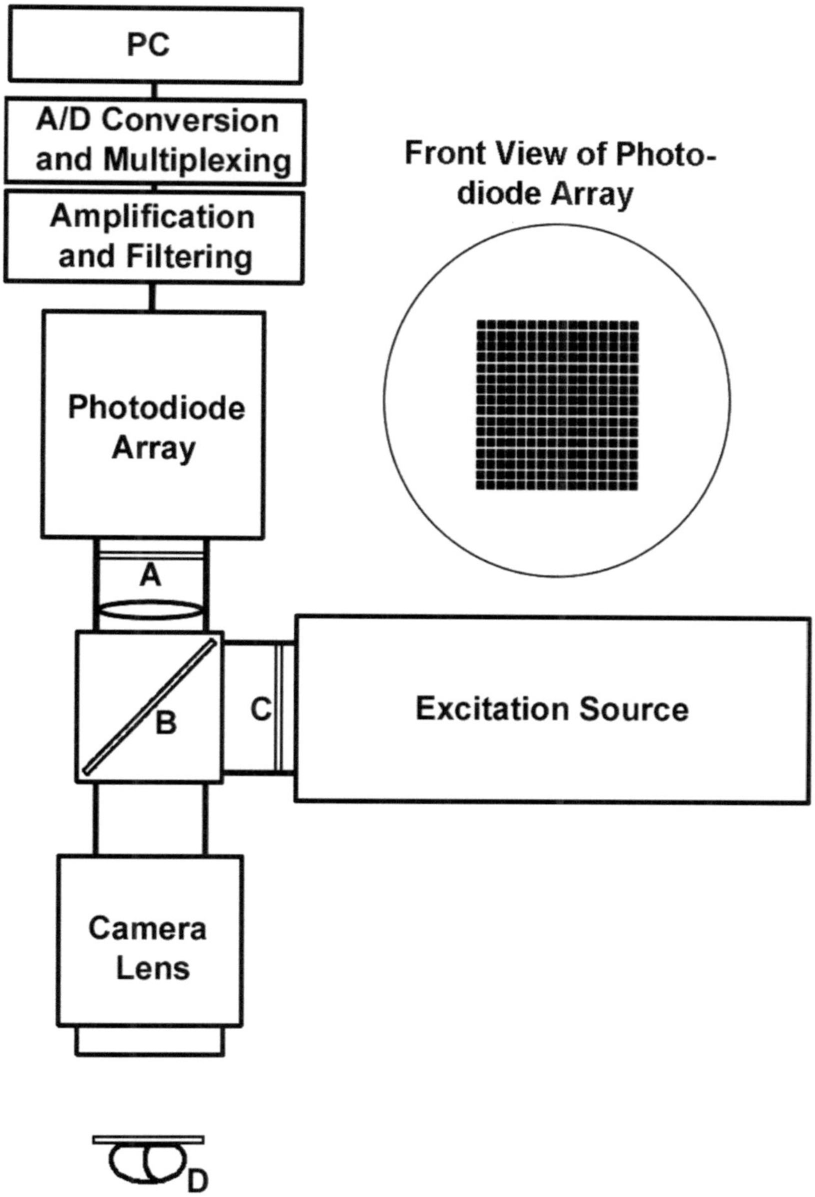

Fig. 2. Top-down view of a typical PDA-based optical mapping system. The thin rectangular boxes marked A, B, and C represent fluorescence band pass, long pass, and excitation band pass filters, respectively. The long pass filter B is housed in an optical cube. The elliptical shape between A and B represents a condensing lens. The front view of the PDA shows the 16x16 element photoactive region of the detector. The oval-like shape marked D represents the heart, which is pressed against a flat plate in order to create a two-dimensional surface so that the entire mapped region lies within the focal plane. Electrically connected components are separated by a thin solid line, while optically connected components are separated by hollow rectangles.

Current alternatives to PDAs include photomultiplier (PMT) systems, charge-coupled device (CCD) cameras, and complimentary metal-oxide semiconductor (CMOS) cameras. PMT systems have an extremely high gain (up to 10^8 increase in intensity), and can even be used in a process call "photon counting" whereby the release of individual photons may be recorded. PMT systems are capable of extremely high sensitivity, and typically have a quicker response time than photodiode systems. However for normal optical mapping applications, the sensitivity and response time of photodiodes is more than sufficient. Typically, PMT systems employ sequential rather than simultaneous recordings of adjacent spots. However, due to their rapid response time they may be coupled with a laser scanner to produce acquisition rates equivalent to those seen in photodiode-based systems. For example, a laser scanner which scans at 256 kHz with a 256-spot grid is essentially equivalent to a PDA-based 256-spot system which scans at a rate of 1 kHz. For this example, there will be a time difference of ~1 ms between the 1st and 256th spots scanned. This difference can often be neglected during normal propagation in cardiac tissue. The basis for the photomultiplier tube is the *photoelectric effect*, whereby an incident photon strikes a photocathode and starts a chain reaction. The photocathode emits an electron, which strikes subsequent dynodes, at each step of the way releasing a greater number of electrons than those which struck the preceding dynode. At the final anode, the signal has been "multiplied" many fold (~10^6) and the resultant electrons result in a change in voltage.

Cameras using CCD technology typically have impressive spatial resolution, however these systems have not been able to achieve the same temporal resolution as PDA-based systems. The CardioCCD-SMQ (RedShirt Imaging, Decatur, GA) currently offers a spatial resolution of 80x80 while achieving a temporal resolution of 2000 frames per second (fps). The CCD camera may achieve higher temporal resolution and/or higher signal-to-noise ratios by a process known as "binning," however this process decreases the spatial resolution of the camera, the major advantage of CCD technology. When using the CCD camera for cardiac mapping experiments, light saturation of the detector may be an issue and thus gain must be carefully controlled. Typically this is less of a problem with PDA systems. The basis of operation of the CCD detector is a shift registration process. Charge which is accumulated due to photons falling upon light-sensitive regions within the detector is transferred along the 2-dimensional detector array until the charge within a given potential well reaches the final readout electrode. The transfer of charge is accomplished by changing the voltage in an adjacent pixel, causing electrical charge to flow in the desired direction.

Although more costly than either CCD cameras or PDAs, the CMOS camera boasts exciting technology that has recently become capable of delivering extremely high spatiotemporal resolution. RedShirt Imaging lists the CardioCMOS-128f as being capable of recording 128x128 spots at an acquisition rate of 10,000 fps (RedShirt Imaging, LLC, Decatur, GA). CMOS technology typically allows for a very large well depth and a large dynamic range, although these cameras still lack the DC coupling ability of the PDA system. The architecture of the CMOS detector is what sets it apart from the CCD detector. The CMOS detector uses specialized manufacturing techniques to create micro-arrays of photodetectors, each with their own dedicated amplifiers and independent circuitry. Thus CMOS detectors are capable of performing signal processing functions on a pixel-by-pixel basis. The ability of the CMOS detector pixel to record and process signals on a pixel-by-pixel basis is in contrast to the CCD detector, which must transfer signals from individual pixels to be processed by downstream electrical circuitry. This fundamental difference in circuitry architecture is reflected in the higher acquisition rates typically observed in CMOS detectors.

The PDA, as well as the individual photodiode, remains a cost-effective and rugged solution to a wide variety of problems within the field of COM. Photodiode arrays boast a wide spectral response, high dynamic range, high temporal resolution, and the largest well depth of the COM detectors. The PDA is typically a rugged device and can operate in high-light conditions typical in most laboratory experiments, while still delivering a high signal-to-noise ratio. Although low-light conditions can be achieved in COM applications, this typically places limitations on $\Delta F/F$. Another interesting and unique feature of the PDA as a COM detector is that they may be AC coupled. This removal of the DC component of the signal allows the entire dynamic range of the detector to be used on the signal itself, rather than on both the signal and the background baseline fluorescence. This is particularly helpful when imaging with dyes that have a large background fluorescence signal (e.g., di-4-ANEPPS). Due to their robust nature and versatility in a wide variety of applications, photodiode-based systems remain the "workhorse" detector in COM. The PDA operates on principles based on the individual photodiode. Currents from individual photodiodes in the PDA are converted into voltages, which are read out of the detector in parallel and processed by a PC.

The typical noise levels, sensitivity, speed, spatial resolution, and ability to respond to high light levels without saturation (i.e. well depth) are summarized in the following table:

	Noise Levels	Sensitivity	Speed	Spatial Resolution	Well Depth
CCD	**	****	**	****	**
CMOS	****	***	****	*****	****
PDA	*****	****	*****	**	*****
PMT	***	*****	*****	*	**

Table.

For the above chart, more asterisks means better performance, which in the case of COM translates into lower noise, greater sensitivity, greater speed, higher spatial resolution, and a larger well depth. A few caveats apply to the above chart. Although the PMT has extraordinary speed and sensitivity, it is limited by the fact that there is typically no spatial resolution, and must be used in a sequential scanning method. Also, the authors have tried to take into account the *useful* range of a particular characteristic of the detector. For example, many CCD cameras are capable of recording at a rate of 2 kHz, however in practice it is currently challenging to obtain useful signals ($S/N > 5$) at this rate and at full spatial resolution. Rates of 200-500 Hz are more typical in practice. Also, it must be mentioned that these technologies are constantly and rapidly evolving, thus these ratings are of course subject to change. Also, note that the above comparisons refer to detectors commonly used in COM, and are not meant to serve as a basis for comparison for these technologies in general.

2.3 Filtration, digitization, and multiplexing

Optical signals are subject to several types of noise which must be removed in order to accurately study the details of the cardiac AP. We will briefly review the types of noise most relevant to COM systems. Various types of white noise are ubiquitous throughout all types of electronics, and are typically of frequencies well above those of cardiac signals. Thus low pass filters are typically used to help remove white noise. Sixty-cyle is another type of noise that is often encountered when collecting optical signals. This noise may contaminate signals by way of electromagnetic waves from nearby power outlets, or may be introduced if equipment used in optical mapping experiments is powered using AC power (i.e., if the equipment is not isolated). It may be alleviated by the use of a Faraday cage and/or the use of a band-stop (i.e., notch) filter centered at 60-Hz. Mechanical vibrations may also affect optical signals, and can range from fluctuations in air current to vibrations due to foot traffic. Sources of mechanical vibrations are highly varied in nature, and must be dealt with on a case-by-case basis. A research-grade optical table with active isolation should be sufficient to suppress most sources of mechanical noise. Optical recordings may also contain drifts in basline voltage due to several sources. These include photobleaching, dye washout, and dye internalization into the inner leaflet of the cell membrane. One way to reduce the impact of these noise sources is to employ the technique of ratiometry (Knisley et al., 2000), which is discussed in the *Pre/Post-Conditioning* section which follows.

Many optical devices, including photodiodes and photomultiplier tubes, record analog signals that must be digitized before being stored on a PC. Digitization equipment must have sufficient speed and throughput in order to follow the high spatiotemporal resolution required for optical mapping applications. A review article by Entcheva et al. summarizes the state of the art in this sub-field of COM (Entcheva & Bien, 2006).

When data is recorded from a two-dimensional grid of sites simultaneously, the most intuitive storage method is a two-dimensional matrix of values. However, prior to storage on a PC, this data must be routed from the digitization equipment to the PC. This requires arranging the data in a sequential fashion, a process known as multiplexing. For a 16x16 element PDA, data for a single millisecond might be arranged from sites 1 to 256 and then be sequentially sent to the PC for storage. Following this, the data could then be demultiplexed and arranged as a 2D matrix, a more logical form for creating activation maps and other graphics to assist with visualization of the data.

2.4 Pre/post-conditioning of optical data

In contrast to electrical signals, optical signals are highly sensitive to heart motion. Various methods have been developed in order to reduce "motion artifacts" which are often present in optical signals. These motion artifacts are thought to be the result of a change in the location of the mapped region on the heart surface, where a fluorescence gradient typically exists (Himel & Knisley, 2006). Methods to reduce motion artifact include physically restraining the heart (Efimov et al., 1996; Girouard et al., 1996), the use of electromechanical uncouplers (Federov et al., 2007; Jalife et al., 1998; Li & Nattel, 2007; Wu et al., 1998), and the technique of ratiometry (Hooks et al., 2001; Knisley et al., 2000; Kong et al., 2003). Physically restraining the heart reduces motion artifacts simply by limiting the extent to which the heart can move during contraction, thus limiting the amount by which the mapped region moves with respect to its original position on the heart. Electromechanical uncouplers work by a variety of methods, but most have an effect upon the actin-myosin cytoskeleton which is responsible for contraction. Electromechanical uncouplers should be used with care, as some studies have shown that these agents can affect various parameters of the cardiac AP and may also affect the dynamics of ventricular fibrillation (VF) (Baker et al., 2004; Biermann et al., 1998; Hyashi et al., 2003; Lee et al., 2001).

Ratiometry is a signal processing method that requires the use of two recorded optical signals, each of a different band of the wavelength spectrum. We will refer to the longer wavelength signal as "red" and the shorter wavelength signal as "green". When the V_m-sensitive dye di-4-ANEPPS is excited, the peak of the emission spectrum of the dye shifts toward shorter wavelengths (green). Thus the green signal would show an increase in fluorescence intensity while the red signal would show a corresponding decrease in intensity. Using this method, an upright cardiac AP would be recorded in the green signal while an inverted (or "upside-down") AP would be recorded in the red signal. The important thing to consider is that the emission signal corresponding to V_m is emitted at a relatively narrow frequency band. Contrastingly, emission due to motion is not heavily wavelength-dependent, and will cause the change in fluorescence signals in the same direction regardless of the wavelength band of the collected signal. Since the motion signals are common to both collected wavelengths, we may reduce motion artifacts by simply taking the ratio of the green signal to the red on a point-by-point basis. This will cause a significant reduction in the motion artifact, and will help us to isolate the electrical signal. This technique could be achieved in the laboratory by using dual PDAs and separating fluorescence emission into two wavelength bands, one above and one below the peak emission wavelength of the dye of interest. In addition to motion artifact removal, ratiometric signals have also been used to study motion artifacts optically. This may be achieved by subtracting the electrical signal from the signal containing both electrical and motion components (Himel et al., 2006).

Since fibrillation (both ventricular and atrial) is a topic of great clinical and theoretical interest, considerable effort has been expended in order to analyze data recorded from the fibrillating heart. This data is challenging to analyze and interpret, since recordings of fibrillation often have a chaotic appearance when viewed with time as the horizontal axis. Thus a variety of alternate methods have been used to gain insight into the nature of fibrillation, including dominant frequency analysis (Caldwell et al., 2007; Choi et al., 2003; Choi et al., 2006; Joel & Hsia, 2005; Moreno et al., 2005; Wu et al., 2004; Wu et al., 2006; Zaitsev et al., 2003) and mutual information (Omichi et al., 2004, Wu et al., 2005). More recently the use of a metric known as spatiotemporal entropy has been used to analyze oscillatory dynamics in cardiac and neural systems (Bub et al., 2005; Jung et al., 2000; Himel et al., 2009).

Dominant frequency analysis involves use of the Fourier transform to examine the frequency content of fibrillation recordings. Some groups have used dominant frequency analysis to support the theory of a "mother rotor" (i.e., a high-frequency region of the heart that drives fibrillatory activity) (Chen et al., 2000; Jalife et al., 1998; Zaitsev et al., 2000), however this issue remains controversial (Berenfeld et al., 2001). Others believe that fibrillation is maintained by the constant creation and annihilation of wavelets which occur due to functional and anatomical heterogeneities (Choi et al., 2002; Lee et al., 1996; Moe, 1962; Rogers et al., 1999; Valderrábano et al., 2002). More recently, some groups have concluded that fibrillatory activity may be driven by both mechanisms, depending on specific conditions in the heart (Chen et al., 2003; Liu et al., 2004; Nash et al., 2006; Wu et al., 2002).

Mutual information techniques have been used to examine the relationship between V_m and Ca_i. Algorithms assign a numerical value for individual signals in order to quantify the degree of similarity between V_m and Ca_i during fibrillation in order to give insight into the mechanisms of arrhythmogenesis and the maintenance of fibrillation; however, mutual information must be calculated for individual signal pairs and by itself does not indicate *spatiotemporal* heterogeneities in V_m/Ca_i relationships.

Spatiotemporal entropy has been used to quantify the degree of uncertainty in both time and space by considering them as lumped parameters, and analyzing activations in the context of space-time cubes (i.e., stacked two-dimensional optical maps with time as the third dimension). Spatiotemporal entropy has been used to analyze neural simulations as well as oscillatory dynamics in cultured cell monolayers (Bub et al., 2005; Jung et al., 2000). Spatiotemporal entropy analysis is appealing when analyzing optical mapping data, since one of the main strengths of COM data lies in its spatial resolution.

3. Studies in cardiac electrophysiology research using photodiode arrays

This section will showcase three recent optical mapping studies from this lab which examine cardiac arrhythmia mechanisms in the context of global ischemia (Himel et al., 2009; Lakireddy et al., 2005; Lakireddy et al., 2006). These studies examine the dynamic relationship between V_m and Ca_i over the course of ischemia/reperfusion injury. These studies used a photodiode-based system which simultaneously records V_m and Ca_i with two separate 16x16 PDAs (figure 2 shows 16x16 element photoactive region of the PDA, see figure 3 for a schematic of the simultaneous dual-measurement system). This system was designed by B.R. Choi and G. Salama, and uses two Hamamatsu C4675-103 detectors. Please see the excellent review by Salama et al. for more details regarding this system (Salama et al., 2009). The whole-heart guinea pig (GP) Langendorff model was used, where the aorta was cannulated and hung vertically within a surrounding bath of physiological solution referred to as Tyrode's solution. Heated (~37° C), oxygenated Tyrode's solution was pumped retrogradely through the aorta in order to nourish and oxygenate the heart via the coronary arteries. These studies simulated the condition of global ischemia (i.e., oxygen deprivation) and reperfusion by temporarily interrupting perfusion into the coronary arteries and then restarting the perfusion. The guinea pig animal model is similar to human physiology in terms of its AP morphology and intracellular calcium handling.

Study 1 (Himel et al., 2009): Ventricular tachycardia is frequently observed in the clinical environment. VT may either be brief (<30 seconds) or may continue indefinitely, and patient survival is critically dependent on the duration of the VT episode. Spontaneous termination of the VT episode within a few seconds leaves the patient unharmed. If the VT episode continues however, the patient is typically dependent upon a defibrillatory shock to terminate the episode, without which the patient may not survive. Although extremely important clinically, the mechanism which determines the duration of the VT episode is not well understood. It is thought that uncoupling between V_m and Ca_i coupling may play a role in determining whether an episode of VT will terminate spontaneously or continue indefinitely. This study sought to use the concept of spatiotemporal entropy (E) as a metric to determine the degree of uncoupling between V_m and Ca_i during the early phases of a VT episode, and to test the predictive power of E for VT duration. During normal sinus rhythm, propagation of excitation waves in the heart is uniform and wavebreak is typically not observed. During episodes of VT, wavebreak is common, and fragmentation of excitation waves in the heart is frequently observed. Moreover, differences between V_m and Ca_i signals are often observed. These differences in V_m/Ca_i coupling may be quantified using the absolute value of the difference in E, symbolized by E_d.

This study examined several groups of VT episodes which were divided according to whether or not they terminated spontaneously. Self-terminating episodes of VT were further classified as short (<5 seconds) or long (>5 seconds). E_d was determined for the first 500 ms of all VT episodes. E_d values for non self-terminating episodes of VT were significantly greater than self-terminating VT episodes. Further, E_d values for long self-terminating episodes of VT were significantly greater than those for short self-terminating episodes.

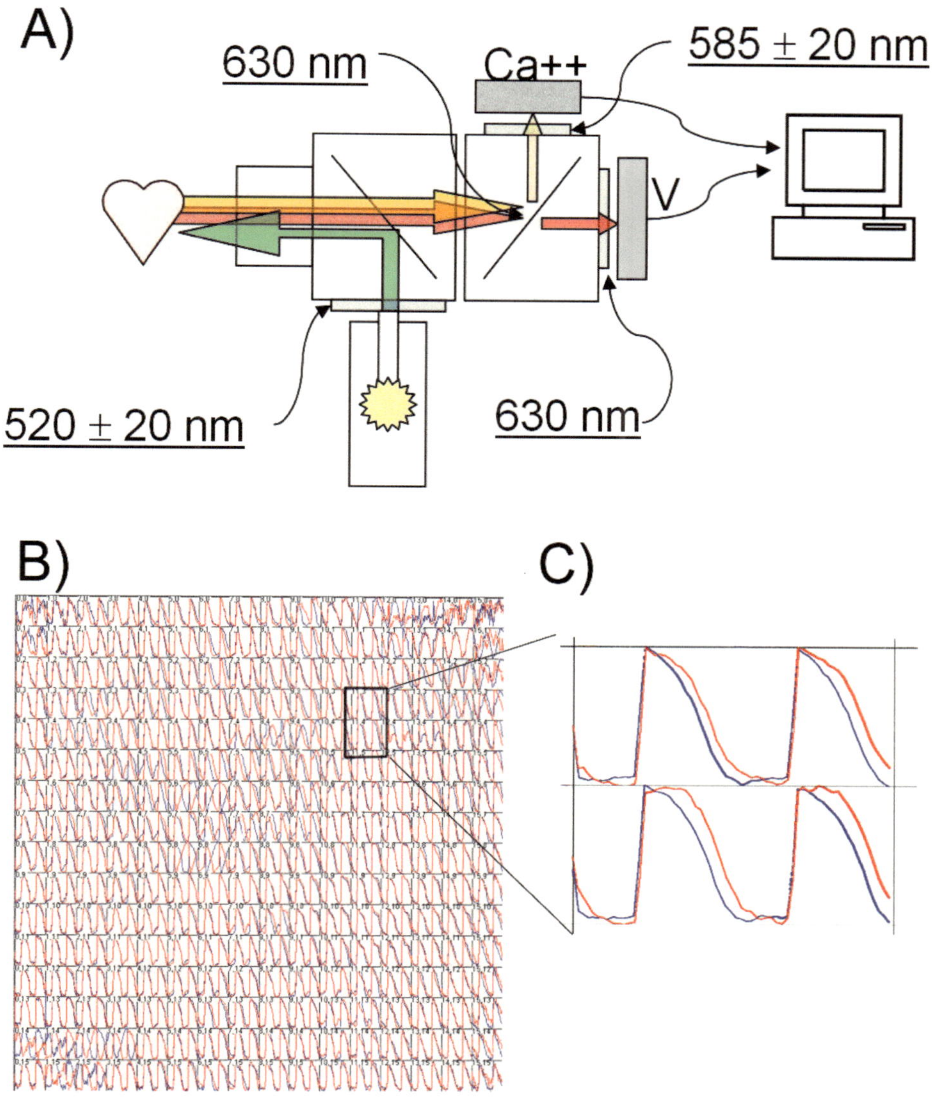

Fig. 3. Optical mapping diagram for simultaneous V_m and Ca_i measurements. For panel A), the position of each filter is indicated by thin curved lines with arrowheads, and the characteristics of the filters are underlined (high pass filters are indicated by a single number, whereas band pass filters have number ± band). The calcium and voltage PDAs are shaded gray. The large color-filled lines indicate the path of the given color of light. The excitation source is the bottom component in panel A). Panel B) shows an map of optical signals, and panel C) shows enlarged traces of the indicated pixels. The blue signal in panel C) represents V_m, whereas the red signal indicates Ca_i.

Wavebreak and differences in V_m and Ca_i spatial fluorescence maps were consistently identified during periods of high E and high E_d. The results are illustrated by figures 4 and 5, which show typical examples of low and high E_d during VT episodes.

High E_d correlated with a greater duration of a VT episode. This may be related to destabilization of propagation and uncoupling between V_m and Ca_i activation wavefronts.

Study 2 (Lakireddy et al., 2005): This study examines spatial dispersion of repolarization in the context of global ischemia, and also the role spatial dispersion plays in the development of electrical alternans. Electrical alternans is a term used to describe beat-to-beat alterations in AP morphology. For example, a one-to-one APD alternans occurs when a normal AP is followed by a short-duration AP, which is then followed by normal AP, short-duration AP, and so on. Electrical alternans are considered to be a strong marker of electrical instability, and often precede malignant arrhythmias such as ventricular tachycardia (VT) and VF (Hohnsloser et al., 1998; Gold et al., 2000; Ikeda et al., 2006; Pastore et al., 1999; Pham et al., 2003; Rashba et al., 2004; Rosenbaum et al., 1994).

In this study by Lakireddy et al., ischemia-induced changes in APD and intracellular calcium transient duration (Ca_iT-D) were determined, and their relationship with electrical alternans was investigated. Recordings show that ischemia resulted in a significant decrease in APD, but resulted in a significant increase in Ca_iT-D. In addition, changes in APD were spatially heterogeneous while changes in Ca_iT-D were relatively homogeneous (see figure 6). Sites with less shortening of APD displayed alternans in both Ca_iT-D and APD, while sites with more shortening of APD displayed Ca_iT-D alternans but little or no APD alternans, leading to a condition of significant spatial dispersion of the APD. The condition of increased spatial dispersion due to ischemia is thought to account for the vulnerability of the heart to alternans.

Study 3 (Lakireddy et al., 2006): In a second study by Lakireddy et al., the association between arrhythmogenesis and spontaneous calcium oscillations (S-CaOs) was examined in the intact heart. It is known that ischemia/reperfusion leads to elevated Ca_i and an alteration in Ca_i kinetics (Bers, 2002; Marban et al., 1990; Steenbergen et al., 1987). This alteration in normal Ca_i kinetics can lead to S-CaOs. Under such conditions, the normal master/slave relationship between V_m/Ca_iT signals is reversed (i.e., calcium signals precede and drive V_m).

The goal of this study was to investigate the correlation between S-CaOs and arrhthmogenesis using an experimental GP model with 15 minutes of no flow ischemia followed by 15 minutes of reperfusion. Changes in Ca_i and Vm in a limited zone of the epicardial surface of the GP heart were simultaneously recorded and carefully examined. The study provided evidence of a linkage between S-CaOs and arrhythmogenesis in the setting of ischemia/reperfusion (I/R). In the intact heart during I/R, spontaneous premature beats (PBs) occurred and were ubiquitous. Some PBs initiated a VT or VF (see figure 7), while others remained confined to their site of origin and did not result in an arrhythmia (see figure 8). Two important observations had to be made in order to link an arrhythmia to S-CaOs in the experimental model. First, the beginning of S-CaOs preceded the onset of the simultaneously recorded membrane depolarization by 2-15 ms at a very restricted site in the optical field. In recordings obtained further away from the focal site of origin, the relative amplitude of the S-CaOs gradually decreased and the start of membrane depolarization preceded the onset of S-CaOs. Second, the presence of some degree of conduction block, which by definition is the failure of S-CaOs to trigger a fully propagated response, was essential for the localization of the focal site of origin. Thus S-CaOs may remain concealed (and hence benign) or may manifest as PBs, VT or VF.

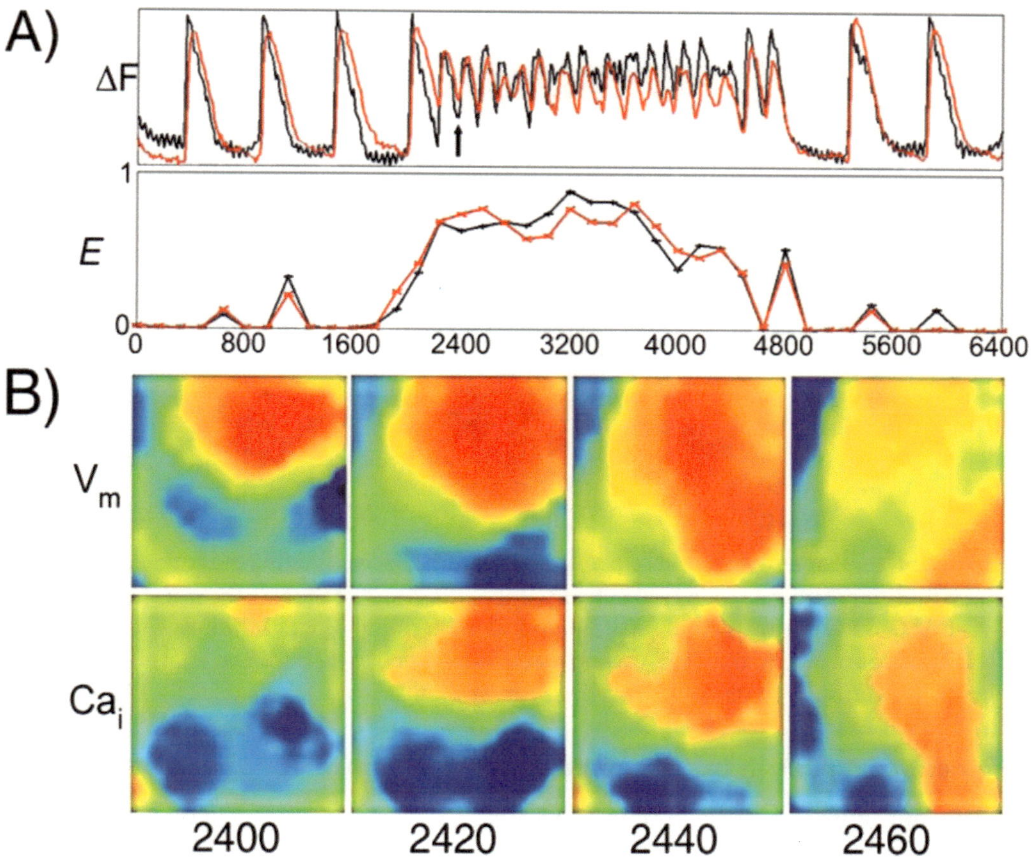

Fig. 4. Example of spatiotemporal entropy from a short self-terminating run of a ventricular tachyarrhythmia. V_m and Ca_i were recorded using a dual PDA system. Entropy traces (2nd row, panel A) show small differences between V_m and Ca_i entropy, indicating a low degree of spatiotemporal uncoupling. Fluorescence maps in panel B show similar, but not identical excitation wavefronts for V_m and Ca_i. Values given for traces and fluorescence maps are given in ms. (*reproduced with permission from Himel et al., 2009*).

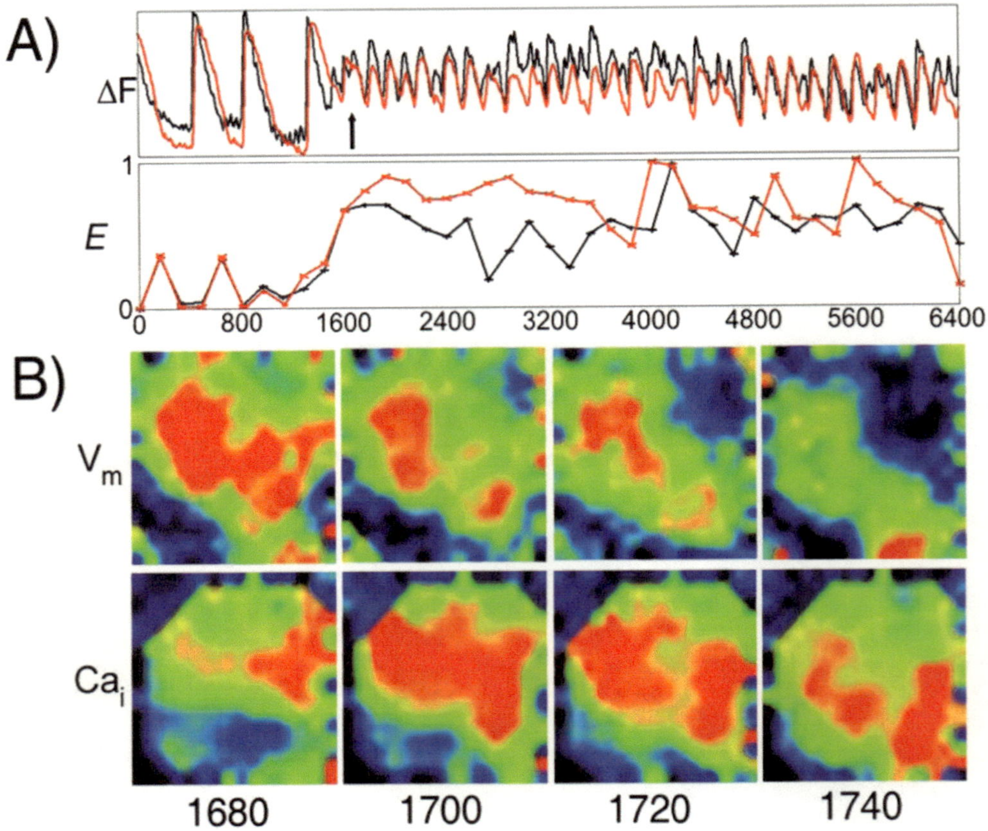

Fig. 5. Example of spatiotemporal entropy from a non self-terminating run of a ventricular tachyarrhythmia. Differences in the entropy traces (2nd row, panel A) show a disparity between V_m and Ca_i entropy and spatiotemporal uncoupling. Spatial fluorescence maps reflect the disparity shown by the V_m and Ca_i traces. Values given for traces and fluorescence maps are in ms. (*reproduced with permission from Himel et al., 2009*).

4. Recent advances in cardiac optical mapping

We will now turn to briefly discuss a few of the important recent advances in COM. Optrodes are bundles of microscopic fiberoptic cables which are inserted into cardiac tissue in order to interrogate intramural activation patterns. They are similar to plunge needle electrodes in their usage; however optrodes are capable of measuring complete APs, including the repolarization phase, whereas plunge electrodes measure extracellular potentials only. It is thought that in the future optrodes will play an important role in more carefully examining transmural dispersion of repolarization, an important factor in arrhythmogenesis in a variety of cardiac diseases (Antzelevitch, 2007; El-Sherif et al., 1996; Milberg et al., 2005; Shimizu et al., 1997). Optrodes may also be important in the study of the dynamics of arrhythmic circuits, since they are often present deeper in the myocardial wall (Allison et al., 2007; Li et al., 2008; Valderrábano et al., 2001). However, like plunge electrodes optrodes must also be inserted into the tissue and therefore cause damage which may by itself alter activation patterns. Thus the effects of this insertion must be carefully

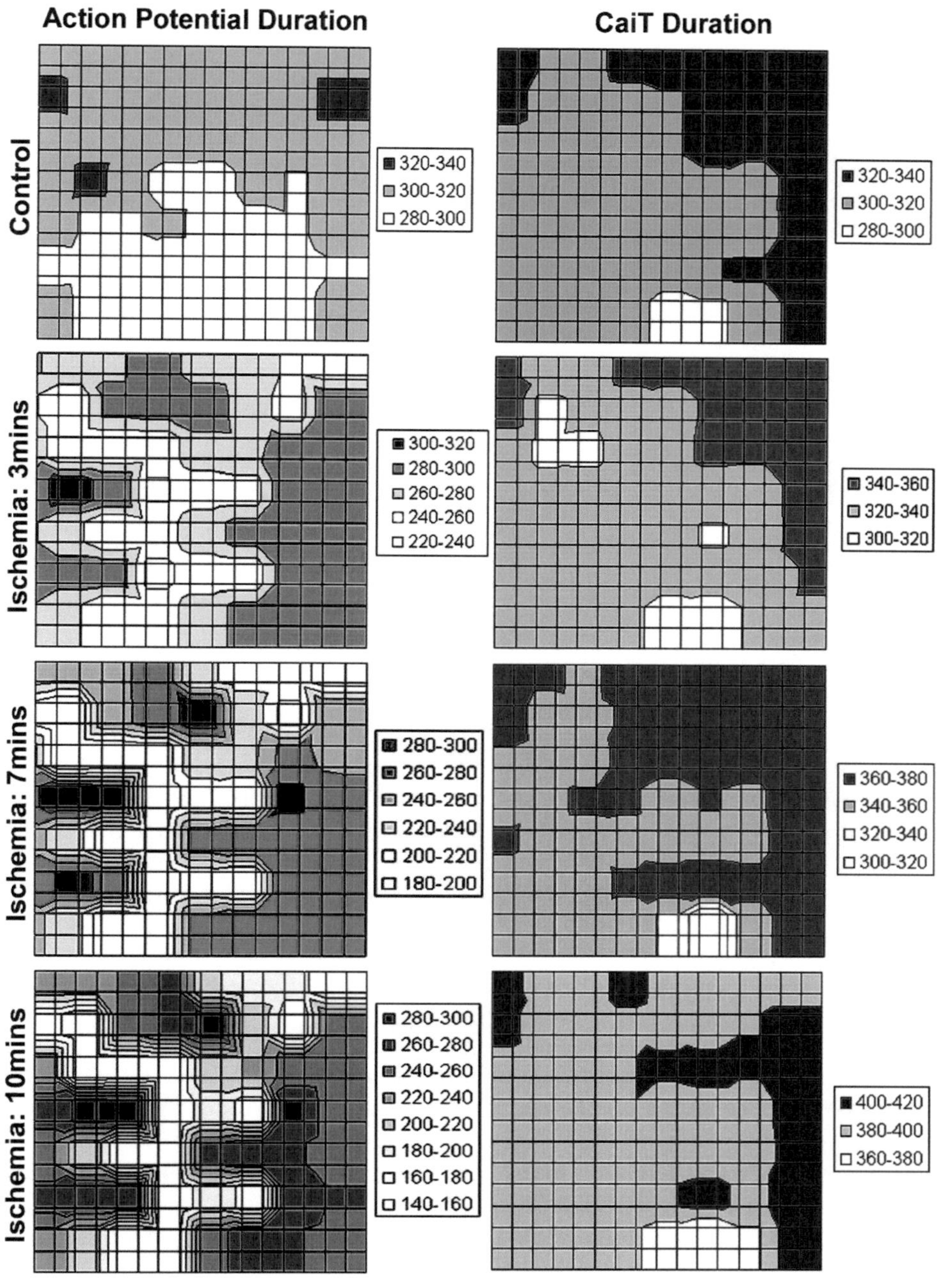

Fig. 6. APD and Ca_iT-D during normal perfusion and into ischemia. Scales to the right indicate the color of a given APD or Ca_iT-D. *(reproduced with permission from Lakireddy et al., 2005).*

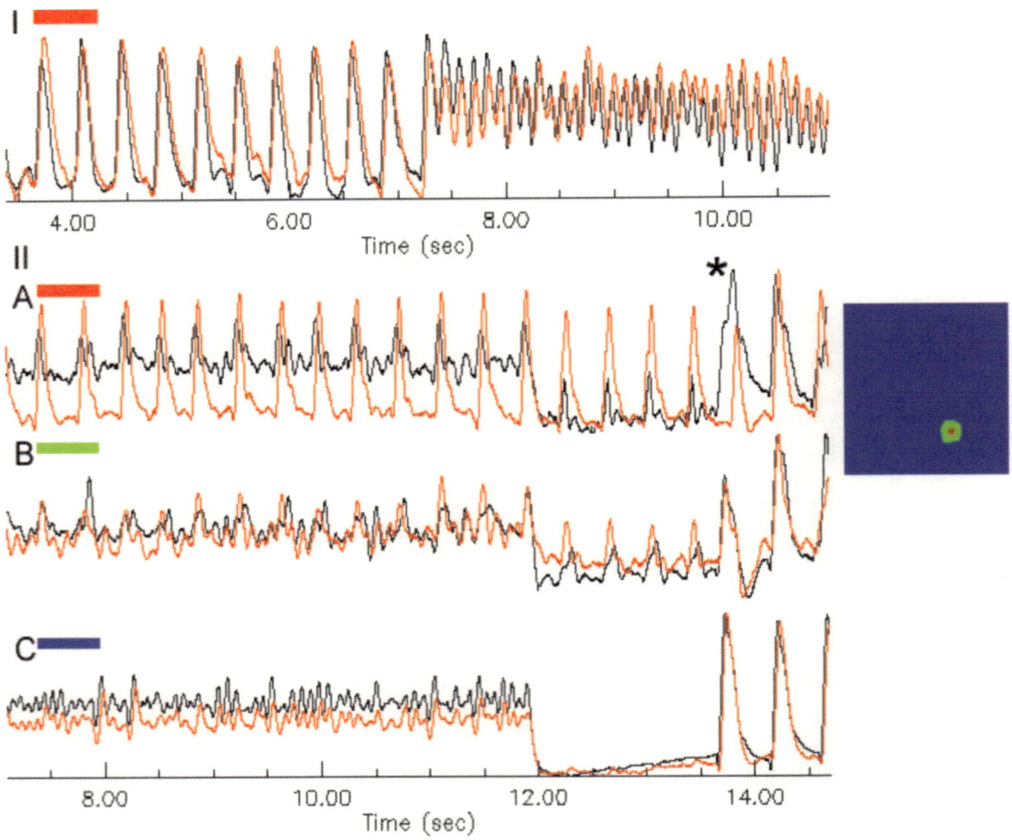

Fig. 7. Concealed spontaneous calcium oscillations (S-CaOs). Recordings were obtained from an experiment in which localized S-CaOs developed during an episode of self-terminating VF and continued uninterrupted after the resumption of spontaneous cardiac rhythm. Panel I illustrates the initiation of VF. Panel II shows recordings from three representative pixels (marked by different colors in the map of the optical field, seen to the right of the traces). After the self-termination of VF (at approximately 12 seconds), the majority of the optical field showed a pause with no electrical activity (trace C of panel II), while the localized S-CaOs continued. *(reproduced with permission from Lakireddy et al., 2006).*

considered when interpreting intramural data (El-Sherif, 2007). Photodiodes have played a dominant role in the construction of optrodes (Caldwell et al., 2005; Kong et al., 2007; Byars et al., 2003).

Several groups have recently begun to use multiple cameras to simultaneously interrogate opposing sides of the ventricular wall (Evertson et al., 2008; Kay & Rogers, 2006; Kay et al., 2004; Kay et al., 2006; Rogers et al., 2007). In addition, some of these groups use additional cameras to recreate the geometry of the heart in order to properly orient optical maps from several cameras on the epicardial surface (Kay et al., 2004; Evertson et al., 2008). Most Panoramic optical mapping systems are based on CCD technology, however systems have also been built using multiple PDAs (Qu et al., 2007). Panoramic optical mapping does not address the problem of lost depth information, but does provide a significant improvement over traditional optical mapping which only maps a limited region on the epicardial surface.

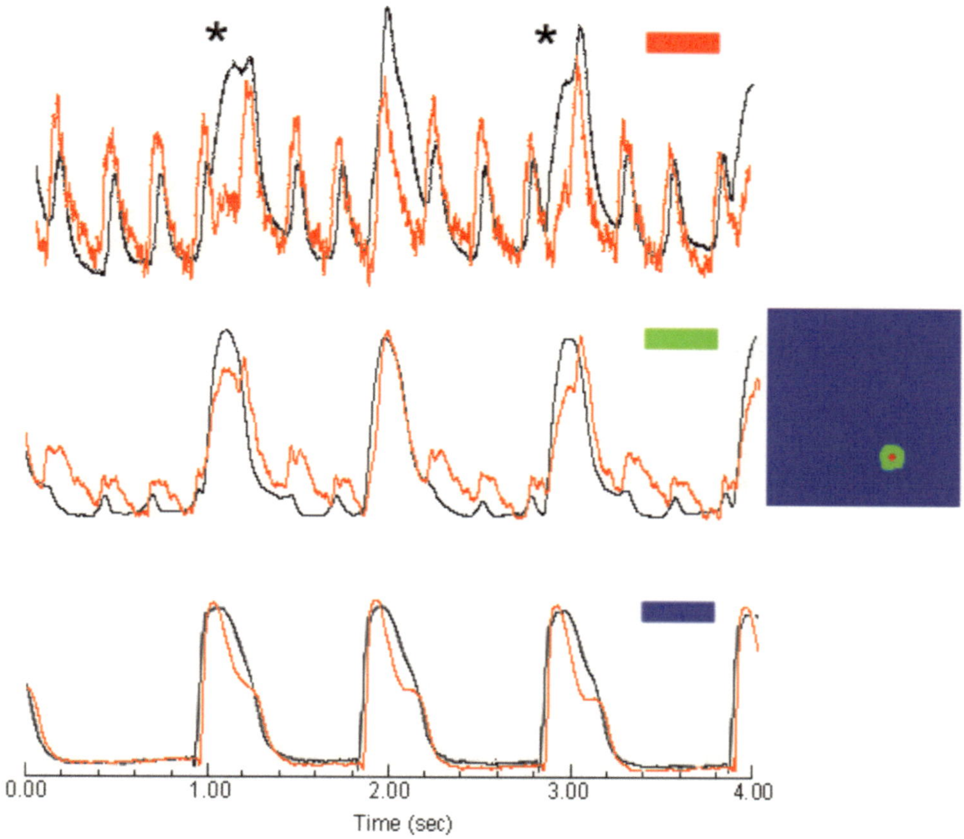

Fig. 8. Calcium oscillations confined to a site within the mapping field. The top, middle, and bottom traces show recordings from the red, green, and blue regions of the mapping field, respectively. The top trace shows regular calcium oscillations driving V_m. The middle trace shows the presence of calcium oscillations which are significantly depressed with respect to those in the top trace, and do not precede V_m. The bottom row shows that the calcium transients are being driven by voltage, implying that the calcium oscillations in the red region of the map have failed to escape the red/green region of the map and propagate through to the blue region. *(reproduced with permission from Lakireddy et al., 2006).*

The use of monolayer cell cultures in COM also represents an important advance, allowing for highly controlled studies of basic conduction as well as studies to elucidate fundamental arrhythmic mechanisms (Bub et al., 1998; Entcheva et al., 2000; Fast et al., 2000; Iravanian et al., 2003; Tung & Cysyk, 2007). An appealing aspect of the cardiac monolayer is that it allows us to study conduction in cardiac tissue without the complexity associated with the three-dimensional whole-heart Langendorff model. Since the cardiac monolayer is essentially two-dimensional (only tens of micrometers thick while being tens of millimeters in diameter), the entire monolayer may be mapped; therefore data interpretation is not complicated by the absence of missing depth information. And although the monolayer is technically three-dimensional, typical optical mapping systems interrogate at sufficient depths so that no information is lost beneath the surface (Ding et al., 2001). Despite being similar to whole-heart mapping in many respects, the actual practice of monolayer mapping carries with it significant challenges, and is in many respects more difficult than whole-heart mapping (Entcheva & Bien, 2006).

5. Conclusion

Photodiodes have played an essential role in the development of the field of COM. They were used in the earliest COM systems and continue to have widespread use today, both in typical applications as well as more modern designs such as optrodes and panoramic systems. Applications for photodiodes within COM continue to emerge, and will likely remain a vital part of this important and ever-expanding branch of cardiac electrophysiology research.

6. List of abbreviations

AP - action potential
AP-A - anthopleurin-A
APD - action potential duration
Ca_i - intracellular calcium
Ca_iT - intracellular calcium transient
Ca_iT-D - intracellular calcium transient duration
CCD - charge-coupled device
CL - cycle length
CMOS - complimentary metal-oxide semiconductor
COM - cardiac optical mapping
GP - guinea pig
I/R - ischemia/reperfusion
LQTS - long QT syndrome
LQT3 - long QT syndrome 3
PB - premature beat
PDA - photodiode array
PMT - photomultiplier tube
TdP - Torsades de Pointes
VF - ventricular fibrillation
V_m - transmembrane voltage
VT - ventricular tachycardia

7. References

Allison JS, Qin H, Dosdall DJ, Huang J, Newton JC, Allred JD, Smith WM, Ideker RE. The transmural activation sequence in porcine and canine left ventricle is markedly different during long-duration ventricular fibrillation. J Cardiovasc Electrophysiol 2007;18:1306-1312.

Al-Khadra A, Nikolski V, Efimov IR. The role of electroporation in defibrillation. Circ Res 2000;87:797-804.

Antzelevitch C. Role of spatial dispersion of repolarization in inherited and acquired sudden cardiac death syndromes. Am J Physiol Heart Circ Physiol 2007;293:H2024–H2038.

Baker LC, Wolk R, Choi BR, Watkins S, Plan P, Shah A, Salama G. Effects of mechanical uncouplers, diacetyl monoxime, and cytochalasin-D on the electrophysiology of perfused mouse hearts. J Physiol Heart Circ Physiol 2004;287:H1771-H1779.

Baxter WT, Mirinov SF, Zaitsev AV, Jalife J, Pertsov AM. Visualizing excitation waves inside cardiac muscle using transillumination. Biophys J 2001;80:516-530.

Berenfeld O, Pertsov AM, Jalife J. Letter to the editor: What is the organization of waves in ventricular fibrillation. Circ Res 2001;89:e22.

Bers DM. Calcium and cardiac rhythms: physiological and pathophysiological. Circ Res 2002;90:14-17.

Biermann M, Rubart M, Moreno A, Wu J, Josiah-Durant A, Zipes DP. Differential effects of cytochalasin D and 2,3 butanedione monoxime on isometric twitch force and transmembrane action potential in isolated ventricular muscle: implications for optical measurements of cardiac repolarization. J Cardiovasc Electrophysiol 1998;9:1348-1357.

Bub G, Glass L, Publicover NG, Shrier A. Bursting calcium rotors in cultured cardiac myocyte monolayers. Proc Natl Acad Sci USA 1998;95:10283-10287.

Bub G, Shrier A, Glass L. Global organization of dynamics in oscillatory heterogeneous excitable media. Physical Review Lett 2005;94:028105.

Byars JL, Smith WM, Ideker RE, Fast VG. Development of an optrode for intramural multisite optical recordings of Vm in the heart. J Cardiovasc Electrophysiol 2003;14:1196-1202.

Cabo C, Pertsov AM, Baxter WT, Davidenko JM, Gray RA, Jalife J. Wave-front curvature as a cause of slow conduction and block in isolated cardiac muscle. Circ Res 1994;75:1014-1028.

Caldwell J, Burton FL, Smith GL, Cobbe ST. Heterogeneity of ventricular fibrillation dominant frequency during global ischemia in isolated rabbit hearts. J Cardiovasc Electrophysiol 2007;18:854-861.

Caldwell BJ, Legrice IJ, Hooks DA, Tai D, Pullan AJ, Smaill BH. Intramural measurement of transmembrane potential in the isolated pig heart: Validation of a novel technique. J Cardiovasc Electrophysiol 2005;16:1001-1010.

Chen J, Mandapati R, Berenfeld O, Skanes AC, Jalife J. High-frequency periodic sources underlie ventricular fibrillation in the isolated rabbit heart. Circ Res 2000;86:86-93.

Chen PS, Wu TJ, Ting CT, Karagueuzian HS, Garfinkel A, Lin SF, Weiss JN. A tale of two fibrillations. Circulation 2003;108:2298-2303.

Cheng Y. Optical mapping of shock-induced arrhythmogenesis in the rabbit heart with healed myocardial infarction: fluorescent imaging with a photodiode array. Methods Mol Med 2006;129:149-61.

Choi BR, Burton F, Salama G. Cytosolic Ca2+ triggers early afterdepolarizations and torsade de pointes in rabbit hearts with type 2 long QT syndrome. J Physiol 2002;543:615-631.

Choi BR, Hatton WJ, Hume JR, Liu T, Salama G. Low osmolarity transforms ventricular fibrillation from complex to highly organized, with a dominant high-frequency source. Heart Rhythm 2006;3:1210-1220.

Choi BR, Jang W, Salama G. Spatially discordant voltage alternans cause wavebreaks in ventricular fibrillation. Heart Rhythm 2007; 4:1057–1068.

Choi BR, Liu T, Lavasani M, Salama G. Fiber orientation and cell-cell coupling influence ventricular fibrillation dynamics. J Cardiovasc Electrophysiol 2003;14:851-860.

Choi BR, Salama G. Simultaneous maps of optical action potentials and calcium transients in guineapig hearts: mechanisms underlying concordant alternans. J Physiol 2000;529:171-188.

Coraboeuf E, Weidmann S. Potentiels d'action du muscle obtenus à l'aide de microélectrodes intracellulaires. Présence d'une inversion de potential. CR Soc Biol (Paris) 1949;143:1360-1360.

Coraboeuf E, Weidmann S. Potentiels de repos et potentials d'action du muscle cardiaque, mesurés á l'aide d'électrodes intracellulaires. CR Soc Biol (Paris) 1949; 143:1329-1331.

Draper MH, Weidmann S. Cardiac resting and action potentials recorded with an intracellular electrode. J Physiol 1951;115:74-94.

Ding L; Splinter, R.; Knisley, S.B. Quantifying spatial localization of optical mapping using Monte Carlo simulations. IEEE Trans Biomed Eng 2001;48:1098-1107.

Efimov IR, Ermentrout B, Huang DT, Salama G. Activation and repolarization patterns are governed by different structural characteristics of ventricular myocardium: Experimental study with voltage-sensitive dyes and numerical simulations. J Cardiovasc Electrophysiol 1996;7:512-530.

El-Sherif N. The challenge of cardiac tridimensional mapping. Heart Rhythm 2007;4:1437-1440.

El-Sherif N, Caref EB, Yin H, Restivo M. The electrophysiological mechanism of ventricular arrhythmias in the long QT syndrome. Circ Res 1996;79:474-492.

Entcheva E, Bien H. Macroscopic optical mapping of excitation in cardiac cell networks with ultra-high spatiotemporal resolution. Progress in Biophysics and Molecular Biology 2006;92:232–257.

Entcheva E, Lu SN, Troppman RH, Sharma V, Tung L. Contact fluorescence imaging of reentry in monolayers of cultured neonatal rat ventricular myocytes. J Cardiovasc Electrophysiol 2000;11:665-676.

Evertson DW, Holcomb MR, Eames MDC, Bray MA, Sidorov VY, Xu J, Wingard H, Dobrovolny HM, Woods MC, Gauthier DJ, Wikswo JP. High-resolution high-speed panoramic cardiac imaging system. IEEE Trans Biomed Eng 2008;55:1241-1243.

Fast VG. Simultaneous optical imaging of membrane potential and intracellular calcium. J Electrocardiol 2005;38:107-112.

Fast VG, Ideker RE. Simultaneous optical mapping of transmembrane potential and intracellular calcium in myocyte cultures. J Cardiovasc Electrophysiol 2000;11:547-556.

Fast VG, Sharifov OF, Cheek ER, Newton JC, Ideker RE. Intramural virtual electrodes during defibrillation shocks in left ventricular wall assessed by optical mapping of membrane potential. Circulation 2002;106:1007-1014.

Fedorov VV, Kostecki G, Hemphill M, Efimov IR. Atria are more susceptible to electroporation than ventricles: Implications for atrial stunning, shock-induced arrhythmia and defibrillation failure. Heart Rhythm 2008;5:593-604.

Fedorov VV, Lozinsky IT, Sosunov EA, Anyukhovsky EP, Rosen MR, Balke W, Efimov IR. Application of blebbistatin as an excitation– contraction uncoupler for electrophysiologic study of rat and rabbit hearts. Heart Rhythm 2007;4:619-626.

Girouard SD, Pastore JM, Laurita KR, Gregory KW, Rosenbaum DS. Optical mapping in a new guinea pig model of ventricular tachycardia reveals mechanisms for multiple wavelengths in a single reentrant circuit. Circulation 1996;93:603-613.

Gold MR, Bloomfield DM, MD, Anderson KP, El-Sherif N, Wilber DJ, Groh WJ, Estes M, Kaufman ES, Greenberg ML, Rosenbaum DS. A comparison of T-wave alternans, signal averaged electrocardiography and programmed ventricular stimulation for arrhythmia risk stratification. JACC 2000;36:2247-2253.

Hayashi H, Miyauchi Y, Chou CC, Karagueuzian HS, Chen PS, Lin SF. Effects of cytochalasin D on electrical restitution and the dynamics of ventricular fibrillation in isolated rabbit heart. J Cardiovasc Electrophysiol 2003;14:1077-1084.

Himel IV, HD, Bub G, Yue Y, and El-Sherif N. Early voltage/calcium uncoupling predestinates the duration of ventricular tachyarrhythmias during ischemia/reperfusion. Heart Rhythm 2009;6:1359-1365.

Himel IV, HD, Dumas JH, Kiser AC, Knisley SB. Translesion stimulus-excitation delay indicates quality of linear lesions produced by radiofrequency ablation in rabbit hearts. Physiol. Meas. 2007;28:611–623.

Himel IV, H.D., Knisley, S.B. Imaging of cardiac movement using ratiometric and nonratiometric optical mapping: effects of ischemia and 2, 3-butaneodione monoxime. IEEE Trans Med Imaging 2006;25:122-127.

Hohnloser SH, Klingenheben T, Li YG, Zabel M, Peetermans J, Cohen RJ. T wave alternans as a predictor of recurrent ventricular tachyarrhythmias in ICD recipients: Prospective comparison with conventional risk markers. J Cardiovasc Electrophysiol 1998;9:1258-1268.

Hooks DA, LeGrice IJ, Harvey JD, Smaill BH. Intramural multisite recording of transmembrane potential in the heart. Biophys J 2001;81:2671-2680.

Hyatt CJ, Mirinov SF, Vetter FJ, Zemlin CW, Pertsov AM. Optical Action Potential Upstroke Morphology Reveals Near-Surface Transmural Propagation Direction. Circ Res 2005;97:277-284.

Ikeda T, Yoshino H, Sugi K, Tanno K, Shimizu H, Watanabe J, Kasamaki Y, Yoshida A, Kato T. Predictive value of microvolt T-wave alternans for sudden cardiac death in patients with preserved cardiac function after acute myocardial infarction. JACC 2006;48:2268-2274.

Iravanian S, Nabutovsky Y, Kong CR, Saha S, Bursac N, Tung L. Functional reentry in cultured monolayers of neonatal rat cardiac cells. Am J Physiol Heart Circ Physiol 2003;285:H449–H456.

Jalife J, Berenfeld O, Skanes A, Mandapati R. Mechanisms of atrial fibrillation: mother rotors or multiple daughter wavelets, or both? J Cardiovasc Electrophysiol 1998;9:S2-12.

Joel SE, Hsia PW. Discovery of gradient pattern in dominant frequency maps during fibrillation: implication of rotor drift and epicardial conduction velocity changes. J Electrocardiol 2005;38:159-165.

Jung P, Wang J, Wackerbauer R, Showalter K. Coherent structure analysis of spatiotemporal chaos. Physical Review E 2000;61:2095-2098.

Kay MW, Amison PM, Rogers JM. Three-dimensional surface reconstruction and panoramic optical mapping of large hearts. IEEE Trans Biomed Eng 2004;51:1219-1229.

Kay MW, Rogers JM. Epicardial rotors in panoramic optical maps of fibrillating swine ventricles. Conf Proc IEEE Med Biol Soc 2006;1:2268-2271.

Kay MW, Walcott GP, Gladden JD, Melnick SB, Rogers JM. Lifetimes of epicardial rotors in panoramic optical maps of fibrillating swine ventricles. Am J Physiol Heart Circ Physiol 2006;291:1935-1941.

Knisley SB, Baynham T. Line stimulation parallel to myofibers enhances regional uniformity of transmembrane voltage changes in rabbit hearts. Circ Res 1997;81:229-241.

Knisley SB, Hill BC. Effects of bipolar point and line stimulation in anisotropic rabbit epicardium: assessment of the critical radius of curvature for longitudinal block. IEEE Trans Biomed Eng 1995;42:957-966.

Knisley SB, Hill BC, Ideker RE. Virtual electrode effects in myocardial fibers. Biophys J 1994;66:719-728.

Knisley SB, Justice RK, Kong W, Johnson PL. Ratiometry of transmembrane voltage-sensitive fluorescent dye emission in hearts. Am J Physiol Heart Circ Physiol 2000; 279: H1421–H1433.

Knisley SB, Trayanova N, Aguel F. Roles of electric field and fiber structure in cardiac electric stimulation. Biophys J 1999;77:1404-1417.

Kong W, Fakhari N, Sharifov OF, Ideker RE, Smith WM, Fast VG. Optical measurements of intramural action potentials in isolated porcine hearts using optrodes. Heart Rhythm 2007;4:1430-1436.

Kong W, Walcott GP, Smith WM, Johnson PL, Knisley SB. Emission ratiometry for simultaneous calcium and action potential measurements with coloaded dyes in rabbit hearts: reduction of motion and drift. J Cardiovasc Electrophysiol 2003;14:76-82.

Lakireddy V, Baweja P, Syed A, Bub G, Boutjdir M, El-Sherif N. Contrasting effects of ischemia on the kinetics of membrane voltage and intracellular calcium transient underlie electrical alternans. Am J Physiol Heart Circ Physiol 2005; 288:400-407.

Lakireddy V, Bub G, Baweja P, Syed A, Boutjdir M, El-Sherif N. The kinetics of spontaneous calcium oscillations and arrhythmogenesis in the in vivo heart during ischemia/reperfusion. Heart Rhythm 2006;3:58-66.

Lan DZ, Pollard AE, Knisley SB. Optical mapping of Vm and Cai 2+ in a model of arrhythmias induced by local catecholamine application in patterned cell cultures. Eur J Physiol 2007;453:871–877.

Laurita KR, Singal A. Mapping action potentials and calcium transients simultaneously from the intact heart. Am J Physiol Heart Circ Physiol 2001;280: H2053–H2060.

Lee JJ, Kamjoo K, Hough D, Hwang C, Fan W, Fishbein MC, Bonometti C, Ikeda T, Karagueuzian HS, Chen PS. Reentrant wave fronts in wiggers' stage II ventricular fibrillation. Circ Res 1996;78:660-675.

Lee MH, Lin SF, Ohara T, Omichi C, Okuyama Y, Chudin E, Garfinkel A, Weiss JN, Karagueuzian HS, Chen PS. Effects of diacetyl monoxime and cytochalasin D on ventricular fibrillation in swine right ventricles. Am J Physiol Heart Circ Physiol 2001;280: H2689–H2696.

Li D, Nattel S. Pharmacological elimination of motion artifacts during optical imaging of cardiac tissues: Is blebbistatin the answer? Heart Rhythm 2007;4:627-628.

Li L, Jin Q, Huang J, Cheng KA, Ideker RE. Intramural foci during long duration fibrillation in the pig ventricle. Circ Res 2008;102:1256-1264.

Liu YB, Pak HN, Lamp ST, Okuyama Y, Hayashi H, Wu TJ, Weiss JN, Chen PS, Lin SF. Coexistence of two types of ventricular fibrillation during acute regional ischemia in rabbit ventricle. J Cardiovasc Electrophysiol 2004;15:1433-1440.

Marban E, Kitakaze M, Koretsune Y, Yue DT, Chacko VP, Pike MM. Quantification of $[Ca2+]_i$ in perfused hearts: critical evaluation of the 5F-BAPTA and nuclear magnetic resonance method as applied to the study of ischemia and reperfusion. Circ Res 1990;66:1255-1267.

Matiukas A, Mitrea BG, Pertsov AM, Wuskell JP, Wei M, Watras J, Millard AC, Loew LM. New near-infrared optical probes of cardiac electrical activity. Am J Physiol Heart Circ Physiol 2006;290: H2633–H2643.

Matiukas, A, Mitrea BG, Qin M, Pertsov AM, Shvedko AG, Warren MD, Zaitsev AV, Wuskell JP, Wei M, Watras J, Loew LM. Near-infrared voltage-sensitive fluorescent dyes optimized for optical mapping in blood-perfused myocardium. Heart Rhythm 2007;4:1441-1451.

Milberg P, Reinsch P, Wasmer K, Mönnig G, Stypmann P, Osada N, Breithardt G, Haverkamp W, Eckardt L. Transmural dispersion of repolarization as a key factor of arrhythmogenicity in a novel intact heart model of LQT3. Circ Res 2005;65:397-404.

Moe GK. On the multiple wavelet hypothesis of atrial fibrillation. Arch Int Pharmacodyn Ther 1962;140:183–188.

Morad M, Dillon S. A new laser scanning system for measuring action potential propagation in the heart. Science 1981; 214:453-456.

Morad M, Salama G. Optical probes of membrane potential in heart muscle. J Physiol 1979;292:267-295.

Moreno J, Zaitsev AV, Warren M, Berenfeld O, Kalifa J, Lucca E, Mironov S, Guha P, Jalife J. Effect of remodelling, stretch and ischaemia on ventricular fibrillation frequency and dynamics in a heart failure model. Cardiovasc Res 2005;65:158-166.

Nash MP, Mourad A, Clayton RH, Sutton PM, Bradley CP, Hayward M, Paterson DJ, Taggart P. Evidence for multiple mechanisms in human ventricular fibrillation. Circulation 2006;114:536-542.

Neunlist M, Tung L. Spatial distribution of cardiac transmembrane potentials around an extracellular electrode: Dependence on fiber orientation. Biophys J 1995;68:2310-2322.

Omichi C, Lamp ST, Lin SF, Yang J, Baher A, Zhou S, Attin M, Lee MH, Karagueuzian HS, Kogan B, Qu Z, Garfinkel A, Chen PS, Weiss JN. Intracellular Ca dynamics in ventricular fibrillation. Am J Physiol Heart Circ Physiol 2004;286: H1836–H1844.

Pastore JM, Girouard SD, Laurita KR, Akar FG, Rosenbaum DS. Mechanism linking T-wave alternans to the genesis of cardiac fibrillation. Circulation 1999;99:1385-1394.

Perez FJ, Wood MA, Schubert CM. Effects of gap geometry on conduction through discontinuous radiofrequency lesions. Circulation 2006;113:1723-1729.

Pham Q, Quan KJ, Rosenbaum DS. T-wave alternans: marker, mechanism, and methodology for predicting sudden cardiac death. J Electrocardiol 2003;36:75-81.

Pruvot EJ, Katra RP, Rosenbaum DS, Laurita KR. Role of calcium cycling versus restitution in the mechanism of repolarization alternans. Circ Res 2004;94:1083-1090.

Qu F, Ripplinger CM, Nikolski VP, Grimm C, Efimov IR. Three-dimensional panoramic imaging of cardiac arrhythmias in rabbit heart. J Biomed Opt. 2007;12:044019.

Rashba EJ, Osman AF, Macmurdy K, Kirk MM, Sarang SE, Peters RW, Shorofsky SR, Gold MR. Enhanced detection of arrhythmia vulnerability using T wave alternans, left ventricular ejection fraction, and programmed ventricular stimulation: A prospective study in subjects with chronic ischemic heart disease. J Cardiovasc Electrophysiol 2004;15:170-176.

Rogers JM, Huang J, Smith WM, Ideker RE. Incidence, evolution, and spatial distribution of functional reentry during ventricular fibrillation in pigs. Circ Res 1999;84:945-954.

Rogers JM, Walcott GP, Gladden JD, Melnick SB, Kay MW. Panoramic optical mapping reveals continuous epicardial reentry during ventricular fibrillation in the isolated swine heart. Biophys J 2007;92:1090-1095.

Rosenbaum DS, Jackson LE, Smith JM, Garan H, Ruskin JN, Cohen RJ. Electrical alternans and vulnerability to ventricular arrhythmias. N Engl J Med 1994;330:235-241.

Sakai T. Optical mapping analysis of the spatiotemporal pattern of experimental tachyarrhythmia in improved isolated rat atrium preparation. J Physiol Sci 2008;58:87-97.

Salama G. Merocyanine 540 as an optical probe of transmembrane electrical activity in the heart. Science 1976;191:485-487.

Salama G, Choi BR, Azour G, Lavasani M, Tumbev V, Salzberg BM, Patrick MJ, Ernst LA, Waggoner AS. Properties of new, long-wavelength, voltage-sensitive dyes in the heart. J. Membrane Biol 2005;208:125–140.

Salama G, Hwang SM. Simultaneous optical mapping of intracellular free calcium and action potentials from langendorff perfused hearts. Curr Protoc Cytom 2009; Chapter 12: Unit 12.17.

Sano T, Ohtsuka E, Shimamoto T. "Unidirectional" atrioventricular conduction studied by microelectrodes. Circ Res 1960;8:600-608.

Sano T, Takayama N, Shimamoto T. Directional difference of conduction velocity in the cardiac ventricular syncytium studied by microelectrodes. Circ Res 1959;7:262-267.

Scherz P. Practical Electronics for inventors. Second edition. New York, NY: McGraw-Hill; 2007.

Shabetai R, Surawicz B, Hammill W. Monophasic Action Potentials in Man. Circulation 1968;38:341-352.

Shimizu W, Antzelevitch C. Sodium channel block with mexiletine is effective in reducing dispersion of repolarization and preventing torsade de pointes in LQT2 and LQT3 models of the long-QT syndrome. Circulation 1997;96:2038-2047.

Steenbergen C, Murphy E, Levy L, London RE. Elevation in cytosolic free calcium concentration early in myocardial ischemia in perfused rat heart. Circ Res 1987;60:700-707.

Tung L, Cysyk J. Imaging fibrillation: defibrillation in a dish. J Electrocardiol 2007;40:S62-S65.

Tung L, Zhang Y. Optical imaging of arrhythmias in tissue culture. J Electrocardiol 2006;39:S2-S6.

Valderrábano M, Lee MH, Ohara T, Lai AC, Fishbein MC, Lin SF, Karagueuzian HS, Chen PS. Dynamics of intramural and transmural reentry during ventricular fibrillation in isolated swine ventricles. Circ Res 2001;88:839-848.

Valderrábano M, Yang J, Omichi C, Kil J, Lamp ST, Qu Z, Lin SF, Karagueuzian HS, Garfinkel A, Chen PS, Weiss JN. Frequency analysis of ventricular fibrillation in swine ventricles. Circ Res 2002;90:213-222.

Weidmann, S. Effect of current flow on the membrane potential of cardiac muscle. J Physiol 1951;115:227-236.

Wu J, Biermann M, Rubart M, Zipes DP. Cytochalasin D as excitation-contraction uncoupler for optically mapping action potentials in wedges of ventricular myocardium. J Cardiovasc Electrophysiol 1998;9:1336-1347.

Wu S, Weiss JN, Chou CC, Attin M, Hayashi H, Lin SF. Dissociation of membrane potential and intracellular calcium during ventricular fibrillation. J Cardiovasc Electrophysiol 2005;16:186-192.

Wu TJ, Lin SF, Baher A, Qu Z, Garfinkel A, Weiss JN, Ting CT, Chen PS. Mother rotors and the mechanisms of D600-induced type 2 ventricular fibrillation. Circulation 2004;110:2110-2118.

Wu TJ, Lin SF, Hsieh YC, Ting CT, Chen PS. Ventricular fibrillation during no-flow global ischemia in isolated rabbit hearts. J Cardiovasc Electrophysiol 2006;17:1112-1120.

Zaitsev AV, Guha PK, Sarmast F, Kolli A, Berenfeld O, Pertsov AM, de Groot JR, Coronel R, Jalife J. Wavebreak formation during ventricular fibrillation in the isolated, regionally ischemic pig heart. Circ Res 2003;92:546-553.

Wu TJ, Lin SF, Weiss JN, Ting CT and Chen PS. Two types of ventricular fibrillation in isolated rabbit hearts: Importance of excitability and action potential duration restitution. Circulation 2002;106:1859-1866.

Zaitsev AV, Berenfeld O, Mirinov SF, Jalife J, Pertsov AM. Distribution of excitation frequencies on the epicardial and endocardial surfaces of fibrillating ventricular wall of the sheep heart. Circ Res 2000;86:408-417.

Zemlin CW, Bernus O, Matiukas A, Hyatt CJ, Pertsov AM. Extracting intramural wavefront orientation from optical upstroke shapes in whole hearts. Biophys J 2008;95:942-950.

9

Photodiode Array Detection in Clinical Applications; Quantitative Analyte Assay Advantages, Limitations and Disadvantages

Zarrin Es'haghi
Department of Chemistry, Payame Noor University, 19395-4697 Tehran, I.R. of IRAN

1. Introduction

1.1 Optical spectroscopy

Study of the electromagnetic radiation by matter, as related to the dependence of these processes on the wavelength of the radiation. More recently, the definition has been expanded to include the study of the interactions between particles such as electrons, protons, and ions, as well as their interaction with other particles as a function of their collision energy. Spectroscopic analysis has been crucial in the development of the most fundamental theories in physics, including quantum mechanics, the special and general theories of relativity, and quantum electrodynamics.

Spectroscopic techniques have been applied in virtually all technical fields of science and technology. One of the most famous kinds of spectroscopy, optical spectroscopy is used routinely to identify the chemical composition of matter and to determine its physical structure.Spectroscopic techniques are extremely sensitive. Single atoms and even different isotopes of the same atom can be detected among 10^{20} or more atoms of a different species. Isotopes are all atoms of an element that have unequal mass but the same atomic number. Isotopes of the same element are virtually identical chemically. Trace amounts of pollutants or contaminants are often detected most effectively by spectroscopic techniques. Because of this sensitivity, the most accurate physical measurements have been frequency measurements.

Spectroscopy now covers a sizable fraction of the electromagnetic spectrum. The table (1) summarizes the electromagnetic spectrum over a frequency range of 16 orders of magnitude. Spectroscopic techniques are not confined to electromagnetic radiation, however. Because the energy E of a photon (a quantum of light) is related to its frequency v by the relation $E = hv$, where h is Planck's constant, spectroscopy is actually the measure of the interaction of photons with matter as a function of the photon energy. In instances where the probe particle is not a photon, spectroscopy refers to the measurement of how the particle interacts with the test particle or material as a function of the energy of the probe particle.

Electromagnetic radiation is composed of oscillating electric and magnetic fields that have the ability to transfer energy through space. The energy propagates as a wave, such that the crests and troughs of the wave move in vacuum at the speed of 299,792,458 metres per second.

Electromagnetic phenomena		
Gamma rays (γ rays)	$<5 \times 10^{-12}$	$>6 \times 10^{19}$
X-rays	5×10^{-12}–1×10^{-8}	3×10^{16}–6×10^{19}
Ultraviolet	1×10^{-8}–4×10^{-7}	7×10^{14}–3×10^{16}
Visible light	4×10^{-7}–7×10^{-7}	4×10^{14}–7×10^{14}
Infrared	8×10^{-7}–1×10^{-3}	3×10^{11}–4×10^{14}
Microwaves, Radar	1×10^{-3}–1	3×10^{8}–3×10^{11}
Television waves	1–10	3×10^{7}–3×10^{8}
Radio waves	10–1,000	3×10^{5}–3×10^{7}

Table 1. Frequency and wavelength domain of electromagnetic radiations

The decomposition of electromagnetic radiation into its component wavelengths is fundamental to spectroscopy. Evolving from the first crude prism spectrographs that separated white light into its constituent colours, modern spectrometers have provided ever-increasing wavelength resolution. Large-grating spectrometers are capable of resolving wavelengths as close as 10^{-3} nanometre, while modern laser techniques can resolve optical wavelengths separated by less than 10^{-10} nanometre. The frequency with which the electromagnetic wave oscillates is also used to characterize the radiation. The product of the frequency (ν) and the wavelength (λ) is equal to the speed of light (c); *i.e.*, $\nu\lambda = c$. The frequency is often expressed as the number of oscillations per second, and the unit of frequency is hertz (Hz), where one hertz is one cycle per second.

Spectroscopy is used as a tool for studying the structures of atoms and molecules. The large number of wavelengths emitted by these systems makes it possible to investigate their structures in detail, including the electron configurations of ground and various excited states.

Spectroscopy also provides a precise analytical method for finding the constituents in material having unknown chemical composition. In a typical spectroscopic analysis, a concentration of a few parts per million of a trace element in a material can be detected through its emission spectrum.

Production and analysis of a spectrum usually require the following: (1) a source of electromagnetic radiation, (2) a disperser to separate the light into its component wavelengths, and (3) a detector to sense the presence of light after dispersion (See Figure 1). The apparatus used to accept light, separate it into its component wavelengths, and detect the spectrum is called a spectrometer. Spectra can be obtained either in the form of emission spectra, which show one or more bright lines or bands on a dark background, or absorption spectra, which have a continuously bright background except for one or more dark lines.

1.1.1 Optical detectors

The principal detection methods used in optical spectroscopy are photographic (*e.g.*, film), photoemissive (photomultipliers), and photoconductive (semiconductor). Prior to about 1940, most spectra were recorded with photographic plates or film, in which the film is placed at the image point of a grating or prism spectrometer. An advantage of this technique is that the entire spectrum of interest can be obtained simultaneously, and low-intensity spectra can be easily taken with sensitive film.

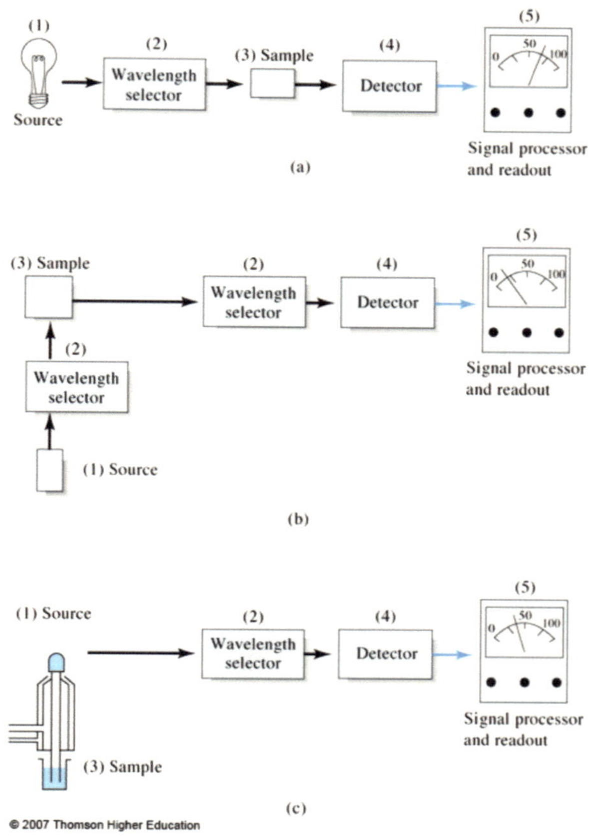

Fig. 1. Components of Optical Instruments. The generic spectrometer, (a) Molecular absorption , (b) Molecular emission and (c) Atomic absorption

Photoemissive detectors have replaced photographic plates in most applications. When a photon with sufficient energy strikes a surface, it can cause the ejection of an electron from the surface into a vacuum. A photoemissive diode consists of a surface (photocathode) appropriately treated to permit the ejection of electrons by low-energy photons and a separate electrode (the anode) on which electrons are collected, both sealed within an evacuated glass envelope. A photomultiplier tube has a cathode, a series of electrodes (dynodes), and an anode sealed within a common evacuated envelope. Appropriate voltages applied to the cathode, dynodes, and anode cause electrons ejected from the cathode to collide with the dynodes in succession. Each electron collision produces several more electrons; after a dozen or more dynodes, a single electron ejected by one photon can be converted into a fast pulse (with a duration of less than 10^{-8} second) of as many as 10^{7} electrons at the anode. In this way, individual photons can be counted with good time resolution.
Other photodetectors include imaging tubes (*e.g.*, television cameras), which can measure a spatial variation of the light across the surface of the photocathode, and microchannel plates, which combine the spatial resolution of an imaging tube with the light sensitivity of a photomultiplier. A night vision device consists of a microchannel plate multiplier in which the electrons at the output are directed onto a phosphor screen and can then be read out with an imaging tube.

Solid-state detectors such as semiconductor photodiodes detect light by causing photons to excite electrons from immobile, bound states of the semiconductor (the valence band) to a state where the electrons are mobile (the conduction band). The mobile electrons in the conduction band and the vacancies, or "holes," in the valence band can be moved through the solid with externally applied electric fields, collected onto a metal electrode, and sensed as a photoinduced current. Microfabrication techniques developed for the integrated-circuit semiconductor industry are used to construct large arrays of individual photodiodes closely spaced together. The device, called a charge-coupled device (CCD), permits the charges that are collected by the individual diodes to be read out separately and displayed as an image.

1.1.2 Multichannel detectors

Multichannel detectors can be used to sense optical and ionizing radiation or convert to an electrical signal an incoming chemical, physical, mechanical, or thermal stimulus. In other words; multichannel detector, can measure all wavelengths dispersed by a dispersing elemnt simultaneously.

The multichannel detector employs a light source that emits light over a wide range of wavelengths. Employing an appropriate optical system (a prism or diffraction grating), light of a specific wavelength can be selected for detection purposes. The specific wavelength might be chosen where a solute has an absorption maximum to provide maximum sensitivity. Alternatively, the absorption spectra of an eluted substances could be obtained for identification purposes by scanning over a range of wavelengths. The latter procedure, however, differs with the type of multichannel detector being used.

There are two basic types of multi–wavelength detector, the *dispersion* detector and the *diode array detector*, the latter being the more popular. In fact, very few dispersion instruments are sold today but many are still used in the field and so their characteristics will be discussed. All multichannel detectors require a broad emission light source such as deuterium or the xenon lamp, the deuterium lamp being the most popular.

The two types of multichannel detectors have important differences. In the dispersive instrument, the light is dispersed before it enters the sensor cell and thus virtually monochromatic light passes through the cell. However, if the incident light is of a wavelength that can excite the solute and cause fluorescence at another wavelength, then the light falling on the photo cell will contain the incident light together with any fluorescent light that may have been generated. It follows, that the light monitored by the photocell may not be monochromatic and light of another wavelength, if present, would impair the linear nature of the response. This effect would be negligible in most cases but with certain fluorescent materials the effect could be significant. The diode array detector operates quite a differently. Light of *all wavelengths* generated by the deuterium lamp is passed through the cell and then dispersed over an array of diodes. Thus, the absorption at discrete groups of wavelengths is continuously monitored at each diode. However, light falling on a discrete diode may not be derived solely from the incident light but may contain light generated by fluorescence excited by light of a shorter wavelength.

The ideal multichannel detector would be a combination of both the dispersion system and the diode array detector. This arrangement would allow a true monochromatic light beam to pass through the detector and then the transmitted beam would itself be dispersed again onto a diode array. Only that diode sensing the wavelength of the incident light would be used for monitoring the transmission. Under some circumstances, measurement of transmitted light may involve fluorescent light and the absorption spectrum obtained for a substance may be a degraded form of the true absorption curve. In this way any fluorescent light would strike other diodes, the true absorption would be measured and accurate monochromatic sensing could be obtained.

In a multichannel dispersive detector light from the deuterium lamp is collimated by two curved mirrors onto a holographic diffraction grating. The dispersed light is then focused by means of a curved mirror, onto a plane mirror and light of a specific wavelength is selected by appropriately positioning the angle of the plane mirror. Light of the selected wavelength is then focused by means of a lens through the flow cell. The exit beam from the cell is then focused by another lens onto a photocell, which gives a response that is some function of the intensity of the transmitted light. The detector is usually fitted with a scanning facility that allows the spectrum of the solute contained in the cell to be obtained. There is an inherent similarity between UV spectra of widely different types of compounds, and so UV spectra are not very reliable for the identification of most solutes.

A usual use of multichannel choice is to enhance the sensitivity of the detector by selecting a wavelength that is characteristically absorbed by the substance of interest. Conversely, a wavelength can be chosen that substances of little interest in the mixture do not adsorb and, thus, make the detector more specific to those substances that do.

Multichannel dispersive detectors provids adequate sensitivity, versatility and a linear response. But, it has mechanically operated wavelength selection and requires a stop/flow procedure to obtain spectra. In contrast, the *diode array detector* has the same advantages but none of these disadvantages.

Find some important multichannel detector on the list below.

- Photodiode Array (PDA)
  - Semiconductors (Silicon and Germanium)_(see Figure 3)
    - Group IV elements
    - Formation of holes (via thermal agitation/excitation)
    - Doping
      - n-type: Si (or Ge) doped with group V element (As, Sb) to add electrons.
        As: $[Ar]4S^2 3d^{10} 4p^3$
- p-type: Doped with group III element (In, Ga) to added holes
  In: $[Kr]5S^2 4d^{10} 5p^1$ (see Figure 4)
- coupled device (CCD)
- vidicon

1.1.3 Photodiode array detectors

A photodiode array is a linear array of several hundred light sensing diodes light ranging from 128 to 1024 – and even up to 4096 having a thousand phototubes, for every different wavelength. The design of this kind of machine is somewhat different and simpler. (Figures 2-4) Light passes through the sample first. Then it hits the monochromator, and then it is dispersed onto the photodiode array.

This multichannel detector makes an ideal sensor for an entire spectrum in a UV-VIS dispersive spectrophotometer. With that application, newer arrays have been made with adjacent diodes 25.6 mm long and spaced 25 mm on centers.

A polychromatic beam from the source is irradiated onto the inlet slit of the polychromator after passing through the sample compartment. The polychromator disperses the narrow band of the spectrum onto the diode array. The photodiode converts light into electrical signals and temporarily stores them. These signals are then read out as time-series signals via the output line by sequentially turning on the switch array connected to each photodiode with address pulses generated from the shift register.

A silicon photodiode consists of a reversed biased *pn* junction formed on a silicon chip. A photon promotes an electron from the valence bond (filled orbitals) to the conduction bond (unfilled orbitals) creating an electron(-) - hole(+) pair. The concentration of these electron-hole pairs is dependent on the amount of light striking the semiconductor. Spectral resolution limited by size of diode.

PDA detectors are useful in both research and quality assurance laboratories. In the research laboratory, the PDA provides the analyst with a variety of approaches to the analysis. In the quality assurance laboratory, the PDA provides several results from a single run, thereby increasing the throughput of the HPLC.

PDA detection offers the following advantages:

- **Peak measurement at all wavelengths:**
  In methods development, detailed information about the detector conditions required for the analysis may not be known. When a variable wavelength detector is used, a sample must often be injected several times, with varying wavelengths, to ensure that all peaks are detected. When a PDA detector is used, a wavelength range can be programmed and all compounds that absorb within this range can be detected in a single run.
- **Determination of the correct wavelengths in one run:**
  After all peaks have been detected, the maximum absorbance wavelength for each peak can be determined. A PDA detector can collect spectra of each peak and calculate the absorbance maximum.
- **Detection of multiple wavelengths:**
  A PDA detector can monitor a sample at more than one wavelength. This is especially useful when the wavelength maxima of the analytes are different. Wavelengths can be selected to analyze each compound at its highest sensitivity.
- **Peak purity analysis:**
  It is difficult to determine component purity from a chromatogram. However, a PDA detector can analyze peak purity by comparing spectra within a peak. A pure peak has matching spectra throughout the peak (at all wavelengths).
- **Positive peak identification:**
  In liquid chromatography, peak identification is usually based on relative retention times. When a PDA detector is used, spectra are automatically collected as each peak elutes. The PDA software compares the spectra with those stored in a library to determine the best fit matches; this method increases the likelihood of correctly identifying peaks.
- **Scan spectrum very quickly:**
  entire spectrum in <1 second
- **Provides single beam.**
- **Powerful tool for studies of transient intermediates in moderately fast reactions.**
- **Useful for kinetic studies.**
- **Useful for qualitative and quantitative determination of the components exiting from a liquid chromatographic column.**

In addition to above points, there are many major advantages of diode array detection. In the first, it allows for the best wavelength(s) to be selected for actual analysis. This is particularly important when no information is available on molar absorptivities at different wavelengths.

The second major advantage is related to the problem of peak purity. Often, the peak shape in itself does not reveal that it actually corresponds to two or even more components. In such a case, absorbance rationing at several wavelengths is particularly helpful in deciding whether the peak represents a single compound or, is in fact, a composite peak.

As already mentioned, a special feature of some variable wavelength UV detectors is the ability to perform spectroscopic scanning and precise absorbance readings at a variety of wavelengths while the peak is passing though the flow cell. Diode array adds a new dimension of analytical capability to liquid chromatography because it permits qualitative information to be obtained beyond simple identification by retention time.

In absorbance rationing, the absorbance is measured at two or more wavelengths and ratios are calculated for two selected wavelengths. Simultaneous measurement at several wavelengths allows one to calculate the absorbance ratio. Evaluation can be carried out in two ways:

In the first case, the ratios at chosen wavelength are continuously monitored during the analysis: if the compound under the peak is pure, the response will be a square wave function (rectangle). If the response is not rectangle, the peak is not pure.

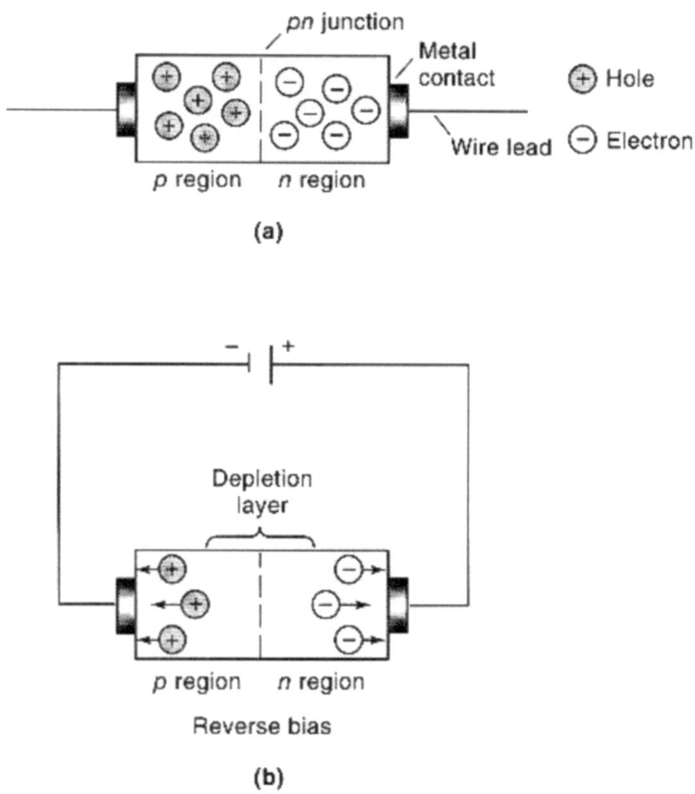

Fig. 2. (a) Schematic of a silicon diode, (b) Formation of depletion layer which prevents of flow of electricity under reverse bias [Skoog & Leary,1992].

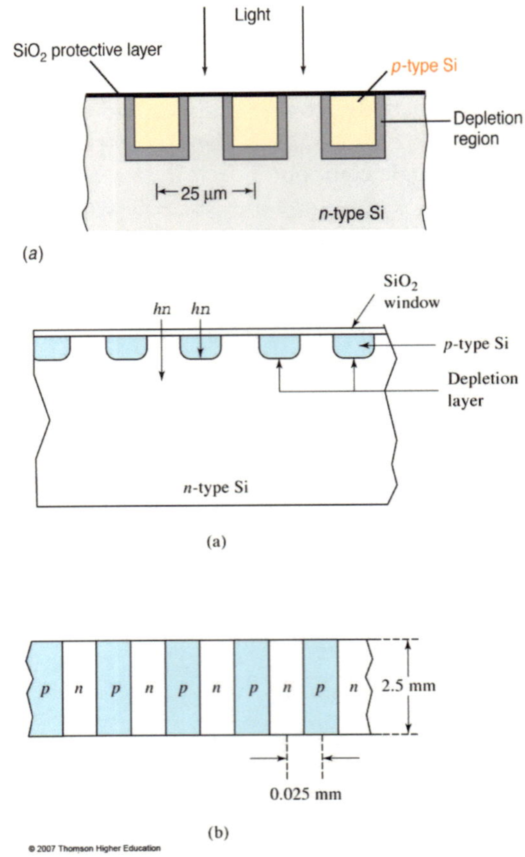

Fig. 3. (a) n-type and (b) p-type photodiode array.

Photodiode array (PDA) detectors scan a range of wavelengths every few milliseconds and continually generate spectral information. Wavelength, time, and absorbance can all be plotted.

In methods development, detailed information about the detector conditions required for the analysis may not be known. When a variable wavelength detector is used, a sample must often be injected several times, with varying wavelengths, to ensure that all peaks are detected. When PDA detectors provide three-dimensional information that allows an accurate assessment of peak identity, purity, and quantitation in a single run. Software support for PDA detectors includes peak purity and spectral library search functions to help determine peak homogeneity and identity.

1.1.4 Photodiode array applications

Spectrometers have developed in many ways since the introduction of simple spectrophotometers which were commercially available from the mid 1950's. Such improvements have enabled us to use PDA type UV-Vis. spectrophotometers.

The scope and performance of conventional single channel detector type UV-Vis spectrophotometers were found to be somewhat limited. This encouraged a search for novel techniques which could be applied to the development of UV-Vis. spectrophotometers.

Dispersed light is focused directly onto the detector array, saving considerable time and greatly reducing instrument complexity. The combination of dispersing element and detector array is employed in most spectrophotometers today.

1.1.4.1 UV-Vis spectroscopy

The introduction of multichannel detectors such as the linear photodiode array (PDA), charge coupled device (CCD) and vidicon enabled new detection systems to be developed for UV-Vis spectrophotometers and encouraged the rapid development of polychromators from the 1970s [Talmi, 1975,1982].

As was expressed earlier a polychromator is an enhanced monochromator which it is accomplished by electronic scanning of the multichannel detector.Multichannel detectors such as the photodiode array, charge coupled device or vidicon are usually flat and are best used with a dispersing arrangement which yields a flat focal plane. Under optimum conditions, they can detect as many wavelengths simultaneously as their number of individual diodes, resolution elements or pixels. Stray light and background per element are negligible because they are arrays and they have very low dark currents.

PDA, on the other hand, is more suited for applications where the light level is relatively high. Because in the PDA the photon saturation charge is greater than CCD so the detection range of PDA is larger than CCD. Furthermore, PDA delivers lower noise than CCD. So it PDA was recommend in applications where higher output accuracy is needed.

This multichannel detector having numbers of elements ranging from 128 to 1024 and even up to 4096. It makes an ideal sensor for an entire spectrum in a UV-VIS dispersive spectrophotometer.

A polychromatic beam from the source is irradiated onto the inlet slit of the polychromator after passing through the sample compartment. The polychromator disperses the narrow band of the spectrum onto the diode array. The photodiode converts light into electrical signals and temporarily stores them. These signals are then read out as time-series signals (see Figure .4).

A spectrum for the whole wavelength range should be acquired for best results. The correlation between wavelengths and particular detector channels in a polychromator facilitates nearly simultaneous measurement of the intensities of the various wavelengths.

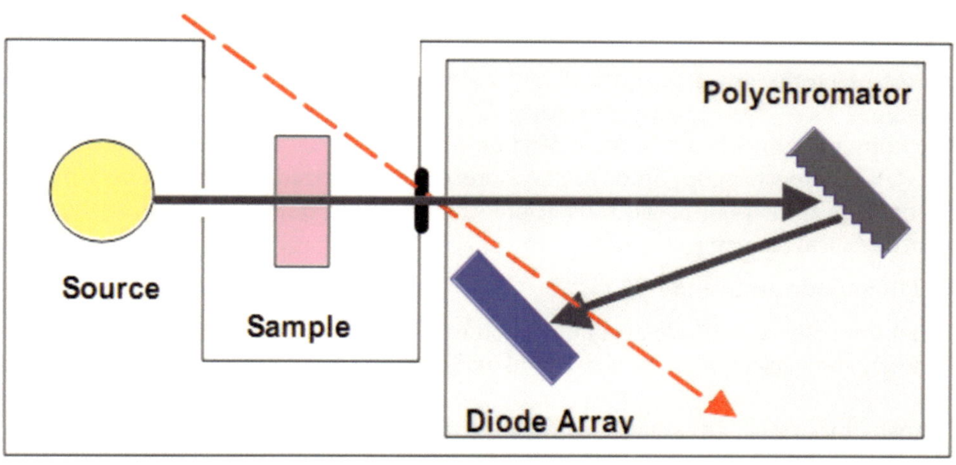

Fig. 4. Schematic of a photodiode array spectrophotometer.

The conventional UV-Vis. spectrophotometer only has one detector. But data for many wavelengths can be acquired with the photodiode array spectrophotometer simultaneously since there are several hundred or a thousand detectors present.Fast spectral acquisition makes diode array spectrophometers the first choice for measurement of fast chemical reactions and kinetics study of materials.

The duration and intensity of illumination determine both the final S/N ratio and the exposure interval needed to acquire a spectrum. This interval is also the integration time for the signal. A longer integration time allows a higher S/N since the signal will be larger and noise averaged more completely towards zero.

There is no Integration function in the conventional UV-Vis spectrophotometer which accumulates the signal. For example, the total required time will be 1000 sec. for 1000 data points and it takes 1 sec. to measure one datum. In this case, all 1000 data have the same signal to noise ratio (S/N). But in a PDA instrument which has a 1000 photodiode array, 1000 data points can be measured in 1 sec. and it would take 1/1000 sec. to achieve the same result obtainable in 1 sec. in a conventional instrument. Therefore, when the same sample is measured for 1000 sec in a PDA instrument, the signal is accumulated and is 1000 times greater than when measuring for 1 sec. The noise will be 1000. This means that the S/N ratio is improved by 1000.

This resulting benefit of fast data acquisition is termed Felgett's Advantage or Multichannel Advantage.

In a conventional UV-Vis spectrophotometer mechanical movement is required to select a specific wavelength. But a photodiode array UV-Vis spectrophotometer acquires data at each wavelength by electrical scanning. In this way, the wavelength reproducibility of a PDA instrument is much better than the conventional mechanical scanning UV-Vis spectrophotometer. In addition, a photodiode array type spectrophotometer has a reversed optic structure which minimizes stray light problems, a serious issue in conventional UV-Vis spectrophotometers.

On the other hand, a PDA is a solid-state device and is more secure and reliable than a PMT (photomultiplier tube). Furthermore, a polychromator avoids the variations in optical performance with wavelength and time that are introduced in a scanning monochromator by moving the grating. Indeed, in a polychromator no mechanical movement is required except perhaps the opening of a shutter at the entrance slit.

The Spectroscopy methods which are used of PDA can be divided into 3 sections: mass spectrometry, atomic spectroscopy and molecular spectroscopy. The applications of PDA for all 3 sections have been growing steadily. UV-Vis, FT-IR, Fluorescence, Raman and NIR spectroscopy instruments are in the molecular group. UV-Vis, is the largest category in this section. UV-Vis spectroscopy finds applications not only in traditional chemistry but also in newer fields such as pharmaceuticals & life science, environment, agriculture, energy and the petrochemical Industry.

1.1.4.2 Photodiode array and HPLC

The great importance of diode-array detection in HPLC can be characterized by the fact that this is solely the subject of an excellent book edited by Huber and George [Huber & George, 1993].

The most important advantage of the diode-array UV detector over conventional multiwavelength UV detectors is the speed of scanning the spectra. Using the reversed optics of the diode-array spectrophotometer enables all points in the spectrum to be measured simultaneously on the array of fixed photodiodes. The speed of scanning the spectrum is thus determined by the speed of data acquisition. In modern diode-array UV detectors equipped with powerful computers the time necessary to take the full spectrum from 190 to 600 nm can be reduced to as short as about 10 msec. This speed is more than sufficient in the overwhelming majority of cases in pharmaceutical analysis when the half-band width of peaks separated by HPLC is usually in the order of 1 min and it is only very rarely in the order of 1-10 sec in fast HPLC systems and especially in capillary electrophoresis where the peaks are in general narrower.

The quality of the UV spectrum of the separated impurities obtained by the diode-array detector is influenced by several of photodiodes. For example, the number of diodes in a DAD of the HPLC instrument is only 205 while in the other it is 1024. If the spectrum has fine structure, better quality spectra are obtainable with the latter. In addition to this the quality of the spectra of especially the low level impurities greatly depends on the baseline noise. This can be reduced by using a light source with high intensity, by selecting a suitable reference wavelength (which is as close to the cut-off wavelength of the separated analyte as possible and a suitable slit width. Generally speaking the sensitivity of the new generation of diode-array detectors is much higher than that of the older ones.

There are three main areas within drug impurity profiling where the advantages of diode-array detectors can contribute to the success of the HPLC (CE) analysis (see Figures 5-7).

(a)Peak purity determination. The determination of peak homogeneity is an integral part of the protocol in the validation of any kind of HPLC (and CE) analysis of pharmaceuticals. In the course of impurity profiling studies it is especially important to check the peak of the main component for its homogeneity from the simple and most widely used absorbance ratio method [Drouen et al.,1984; Wilson et al.,1989] to more sophisticated deconvolution, spectral suppression, spectrum subtraction and other chemometric methods[Huber & George, 1993]. If any kind of peak in-homogeneity is found (impurity on the leading or tailing edges of the main peak or fused impurity peaks, conveniently demonstrated in the three-dimensional mode) the diode-array spectra themselves furnish further information for the identification of the unresolved impurities.

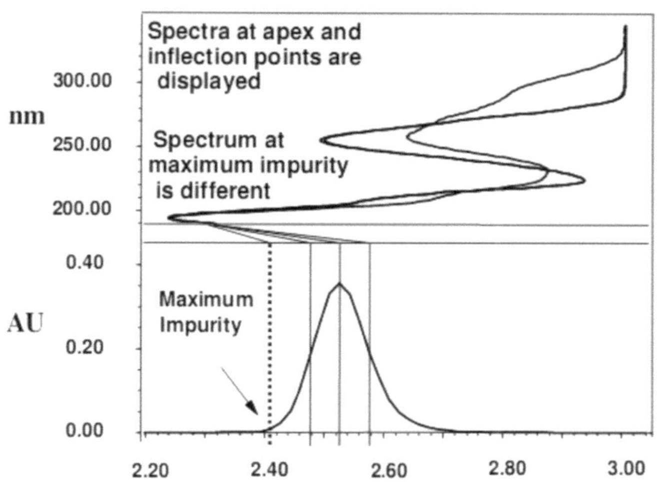

Fig. 5. Peak purity measurement

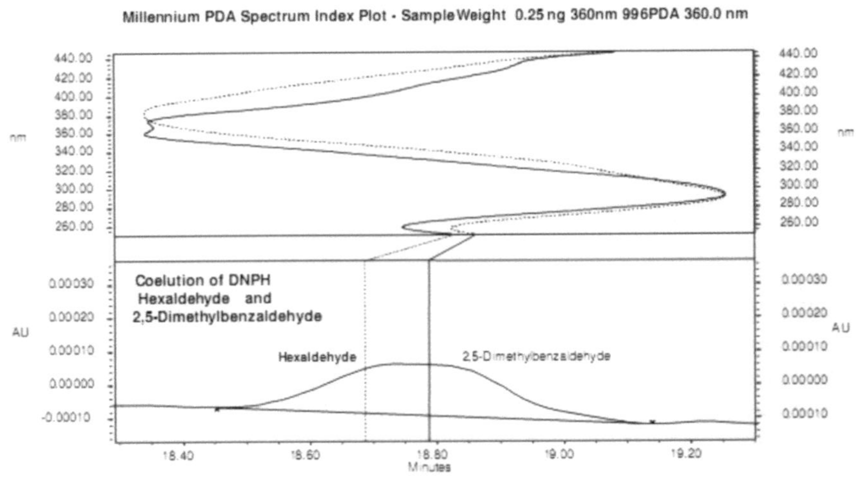

Fig. 6. Maximum impurity detection

(b) Spectral matching. Matching the diode-array spectra of components separated by HPLC with those taken by computer search from spectral libraries is a widely used method [Huber & George, 1993] especially in toxicological analysis . This approach is of limited value in drug impurity profiling since it is unlikely that impurities of especially new drugs are included in spectrum libraries. However, matching the diode-array spectra of the separated impurities with standard materials can greatly support the identification of the impurities on the basis of retention matching.

(c) Structure elucidation of the separated impurities. It is reasonable to begin the search for the structure unknown impurity separated by HPLC or CE with drawing as many conclusions from its diode-array UV spectrum as possible.

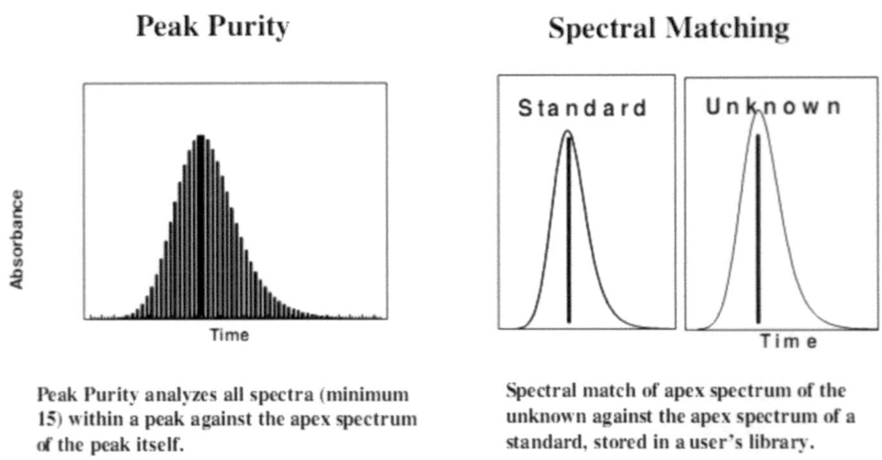

Fig. 7. Determination of peak purity

The short-wavelength parts of the (diode-array) UV spectra can be subject of several distorting effects, moreover even false maxima can occur. In addition to this, short-

wavelength UV bands can originate from different chromophoric functional groups and for this reason they are of limited value in the structure elucidation of organic compounds. As a consequence of these factors it is a prerequisite of drawing useful conclusions from the UV spectrum of an impurity that it should have at least one maximum above 210-220 nm.
Another limitation is that the difference between the structures of the drug material and the impurity should be at or near the chromophoric part of the molecule in order that the difference between their spectra can be of diagnostic value in the structure elucidation of the impurity. For example, the chromophoric group of various steroids is the 4-ene-3-oxo group with an absorption maximum around 240 nm. As it will be shown later, the position of this band is influenced by substituents in the B and C ring of the steroid nucleus but by no means by substituents at C-17. For this reason various esters of 17-hydroxy-4-ene-3-oxo steroids (testosterone, 19-nortestosterone, 17-hydroxyprogesterone, etc.) cannot be differentiated on the basis of their UV spectra.
HPLC with photodiode array detection (HPLC-PDA or HPLC-DAD) is regularly employed for substance identification in the context of Systematic Toxicological Analysis [Koves,1995; Gaillard & Pépin,1997; Herre & Pragst,1997]. With HPLC-PDA the most important parameters in identifying a compound are its retention time and its UV spectrum. Critics of the method often question the specificity of UV detection because of poorly structured spectra and broad absorption bands. Therefore a systematic investigation into the selectivity of PDA detection was carried out by analyzing large numbers of UV spectra with respect to their correlation with chemical structure.
For data analysis the following tools are needed:

1. A spectra library ; the library is embedded into the chromatography software in a way that spectral similarity is compared nm by nm and a "hit list" is returned to the operator.
2. A database of retention times and specific peak areas.
3. A database of all molecular structures with an ability for substructure searches.
4. A structural database of all registered chromophores.

As an alternative to Mass Spectrometers, absorbance detectors (including PDA) are much less expensive and relatively simple to use. LC-DAD is a fast and robust method for screening biological samples in conjunction with a library search algorithm to quickly identify those samples that require confirmatory testing. Numerous methods for using LC-PDA as a screening method have been published and were recently reviewed by Pragst et al. [Pragst et al.,2004]. Because a PDA detector can collect an entire spectrum at each time point in a chromatogram, the data are information rich and more selective than single wavelength chromatograms. Herzler et al. [Herzler et al.,2003] showed that PDA data could be used to selectively identify abused substances in spectrochromatograms based on comparison to a library of over 2500 "toxicologically relevant" substances. Their method relied on the calculation of a 'similarity index' (related to the correlation coefficient) to determine the similarity between a spectrum in an unknown chromatogram and a library spectrum. In addition to spectral matching, a relative retention time was also used to identify the substances of interest.

1.1.4.3 Medical chemistry applications of HPLC-PDA

High performance liquid chromatography (HPLC) with photodiode array detection has been proved to be the demanded method of systematic analysis for unknown drugs in biological sample because of separation efficiency, sensitivity, flexibility and identification potential. HPLC can be an easy way of quantitation as well. Ultraviolet spectra acquired with photodiode array detector together with retention data are used to identify unknown or suspected drugs and metabolites in various biological material. These analytical systems are suitable for toxicological examinations of forensic cases, acute poisonings, drug abuse. They are convenient to subsequent monitoring of serum drug levels during treatment of intoxication as well.

High-performance liquid chromatography coupled with diode array detection (HPLC-DAD) has been widely used as a powerful means for the analysis of multi-component medicines, which can provide a UV chromatogram and comprehensive data about the compounds in complex mixtures [Han et al.,2007; Su etal.,2010; Wei et al.,2010; Zhang et al.,2010]. This technology facilitates identification of unknown components in the matrices system remarkably with high sensitivity and accuracy.

Photodiode array (PDA) detectors record light absorption at different wavelengths and can provide spectra of the analytes. This is useful in identifying unknowns. Mass spectrometry (MS) is a better detector for unknowns. It gives an unambiguous molecular weight of an analyte and provides structural information. When coupled with CE or HPLC, MS can separate co-eluting analytes with different mass to charge ratios. But the Mass spectrometer is an expensive instrument and the possibility of using it is not available in all laboratories. Of course, if possible HPLC/ESI-MS/UV-DAD analysis gives the best sensitivity [Cuyckens& Claeys,2002; Beretta et al.,2009; Christiansen et al.,2011].

The potentials and limitations of high-performance liquid chromatography-photodiode array detection are highlighted in respect to its use in the analysis of different biological matrices followed by the identification of unknowns. The logical analytical approach used in clinical and forensic toxicology, vital for the identification of one or more toxic substances as a cause of intoxication, is largely based on both simple and fast "general unknown screening" methods which cover most relevant drugs and potentially hazardous chemicals. In this field of systematic toxicological analysis, a literature overview shows that HPLC can play a substantial role. Both column packing material and eluent composition have their impact on intra- and inter laboratory reproducibility. In view of the sometimes different retention characteristics of various HPLC columns, several possibilities are addressed to enhance the discriminating power of primary retention parameters. The advantages of photodiode array detection as compared to UV detection have been of paramount importance to the success of HPLC in toxicological analysis. Dedicated libraries with spectral information and searching software are powerful tools in the process of identification of an unknown substance. In the present section, these aspects are also verified in a number of real cases.

HPLC-DAD used as a general unknown screening tool should cover as many drugs and toxicants as possible, but should be also very selective, sensitive and reliable. Liquid chromatography is used in forensic laboratories for numerous applications including examination of drugs. LC with photodiode array detection (PDA) is a hybrid technique which can provide complete UV-visible spectral information on a given peak in a chromatogram, enabling determinations of peak purity to be made, and identification of unknown peaks to be assigned by library searches of spectral information in combination with retention behavior. These are valuable features normally associated with gas chromatography-mass spectrometry. The additional information available on each peak makes LC-PDA a particularly attractive technique for the forensic laboratory where higher levels of certainty are often demanded in test results. This paper reviews some of those applications for LC-PDA in the forensic sciences, including drug screening, drug and pharmaceutical analysis, idenfication of pesticides, fungi, quality control testing and profiling of cosmetics, street drugs and profiling of other complex mixtures. The practical and technical limitations of the technique are explored and its place in the hierarchy of methods available in forensic laboratories is evaluated [Proença et al.,2003; Madej et al.,2003; Proenc et al.,2004; Nieddu et al., 2007; Es'haghi et al.,2010; ; Vosough et al.,2010].

HPLC-DAD offers many advantages in terms of specificity, sensitivity, speed and ruggedness. The data produced, comprising both retention behavior and absorption spectra of eluting chemical entities, result in an identification power at low cost and with widened availability through many laboratories. In addition, the examples showed a great versatility in application fields and excellent quantitative potential. The fast progress in DAD detector technology, computer and software power and HPLC packing material quality have led to an exponential rise of the number of reports on the use of HPLC-DAD. The advent of routine use of HPLC-MS will probably promote HPLC as a viable if not better alternative to GC-MS.

We examined that combined with a sample preparation method; HPLC-PDA can be easy achieved to very low detection limits [Es'haghi et al., 2009, 2010]. In a research, we used of direct suspended droplet microextraction (DSDME) method, based on a three-phase extraction system which is compatible with HPLC-PDA for determination of ecstasy; MDMA (3,4methylendioxy-N-methylamphetamine) in human hair samples. After the extraction, pre-concentrated analyte was directly introduced into HPLC for further analysis. In concentration range between 1.0 and 15,000 ng mL^{-1} calibration curve is drowned. Linearity was observed with r = 0.9921 for analyte. Limit of detection (LOD) were calculated as the minimum concentration providing chromatographic signals three times higher than background noise. Limit of quantification (LOQ) was estimated as the minimum concentration preparing chromatographic signals ten times higher than background noise. Thus, LOD obtained was 0.1 and LOQ was 1.0 ng mL^{-1} too [Es'haghi et al., 2010].

In the other work we successfully used of DSDME method combined with HPLC-PDA for determination of low-residue benzodiazepine, diazepam and lorazepam, in the environmental water samples [Es'haghi et al., 2009, 2009]. After the optimized extraction conditions, the suspended micro-droplet is withdrawn by a HPLC microsyringe, injected to and analyzed by HPLC-DAD. Method was evaluated and enrichment factor 839.8, linearity range from 25 to 5000 ng mL^{-1} with an average of relative standard deviation (n=5) 5.62% for diazepam using a photodiode array detector were determined. HPLC-PDA has good matches with complex matrices such as hair.

A method combining liquid–liquid–liquid microextraction and automated movement of the acceptor and donor phases (LLLME/AMADP) with ion-pair HPLC/DAD has been developed to detect trace levels of chlorophenols in water [Lin etal.,2008] . The extracted chlorophenols, present in anionic form, were then separated, identified, and quantitated by ion-pair high-performance liquid chromatography with photodiode array detection (HPLC/DAD). For trace chlorophenol determination using HPLC/DAD, the chlorophenolate anion provides a better ultraviolet spectrum for quantitative and qualitative analyses than does uncharged chlorophenol. The proposed method was capable of identifying and quantitating each analyte to 0.5 ng mL^{-1} , confirming the HPLC/DAD technique to be quite robust for monitoring trace levels of chlorophenols in water samples.

HPLC/DAD could simultaneously detect UV absorptions at multiple wavelengths and extract the UV spectra of separated analytes in a chromatogram. Absorbance measurements at the band maxima of UV spectra obey the linear Beer's law more accurately than measurements off the band maxima, and UV spectra of the separated analytes can be utilized to identify target analytes in HPLC/DAD. Accordingly, each extracted chlorophenolate anion after ion-pair liquid chromatography separation was quantitated by the maximum adsorption of its own red shift characteristic band, and each target chlorophenlate anion was identified by its own red shift characteristic band as well as its enhanced B band. The chlorophenols were determined under selected experimental conditions to assess repeatability, linearity, coefficient of determination, and detection limit.

A HPLC-DAD method for drug screening in plasma were developed by M. A. Alabdalla [Alabdalla,2005]. This analytical method extracted and tested a number of drugs of different classes. The method included; an acidic and basic Solid Phase Extraction (SPE) of plasma with C18 cartridges, a gradient elution of a modified cyano column with acidic buffer/acetonitrile eluent and a photodiode array ultraviolet (UV) detection. The drug screening procedure applied used retention index and UV spectral data for the identification of compounds, may be appropriate in particular laboratory settings.

Continuous administration of polyphenols from aqueous rooibos (Aspalathus linearis) extract ameliorates dietary-induced metabolic disturbances in hyperlipidemic mice was studied by HPLC-DAD and introduced by R. Beltrán-Debón et al. [Beltrán-Debón et al., 2011]. In this biological matrices and they could find good results.

In a recent study neurons from the olfactory system of the fish crucian carp, Carassius carassius L. were used as components in an in-line neurophysiologic detector (NPD) to measure physiological activities following the separation of substances by high-performance liquid chromatography (HPLC). The skin of crucian carp, C. carassius L. contains pheromones that induce an alarm reaction in conspecifics. Extra-cellular recordings were made from neurons situated in the posterior part of the medial region of the olfactory bulb known to mediate this alarm reaction. The nervous activity of these specific neurons in the olfactory bulb of crucian carp was used as an in-line neurophysiologic detector. HPLC was performed with a diode array detector (DAD) [Brondz et al.,2004].

UV spectral detection was performed at 214, 254 and 345 nm, and scans (190–400 nm) were collected continuously. This system enabled the selection of peaks in the chromatogram with fish alarm pheromone activity. Neurophysiologic detectors (NPDs) in-line with diode array detectors (DADs) are able to provide the physiologically active substances and their spectral characteristics.

Li-wei Yang et al. were developed a method using high-performance liquid chromatography–photodiode array detection (HPLC–DAD) for the quality control of Hypericum japonicum thunb (Tianjihuang), a Chinese herbal medicine. For the first time, the feasibility and advantages of employing chromatographic fingerprint were investigated for the evaluation of Tianjihuang by systematically comparing chromatograms with a professional analytical. The results revealed that the chromatographic fingerprint combining similarity evaluation could efficiently identify and distinguish raw herbs of Tianjihuang from different sources. The effects resulted from collecting locations; harvesting time and storage time on herbal chromatographic fingerprints were also examined [Yang et al.,2005].

1.1.4.4 Photo diode array detector in kinetic study

In kinetic experiments, transient optical absorption is recorded versus time to evaluate rate constants related to the species under investigation. In addition, the recording of a spectrum sometimes becomes necessary in order to identify the species. In most cases, the spectrum is constructed from point-to-point recordings of kinetic curves at selected wavelengths. This procedure is time consuming, and becomes boring especially at long recording times in the second and minute time domain. The use of a device, which enables the recording of a complete spectrum, can be very helpful as it reduces experiment time remarkably. Unwanted side effects, such as photolysis during long recording times, can also be prevented. The application of optical multichannel analyzers which use either a linear charge coupled device (CCD) or a linear photodiode array (PDA) in kinetic experiments was reported by some laboratories [Hunter et al .,1985; Sedlmair et al.,1986; Johnson et al.,1994]. The advantage of using such a detector is the ability to immediately record a complete spectrum from UV to IR with one measurement.

The PDA detector has the ability to record a spectrum over a large range of wavelengths. The uniformity of the analyzing light intensity over the whole range is important because the dynamics and the sensitivity of the measurements depend largely on the intensity. The spectral distribution of the analyzing light, as recorded by the multichannel detector is shown in Figure .8.

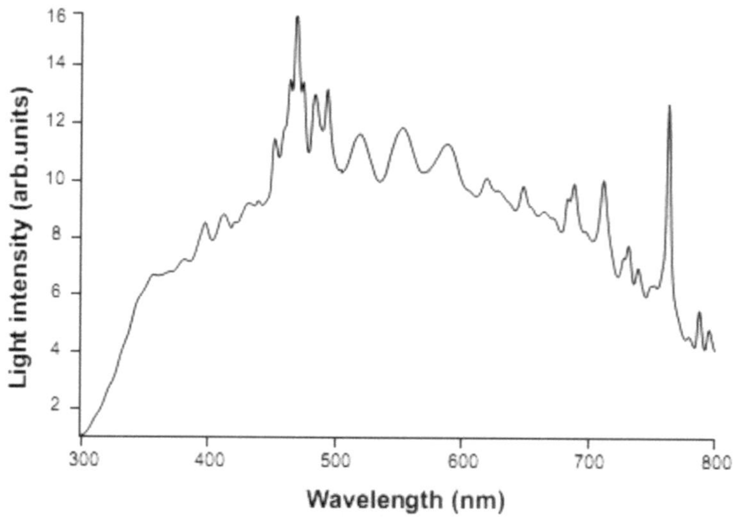

Fig. 8. Light intensity vs. wavelength of an xenon lamp, recorded by the multichannel detector.

The source of the analyzing light is an xenon lamp. The light intensity is attenuated tenfold as compared to kinetic experiments. Although, the recorded intensity of the analyzing light decreases drastically below 350 nm, a spectral range from 300 to 800 nm can be covered. Below 300 nm, recording should be accomplished in small segments and with the help of band-pass filters in order to adjust for the reduced level of analyzing light and for the decreased sensitivity of the detector, and, in addition, to avoid scattered light effects. The measurement depends largely on proper focusing of the light path, i.e., how well the lamp arc is imaged onto the diode array.

Each spectrum is the average of some (for example five) individual measurements; each irradiation consists of a train of ten pulses. The interval between the recordings of the individual spectra or between the pulses in each pulse train was set to zero. The recording at time zero, i.e. before irradiation, shows a straight line. The change in absorption increases with increasing irradiation. In general, kinetic trace scan be constructed from the recorded spectra at selected wavelengths. Similar to the construction of spectra from kinetic traces [Janata,1994].

At measurements in the UV region, Cerenkov emission is a common problem at short measuring times. The intensity of the Cerenkov emission increases with decreasing wavelength and can be much larger than the kinetic signal itself, but probably will not exceed the intensity of the analyzing light. Although this apparatus makes data at longer time scale available, overdriving of the photodiodes and long recovery times are conceivable.
The use of an optical multichannel detector consisting of a linear diode array embedded in the instrumentation for kinetic spectroscopy, as well as the highlights of the computer program used for controlling the gathering and the evaluation of data are described. Complete spectra can be recorded and irradiation can be triggered according to a preset timetable. Due to the read-out time of the photodiode array and the time required by the computer to control the experiment, this apparatus is suitable for application starting in the millisecond time domain and extending up to very long time periods.

1.1.4.5 Chemometrics investigations using photo diode array detection

Chemometrics is a statistical approach to the interpretation of patterns in multivariate data. When used to analyze instrument data, chemometrics often results in a faster and more precise Assessment of composition of a product or even physical or sensory properties. For example, composition of drugs can be quickly measured using LC and chemometrics. Food properties can also be monitored on a continuous basis. In all cases, the data patterns are used to develop a model with the goal of predicting quality parameters for future data. The two general applications of chemometrics technology to predict a property of interest; and to classify the sample into one of several categories (e.g., good versus bad, Type A versus Type B versus Type C etc.). Chemometrics can be used to speed methods development and make routine the use of statistical models for data analysis. Keeping in view of the complexity of the chromatographic fingerprint and the irreproducibility of chromatographic and spectral instruments and experimental conditions, several chemometric approaches such as variance analysis, peak alignment, correlation analysis and pattern recognition were employed to deal with the chromatographic fingerprint. Many mathematical algorithms are used for data processing in chemometric approaches. The basic principles for this approach are variation determination of common peaks/regions and similarity comparison with similarity index and linear correlation coefficient. Similarity index and linear correlation coefficient can be used to compare common pattern of the chromatographic fingerprints obtained. In general, the mean or median of the chromatographic fingerprints under study is taken as the target and both are considered to be reliable [Brereton,1987].
The rapid scanning detectors, as diode array detection, present an alternative technology for rapid, multi-wavelength detection in HPLC. If hyphenated chromatography is further combined with chemometric approaches, clear pictures might be developed for chromatographic fingerprints obtained. A chemical fingerprint obtained by hyphenated chromatography, out of question, will become the primary tool for quality control of medicines.
The full UV-Vis spectrum became accessible as a three-dimensional (3D) data matrix (A, A, t). Data are available in the time, concentration and wavelength domains. This allows the simultaneous use of more than two wavelengths for detection or for the full application of detector information to the analytical problem by means of available chemometric techniques to data from second-order bilinear instruments, as chromatographic and excitation-emission data.
As an alternative to MS, absorbance detectors (including PDA) are much less expensive and relatively simple to use. LC-DAD is a fast and robust method for screening biological samples

in conjunction with a library search algorithm to quickly identify those samples that require confirmatory testing. Numerous methods for using LC-DAD as a screening method have been published and were recently reviewed by Pragst et al. [Pragst et al., 2004]. Because a DAD can collect an entire spectrum at each time point in a chromatogram, the resultant data are information rich and more selective than single wavelength chromatograms.

For the above reasons could be adopted PDA detectors with the various chemometric methods to match spectra contained within a spectrochromatogram to a library.

In a research, triply coupled diode array detection high performance liquid chromatography mass spectroscopy was applied to a complex mixture of at least eight chlorophyll degradation products. Derivatives were employed to determine parts of the chromatogram of composition one. Mass selection was performed on the mass spectroscopic data. Principal components analysis was performed on both the raw and simultaneously normalised/standardised data; three dimensional projections of the data were obtained and compared to conventional two dimensional graphs. Angular plots between diode array loadings characteristic of individual compounds and scores of the diode array data were described. In mass spectra, angular plots between loadings characteristic of individual compounds and the remaining diagnostic masses revealed further mass spectral structure [Zissis et al.,1999].

Liquid chromatography–chemometric methods [LC-Partial least squares (LC-PLS), LC-principle component regression (LC-PCR) and LC-artificial neural network (LC-ANN)] were developed for the determination of anomalin (ANO) and deltoin (DEL) in the root by Alev Tosun et al.[Tosun et al.,2007]. Firstly, chemometric conditions were optimized by testing different mobile phases at various proportions of solvents with various flow rates in different wavelengths by using a normal phase column to obtain the best separation and recovery results. As a result, a mobile phase consisting of n-hexane and ethyl acetate (75:25 v/v) at a constant flow rate of 0.8 mL min $^{-1}$ on the at ambient temperature were found to be the optimal chromatographic conditions for good separation and determination of ANO and DEL in samples. Multi-chromatograms for the concentration set containing ANO and DEL compounds in the concentration range of 50–400 ng mL^{-1} were obtained by using a diode array detector (DAD) system at selected wavelength sets, 300 (A), 310 (B), 320 (C), 330 (D) and 340 (E). Three LC-chemometric approaches were applied to the multichromatographic data to construct chemometric calibrations. As an alternative method, traditional LC at single wavelength was used for the analysis of the related compounds in the plant extracts. All of the methods were validated by analyzing various synthetic ANO–DEL mixtures. After the above step, traditional and chemometric LC methods were applied to the real samples consisting of extracts from roots and aerial parts of analytes.

In a recent research, metabolism disorders in Kunming mice induced by two tumor cells were characterized. Metabolic fingerprint based on high performance liquid chromatography-diode array detector (HPLC-DAD) was developed to map the disturbed metabolic responses. Based on 27 common peaks, principal component analysis (PCA) and partial least squares-discriminant analysis (PLS-DA) were used to distinguish the abnormal from control and to find significant endogenous compounds which have significant contributions to classification. The tumor growth inhibition ratios of Taxol groups were used to validate the predictive accuracies of the PLS-DA models. The predictive accuracies of PLS-DA models for tumors model groups were 97.6 and 100%, respectively. Nine and seven of two models tumors were discovered, including uric acid and cytidine. In addition, the correlations between relative tumor weights and chromatographic data were significant ($p < 0.05$). Investigations on the stability and precision of the established metabolic fingerprints demonstrate that the experiment is well controlled and reliable. This work was shown that the platform of HPLC-DAD coupled with chemometric methods provides a promising method for the study of metabolism disorders [Sun et al., 2011].

2. References

Alabdalla, M.A.,(2005),HPLC-DAD for analysis of different classes of drugs in plasma,*J. Clinical Forensic Med.*, Vol.12, No.6, (December 2005), pp.310-315, ISSN: 13531131

Beltrán-Debón, R., Rull, A., Rodríguez-Sanabria, F., Iswaldi, I., Herranz-López, M., Aragonès, G., Camps, J., Alonso-Villaverde, C., Menéndez, J.A., Micol,V., Segura-Carretero, A., Aragonès,G., Joven, J.(2011), Continuous administration of polyphenols from aqueous rooibos (Aspalathus linearis) extract ameliorates dietary-induced metabolic disturbances in hyperlipidemic mice, *Phytomedicine* ,Article in press,ISSN: 09447113

Beretta, G., Artali, R., Caneva, E., Orlandini, S., Centini, M., Facino, R.M.(2009), Quinoline alkaloids in honey: Further analytical (HPLC-DAD-ESI-MS, multidimensional diffusion-ordered NMR spectroscopy), theoretical and chemometric studies, *J. Pharm. Biomed. Anal.*, Vol.50,No. 3,(15 October 2009), pp. 432-439, ISSN: 07317085

Bothe, E., Janata, E.(1994), Instrumentation of kinetic spectroscopy-12. Software for data acquisition in kinetic experiments, *Radiat. Phys. Chem.* Vol.44, No. 4, (November 1994), pp. 449–454, ISSN: 0969806X

Brereton, R.G.,(1987), Chemometrics in analytical chemistry :a review, *Analyst*, Vol. 112, No.12, (December 1987), pp. 1635-1657, ISSN: 00032654

Brondz, I., Hamdani, E.H., Døving, K.(2004), Neurophysiologic detector - A selective and sensitive tool in high-performance liquid chromatography, *J.Chromatogr. B,* Vol.800, No.1-2,(5 February 2004),pp. 41-47, ISSN: 15700232

Christiansen, A., Backensfeld, T., Kühn, S., Weitschies, W.(2011), Investigating the stability of the nonionic surfactants tocopheryl polyethylene glycol succinate and sucrose laurate by HPLC-MS, DAD, and CAD,*J.Pharm. Sci,* Vol.100, No.5,(May 2011),pp. 1773-1782 , ISSN: 00223549

Cuyckens, F., Claeys, M.(2002), Optimization of a liquid chromatography method based on simultaneous electrospray ionization mass spectrometric and ultraviolet photodiode array detection for analysis of flavonoid glycosides, *Rapid. Comm. Mass. Spectrom.*, Vol.16,No. 24,(nd), pp.2341-2348, ISSN: 09514198

Drouen, A.C.J.H. , Billiet, H.A.H. , De Galan,L. (1984), Dual-wavelength absorbance ratio for solute recognition in liquid chromatography, *Anal. Chem.*, Vol.56, No.6, (May 1984), pp. 971–978, ISSN: 00032700

Es'Haghi, Z., Bandegi, S., Daneshvar, L., Salari, P.(2009),Analysis of diazepam residue from water samples by triple phase-suspended droplet microextraction coupled to high performance liquid chromatography and diode array detection. *Asian J. Chem.* , Vol. 21, No. 8, (October 2009), pp. 6392-6402, ISSN: 09707077

Es'Haghi, Z., Daneshvar, L., Salari, P., Bandegi, S. (2009), Determination of low-residue benzodiazepine, lorazepam, in the environmental water samples by suspended droplet icroextraction and high performance liquid hromatography-diod array detector. *Chemija*, Vol.20, No.3, (nd), pp.181-186, ISSN: 0235-7216

Es'haghi, Z., Mohtaji, M., Hasanzade-Meidani, M., Masrournia, M.(2010), The measurement of ecstasy in human hair by triple phase directly suspended droplet microextraction prior to HPLC-DAD analysis, *J. Chromatogr. B*, Vol.878, No. 1, (1 April 2010),pp. 903–908, ISSN: 15700232

Gaillard, Y. , Pépin, G. (1997), Use of high-performance liquid chromatography with photodio de-array UV detection for the creation of a 600-compound library. Application to forensic toxicology, *J. Chromatogr. A,* vol. 763,No. 1-2,(28 February 1997), pp. 149–163, ISSN: 00219673

Han, J., Ye, M., Guo, H., Yang, M., Wang, B.-r., Guo, D.-a.(2007), Analysis of multiple constituents in a Chinese herbal preparation Shuang-Huang-Lian oral liquid by HPLC-DAD-ESI-MS, *J. Pharm. Biomed. Anal.* Vol.44, No.2, (28 June 2007), pp. 430–438, ISSN: 07317085

Herre, S . , Pragst, F. (1997), Shift of the high performance liquid chromatographic retention ti mes of metabolites in relation to the original drug at an RP-8 column with acidic mobile phase, *J. Chromatogr. B,* vol. 692, No. 1, (25 April 1997), pp. 111–126, ISSN: 15726495

Herzler,M., Herre,S., Pragst,F. (2003), Selectivity of substance identification by HPLC-DAD in toxicological analysis using a UV spectra library of 2682 compounds, *J. Anal. Toxicol.* Vol.27, No. 4, (May 2003),pp. 233-242, ISSN: 01464760

Huber, L. ; George, S.A.(2003). *Diode Array Detection in HPLC* (Second edition), Marcel Dekker, , ISBN: 0-8247-4 ,New York

Hunter, E.P.L., Simic, M.G., Michael, B.D.,(1985), Use of an optical multichannel analyzer for recording absorption spectra of short-lived transients, *Rev. Sci. Instrum.* Vol.56, No.12 (1985), pp. 2199–2204, ISSN: 00346748

Johnson, J.B., Edwards, G., Mendenhall, M.,(1994), A low-cost, high-performance array detector for spectroscopy based on a charged-coupled photodiode, *Rev. Sci. Instrum.,*Vol.65, No.5,(1994), pp. 1782-1783 , ISSN: 00346748

Koves, E. M.(1995), Use of high-performance liquid chromatography-diode array detection in forensic toxicology, *J. Chromatogr. A,* vol. 692, No. 1-2,(nd) , pp. 103–119, ISSN: 00219673

Lambert, W. E., Van-Bocxlaer, J.F. , De Leenheer, A. P .(1997), Potential of high-performance liquid chromatography with photodiode array detection in forensic toxicology, *J. Chromatogr. B,* vol. 689,No.1, (7 February 1997), pp. 45–53, ISSN: 15726495

Lin, C.-Y., Huang, S.-D.(2008),Application of liquid–liquid–liquid microextraction and ion-pair liquid chromatography coupled with photodiode array detection for the determination of chlorophenols in water, *J. Chromatogr. A,* vol. 1193, No. 1-2, (6 June 2008) , pp. 79–84, ISSN: 00219673

Madej, K., Parczewski, A., Kała, M.(2003), HPLC/DAD Screening Method for Selected Psychotropic Drugs in Blood, *Toxicol. Mech. Meth.,* Vol.13, No.2,(April 2003),pp. 121-127, ISSN: 15376524

Nieddu, M., Boatto, G., Serra, D., Soro, A., Lorenzoni, S., Lubinu, F. (2007), HPLC-DAD determination of mepivacaine in cerebrospinal fluid from a fatal case. *J. Forensic Sci.,* Vol.52, No.5, (September 2007), pp.1223-1224, ISSN: 00221198

Pragst, F., Herzler, M., Erxleben, B.-T.,(2004), Systematic toxicological analysis by high-performance liquid chromatography with diode array detection (HPLC-DAD), *Clin. Chem. Lab. Med.* Vol. 42 , No.11,(nd), pp. 1325-1340,ISSN:14346621

Proença, P., Pinho Marques, E., Teixeira, H., Castanheira, F., Barroso, M., Ávila, S., Vieira, D.N.(2003), A fatal forensic intoxication with fenarimol: Analysis by HPLC/DAD/MSD, *Forensic Sci. Int.,*Vol.133, No. 1-2,(23 April 2003),pp. 95-100, ISSN: 03790738

Proença, P., Teixeira, H., Pinheiro, J., Marques, E.P., Vieira, D.N. (2004), Forensic intoxication with clobazam: HPLC/DAD/MSD analysis, *Forensic Sci. Int.,* Vol.143, No.2-3, (16 July 2004), pp. 205–209, ISSN: 03790738

Sedlmair, J., Ballard, S.G., Mauzerall, D.C.,(1986), Diode-array spectrometer (DAPS) for visible and near-IR absorption measurements with 10-ns time resolution, *Rev. Sci. Instrum,* Vol.57, No.12, (nd), pp. 2995–3003, ISSN: 00346748

Skoog, D. A. ; Leary, J. J. (1992). *Principles of Instrumental Analysis* (third edition), Saunders College Publishing, ISBN: 4833702827 , Fort Worth

Su, S., Guo, J., Duan, J.-a., Wang, T., Qian, D., Shang, E., Tang, Y. .(2010), Ultra-performance liquid chromatography-tandem mass spectrometry analysis of the bioactive components and their metabolites of Shaofu Zhuyu decoction active extract in rat plasma, *J. Chromatogr. B,* vol. 878,No. 3-4,(1 February 2010), pp. 355–362, ISSN: 15700232

Sun,X., Liu,Y., Di,D., Wu,G., Guo,H. (2011), Chemometric analysis of metabolism disorders in blood plasma of S180 and H22 tumor-bearing mice by high performance liquid chromatography-diode array detection, *J.Chemometrics,* Article in press, DOI: 10.1002/cem.1387, SSN:1099-128X

Talmi, Y.,(1975), Applicability of TV-type multichannel detectors to spectroscopy, *Anal.Chem.,* Vol.47, No.7, (nd), pp. 658A-670A, ISSN: 00032700

Talmi, Y.,(1982), Spectrophotometry and spectrofluorometry with the self-scanned photodiode array, *App. Spectrosc.,* Vol.36, No.1,(January 1982), pp.1-19 , ISSN: 00037028

Tosun, A., Bahadir, Ö., Dinç, E.(2007), Determination of anomalin and deltoin in Seseli resinosum by LC combined with chemometric methods, *Chromatographia ,* Vol.66, No.9-10, (November 2007),pp. 677-683, ISSN: 00095893

Vosough, M., Bayat, M., Salemi, A.(2010), Matrix-free analysis of aflatoxins in pistachio nuts using parallel factor modeling of liquid chromatography diode-array detection data, *Anal.Chim.Acta,* Vol. 663, No.1, (March 2010),pp.11-18, ISSN: 00032670

Wei, H., Sun, L., Tai, Z., Gao, S., Xu, W., Chen, W. (2010), A simple and sensitive HPLC method for the simultaneous determination of eight bioactive components and fingerprint analysis of Schisandra sphenanthera, *Anal. Chim. Acta,* Vol.662, No. 1, (3 March 2010),pp. 97–104, ISSN: 00032670

Wilson, T.D., Trompeter, W.F., Gartelman, H.F.,(1989), Analysis of barbiturate mixtures using HPLC with diode array detection, *J. Liq. Chromatogr.*Vol.12, No. 7, (nd), pp. 1231-1251, ISSN: 01483919

Yang, L.-W., Wu, D.-H., Tang, X., Peng, W., Wang, X.-R., Ma, Y., Su, W.-W.,(2005), Fingerprint quality control of Tianjihuang by high-performance liquid chromatography–photodiode array detection, , *J.Chromatogr. A,* Vol. 1070, No.1-2,(8 April 2005),pp. 35-42, ISSN: 00219673

Zhang, W., Saif, M.W., Dutschman, G.E., Li, X., Lam, W., Bussom, S., Jiang, Z., (...), Cheng, Y.-C.(2010), Identification of chemicals and their metabolites from PHY906, a Chinese medicine formulation, in the plasma of a patient treated with irinotecan and PHY906 using liquid chromatography/tandem mass spectrometry (LC/MS/MS, *J. Chromatogr. A,* vol. 1217, No. 37,(September 2010), pp. 5785–5793, ISSN: 00219673

Zissis, K.D., Dunkerley, S., Brereton, R.G.(1999),, Chemometric techniques for exploring complex chromatograms: application of diode array detection highperformance liquid chromatography electrospray ionization mass spectrometry to chlorophyll a allomers, *Analyst,* Vol. 124,No.7, (nd) , pp. 971–979, ISSN: 00032654

Part 4

Photodiode for UV-Light Detection

10

UV Photodiodes Response to Non-Normal, Non-Colimated and Diffusive Sources of Irradiance

María-Paz Zorzano, Javier Martín-Soler and Javier Gómez-Elvira
Centro de Astrobiología (INTA-CSIC), Carretera de Ajalvir km 4
Torrejón de Ardoz, Madrid,
Spain

1. Introduction

A photodiode is a type of photodetector capable of converting incident light into either current or voltage, depending upon the mode of operation. The substrate used to make the photodiode sensing part is critical to defining its properties. In particular, only photons with sufficient energy to excite electrons across the material's bandgap will induce photocurrents. Materials commonly used to produce photodiodes include: Silicon (for the wavelength range 190-1100 nm), Germanium (for 400-1700 nm), Indium-Gallium arsenide (for 800-2600 nm) and Lead (II) sulphide (1000-3500 nm), among others. In many applications the semiconductor diode is placed within an opaque housing with different possible sizes (such as TO18, TO5, TO3..) and hermetically sealed with a cover of a material that allows transmission of the desired wavelength range of observation and permits radiation to reach the sensitive part of the device. Some applications require measurement of one specific wavelength range and it is essential that other radiation wavelengths of the range where the substrate shows responsivity do not contribute to the photodiode's current. For this purpose a filter is mounted above the detector to select the wavelength range of responsivity. Common configurations of filtered photodiodes include a metallic platform to hold the filter at an intermediate height between the cover window and the detector, see Figure 1. This mechanical configuration allows the implementation of different filters preserving the external packing dimensions and physical properties of the photodiode sealing with independence of the filter type.

The geometry of arrangement of these elements (cover, housing, and sensing device) defines the nominal field of view (FOV) or geometrical FOV of the sensor. A careful look at this configuration shows that, because of secondary internal reflections against the inner walls of the sensor housing, the sensor may also be excited by photons from trajectories with incident angles greater than this nominal geometrical FOV. Notice that the filter crystal is thick, flat and cut in square shape. Its implementation on top of the sensing dice covers the nominal FOV but leaves a fraction of the photodiode horizontal cross-section uncovered, see Figure 2. As we will explain later, this will have implications for non-normal, non-collimated and diffusive irradiance sources.

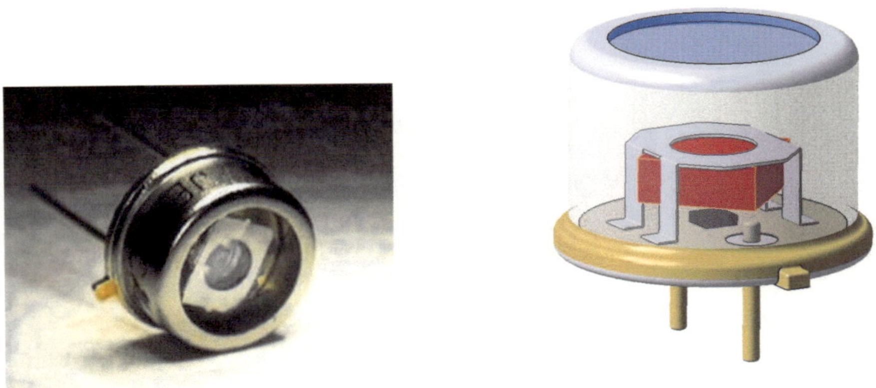

Fig. 1. (Left) Commercial hermetically sealed filtered ultraviolet (UV) photodiode with TO5 housing and the filter mounted on a metallic platform. (Right) Geometrical configuration of the photodiode. The filter crystal is here represented in red and the SiC detector dice in black. The filter is mounted on the metallic structure but does not reach the side of the housing.

Because of changes in the refractive index as the light beam crosses the boundary between two media there is an angle variation which is given by Snell law. The light ray bends toward the normal when the light enters a medium of greater refractive index, and away from the normal when entering a medium of lesser refractive index. For this later case there is a limiting angle beyond which all the rays should be reflected that is called the critical angle of incidence. For instance, in the case of a transition from a quartz crystal to gaseous nitrogen or to air the nominal critical angle is roughly 40°. In theory, for incident angles greater than this angle the light rays should be totally reflected by the cover crystal. However in most cases there is a range of angles where a fraction is reflected and a fraction transmitted. In addition in some implementations, such as the one analyzed here, the covering window is slightly curved. This induces additional angular distortions for non-normal incidence and in practice shifts the critical angle of incidence towards greater values. In summary for the case considered here, beyond the nominal FOV there is a wider angle of allowed incoming rays, what we will here call the critical angle FOV, where incident rays are allowed to pass inside the photodiode housing. Photon rays contained between the nominal FOV and the critical angle FOV may get inside the photodiode housing, be reflected in the walls, avoid the filter and be trapped between the filter and the sensing substrate dice. These photons may thus excite the dice and contribute to the total photodiode output current avoiding the filter selective bandpass action and producing an unexpected current leak. This is schematically represented in Figure 2.

In summary, for a given arbitrary configuration where the radiation from a complex radiating environment excites a photodiode, part of the radiation may be reflected by the cover window and part may pass with slightly different exit angle, part may be contained within the FOV and traverse the filter and part may bounce within the inner part of the sensor housing and hit the dice with a new incidence angle (and here again it may be absorbed, or reflected). When these effects are present the response of the dice to the incident irradiance is neither the one tabulated in the device specifications by the manufacturer (which is usually defined for normal incidence) nor the one obtained in standard laboratory calibrations. For any practical photodiode use, where the incident light is non-normal and non-collimated, these deviations should be explored under representative operation conditions.

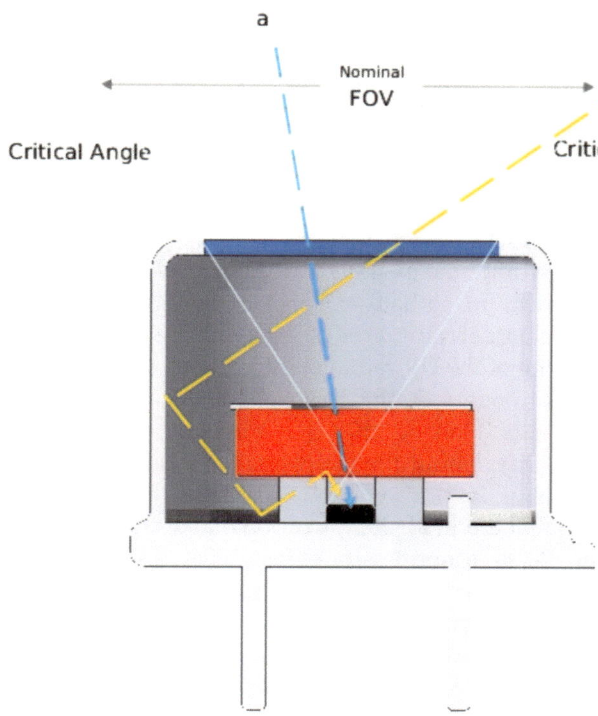

Fig. 2. Nominal field of view (FOV) and critical angle conceptual representation: All ray beams, such as (a), within the FOV solid angle hit the sensing dice after traversing the filter. Photon rays that traverse the window cover with angles of incidence such as (b) hit the inner housing walls and reach the sensing dice after multiple reflections, avoiding the filter selective action.

It is the aim of this chapter to illustrate these effects experimentally with a specific study case and to describe an experimental setup that can be used to quantify the relevance of each contribution. In particular, we describe here the calibration and operation studies of the (ultraviolet) UV photodiodes of the UV sensor for REMS (Rover Environmental Monitoring Station) instrument on-board the MSL rover to Mars (NASA 2012) [Gómez-Elvira et al., 2008]. This sensor consists of a set of 6 photodiodes with different responsivity spectral ranges. One of the photodiodes has no filter and is sensitive in the total SiC responsivity range (from here on named ABC). The other 5 photodiodes correspond to filtered bands named A,B,C,D and E, see Figure 4. Each broadband measurement provides a crude evaluation of the incident irradiance in its relevant spectral range: photodiode C provides a first order estimate of the level of biologically damaging irradiance; photodiodes A and B provide an estimate that may be comparable with terrestrial irradiances while photodiode ABC gives an estimate of the total UV irradiance, and photodiodes D and E are designed to match two UV channels of the MARCI instrument, on-board the Mars Reconnaissance

Orbiter (MRO) satellite to enable direct comparison with reflectance measurements from the top of the atmosphere. The chosen UV photodiodes have a nominal FOV of ± 30º. This is the solid angle contained between the normal to the dice and the imaginary line connecting the sensing dice and the opaque border of the top of the caging, as shown in Figure 2. The photodiodes are mounted in a circular pattern within a metallic box on the rover deck, facing the sky. Each one is embedded within a magnet (to deflect the trajectories of in-falling magnetic dust and protect the window from Martian dust deposition, mimicking the effect of the magnet experiment of the MER rovers) [Kinch et al. 2006]. A scheme of this setup with the nominal field of view of ± 30 º of a photodiode is shown in Figure 3. The whole REMS-UV setup, in an anodized aluminium box of 55 mm x 68 mm x 16 mm with a D25 connector, weights only 72 g. Photodiodes have the advantage of being small in size, light and robust for operation under harsh conditions such as those expected for the MSL rover.

This sensor will deliver for the first time in-situ surface ultraviolet irradiance measurements that will provide ground-truth to radiative transfer models and satellite reflectance measurements as well as first order estimates of biological and chemical doses and UV opacities. A solid understanding of the UV radiation behaviour of the Martian atmosphere is important for photochemical models of the atmosphere [Rodrigo et al., 1990], for the chemistry of the surface minerals [Holland 1978, Mukhin et al., 1996, Quin et al. 2001], has biological implications and is paramount for the assessment of the possible habitability of the Martian surface [Cockell et al. 2000, Patel et al. 2002, Patel et al. 2003, Patel et al. 2004a, Patel et al. 2004B, Cordoba-Jabonero et al. 2003, Cordoba-Jabonero et al. 2005]. In addition, satellite [Mateshvili et al. 2006, Montmessin et al. 2006] and Martian-ground based measurements of UV radiation [Zorzano and Córdoba-Jabonero 2007], are important to retrieve, through radiative transfer studies, [Zorzano et al. 2005], accurate information on atmosphere aerosols, in particular aerosol size distribution, load, dust and cloud dynamics. It is also relevant to estimate the ozone content which in turn serves as a proxy for Mars atmospheric water vapour.

Previous space missions designed to explore the UV Martian conditions, such as the failed Beagle lander mission, also considered a similar UV sensor concept based on photodiodes [Patel et al. 2002]. However, because of their simplicity they also have certain limitations. This study summarizes the evaluation of the response of this setup to representative operation conditions.

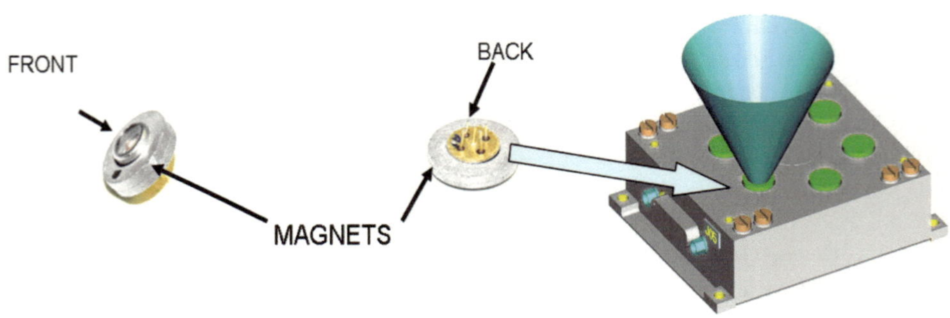

Fig. 3. (Left) Front view of the photodiode and magnet. (Centre) Back view of the photodiode with pins and magnet. (Right) Schematic representation of the nominal FOV of the photodiode within REMS-UV box.

In this space application, the direct radiation source is the Sun. The incident UV radiation comes both as a direct beam, with incident direction according to the solar zenithal angle at the moment of observation, and as diffuse UV irradiance. This diffuse component is the product of the scattering interaction between the incident solar radiation with the dust aerosols and molecules of the thin Martian atmosphere [Zorzano et al. 2005, Zorzano and Córdoba-Jabonero 2007, Zorzano et al. 2009]. The response of the photodiodes in this operation environment shall be investigated.

In summary, we have three scenarios to consider. When the photon ray is within the geometrical FOV, the direct beam is expected to be filtered. When the direct beam is between the geometrical FOV and the critical angle FOV, secondary reflections against the wall may allow extra photons to reach the dice avoiding the filter and thus inducing a current leakage produced by an unfiltered contribution. Finally we shall consider the response to the background diffuse irradiance, i.e. the radiation that has suffered scattering with the atmosphere and reaches the sensor window from almost any direction. In this case the sensor is excited by the diffuse irradiance contained within the solid angle of FOV, which is a significant fraction of the sky diffused irradiance, and shall be filtered. The extra diffuse radiation coming from rays with angles greater than this FOV, but still within the critical angle FOV, can also excite the SiC dice through secondary reflections and, for some photodetectors such as the one considered here, avoiding the filter action. The fraction of diffuse radiation that gets to the dice not being filtered is proportional to the difference between the nominal FOV and the critical angle FOV. If the downwelling diffuse irradiance is uniform then this is a pure geometrical factor. There are second order corrections to this due to the specific reflective, absorption and transmission characteristics of each filter that will be also experimentally observed.

2. Characterization of the response under laboratory conditions

2.1 Spectral calibration of the response with a direct collimated beam of normal or inclined incidence

The response of the sensor to a direct beam of collimated light can be calibrated under controlled operation conditions. This has been done to characterize the spectral responsivity of each photodiode and its dependence with angle of incidence. Its dependence with temperature, the linearity of the response, the degradation with aging and thermal cycling were also analyzed for this specific application in space instrumentation but the results of these tests are beyond the scope of this chapter and shall not be discussed here.

A UV source, focusing optics, a monochromator, a calibrated beam splitter, a detector and a multimeter have been used to calibrate spectrally the response of the photodiodes. One of the photodiodes was sent to The National Physical Laboratory (NPL) (UK's National Measurement Institute) for reference calibration and the results of this calibration setup were referenced with this measurement. The spectral responsivity of each of the 6 chosen photodiodes to a collimated direct beam at normal incidence and ambient temperature is shown in Figure 4. The spectral response of the SiC sensing dice is the one labelled as ABC, namely the one of the photodiode without filter. The rest of photodiodes have a filter and the spectral responsivity is the result of the combined effect of this filter with the underlying SiC responsivity.

The same study has been performed for inclined incidence. Figure 5 shows the measured decay of responsivity (as percentage with respect to the one at normal incidence) for the peak wavelength (the wavelength of maximal responsivity) for all chosen photodiodes together with a 7th order polynomial fit (the r.m.s. error of the fit is 5,3 %) . We observe in this graph both the departure from cosine law and the response of the dices to radiation for incident angles beyond the nominal FOV. This study has been performed for different polar angles and the result is qualitatively similar (not shown), concluding that the response of the photodiode to a direct collimated beam depends mostly on the azimuth angle. At the edge of the nominal FOV (±30º) the response has decayed to a 40% with respect to the normal (and not to 85% as would be expected by a pure cosine like response). Beyond this point there is still significant signal (a 20% of the maximal). The difference between the slow decay of ABC, A and E and the quick decay of B,C, and D is because the maximum of the spectral response of the later ones shows a shift of about 20 nm towards lower wavelength ranges. Furthermore, it is clearly observed in this graph that all the photodiodes show significant responsivity beyond the nominal FOV of ± 30º.

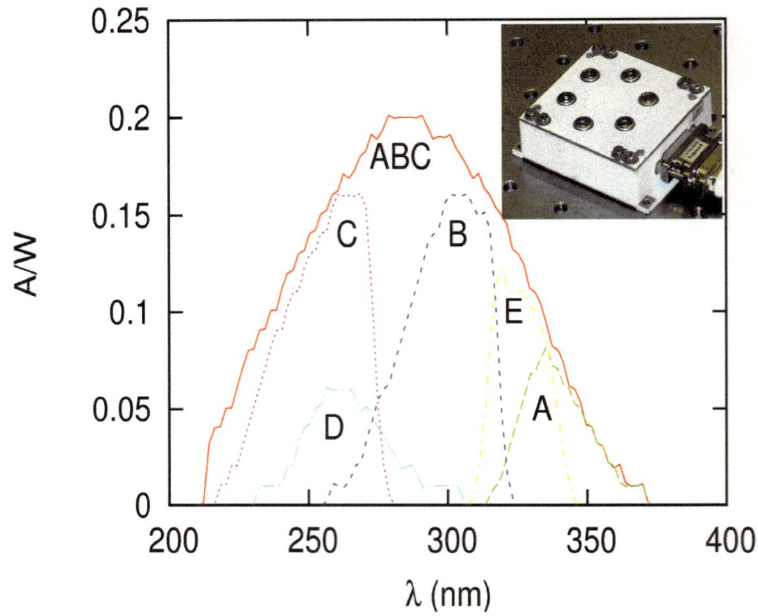

Fig. 4. (Insert) View of the REMS UV box with 6 photodiodes. (Graph) Spectral responsivity, calibrated at ambient temperature with a collimated beam at normal incidence.

2.2 Spectral characterization of the response of the non-filtered contribution with a direct collimated beam at normal incidence

To evaluate qualitatively the spectral weight of the unfiltered contribution, a photodiode was manipulated to separate, in the total current signal, the contribution from the filtered signal and the unfiltered contribution. The same setup used for the spectral calibration of flight model units was used here.

A C type photodiode was opened (by cutting the TO5 housing); an opaque element (a small aluminium plate) was placed on top of the filter, blocking the passage of light rays through this path. Photodiodes that have suffered this manipulation are here named *"ob"*. These photodiodes deliver a current only when photons hit the SiC sensing dice avoiding the filter (which is in this case blocked). By doing this we evaluate the unfiltered contribution of a direct beam, i.e. the direct beam photons that reach and excite the sensing dice avoiding the nominal FOV and thus avoiding the filter. The photodiode base was glued again to the can. Notice that after this manipulation the TO5 cage is filled in by ambient air instead of the original nitrogen encapsulation.

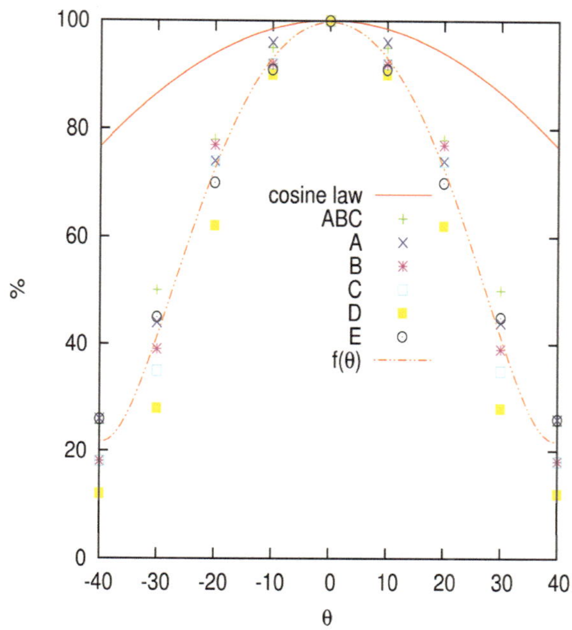

Fig. 5. Zenithal angle dependence of the responsivity to an inclined collimated beam for the SiC unfiltered photodiode (ABC) and 5 filtered photodiodes (A, B, C, D and E). Comparison of the angle dependence between cosine-like decay and a polynomial fit of the measured data with angle. Notice that for a collimated source the photodiodes show significant responsivity beyond the nominal FOV of ± 30º.

In Figure 6 we show the spectral response of this open-blocked photodiode and compare it with the one of a normal C photodiode. Be aware that the filter partially transmits, partially reflects, and partially absorbs. Each specific filter has a different behaviour with respect to transmission and absorption, tuned to select the transmissions properties only. When photons are trapped below the filter, bouncing back and forth until they hit the dice, the filter may also absorb a spectral part of the irradiance. In this specific example a relevant part of the signal of the UVA range seems to be partially blocked, but most of the UVB range radiation is hitting the SiC dice through secondary wall reflections creating a current leak. Thus we may expect different responses for blocked photodiodes of different filters. As we will see later this is indeed observed.

2.3 Characterization of the angular response of the non-filtered contribution with a divergent beam

To qualitatively evaluate under laboratory conditions the angular response of the unfiltered contribution, a new photodiode was manipulated to separate, in the total current signal, the contribution from the filtered signal and the unfiltered contribution. This time a D type photodiode was opened and blocked.

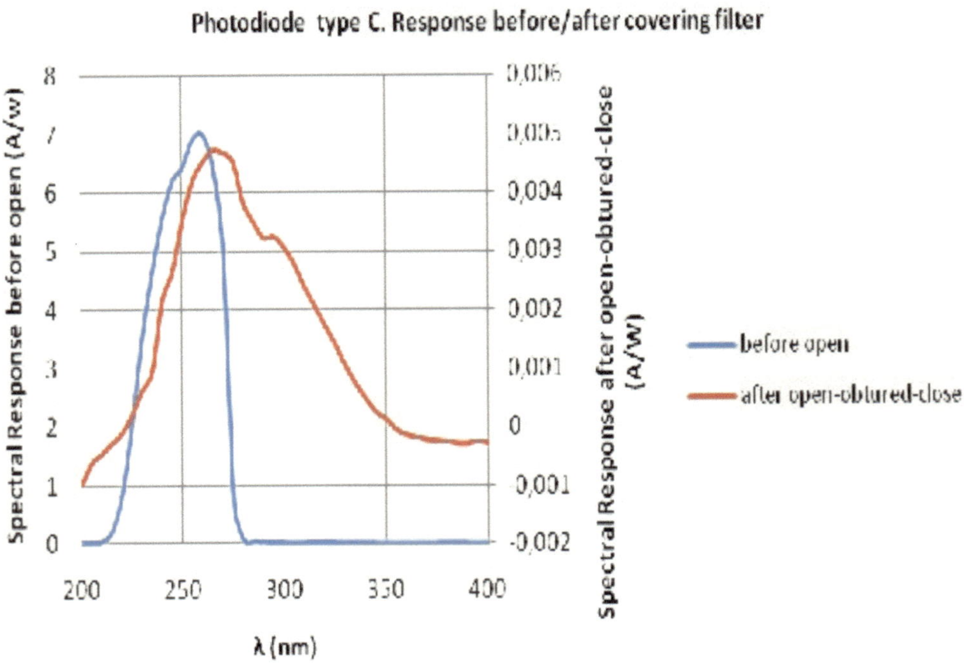

Fig. 6. Spectral response of the open-blocked C photodiode and comparison with original nominal spectral response. The open-blocked responsivity shows the spectral signature of the non-filtered contribution in a C photodiode and extends beyond the nominal UVC range.

A xenon (Xe) light source was placed at a distance of 24 cm to the photodiode (such that the cone of collimated light diverges) and its inclination angle with respect to the diode was varied, see Figure 7. This setup was hard to control but the results give significant qualitative information. The maximal *unfiltered* response is at normal incidence, when the divergent beam hits the photodiode from above. A fraction of the divergent light beam intersects the housing walls; the light in these rays is internally reflected and reaches the dice. As the source goes to lower azimuthal positions, i.e. greater zenithal angle, the signal is reduced up to an angle of roughly 15-20° where it has a minimum. At this point the light beam is half-way to the nominal field of view limit (30°) which is given by the mechanical obstruction of the upper edge in the housing wall. From this point on most of the direct rays hit the side walls. They are then scattered and reflected and reach the dice avoiding the filter. The signal increases then up to roughly 40°-45°. Beyond this point there is a rapid decay and light is partially reflected back at the crystal interface because of the critical angle limit, see Figures 7 and 8. In view of this, we conclude that for field campaign observations, when the Sun direct light beam has a zenithal angle close to 40°-45° (or equivalently an altitude angle between 50°-45°), the measurements of photodiodes where the unfiltered contribution may be comparable to the filtered one shall be discarded.

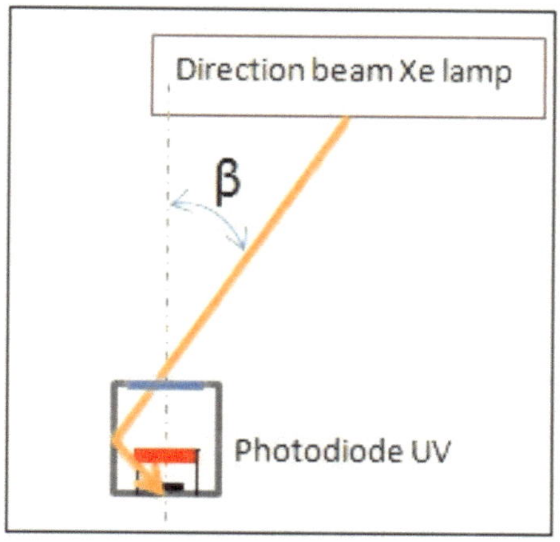

Fig. 7. Schematic representation of the laboratory setup. The beam of the Xe lamp was inclined with respect to the norm to the photodiode. The D photodiode was open, its filter was blocked, and the sensor housing was glued again, to quantify the angular dependence of the non-filtered contribution.

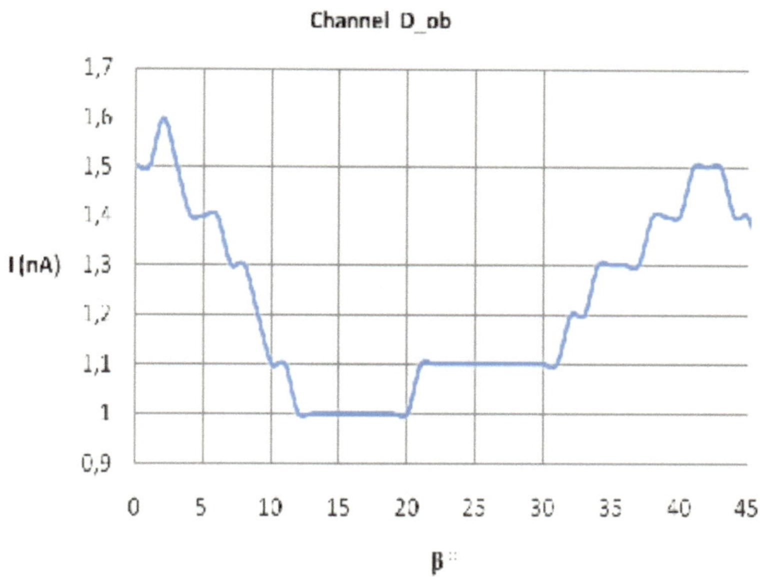

Fig. 8. Non-filtered photodiode's zenithal angle response with an inclined divergent light source exciting a blocked D type photodiode. There is a maximum in the induced current at normal incidence, when a fraction of the cone of the divergent beam hits the inner walls of the caging. As the beam is inclined there is a secondary maximum at 40°-45°, when the centre of the direct divergent beam hits the walls, and is reflected downwards to the sensing dice (avoiding the blocked filter).

Through these experiments we have illustrated the existence of internal reflections for direct beam sources. We therefore redefine a REMS-UV operational strategy that discards observations when the Sun direct beam is in the vicinity of 40-45° w.r.t. the norm. We will evaluate how relevant this internal reflection contribution is for the unavoidable and continuous background of diffusive irradiance.

3. Response characterization under representative operation conditions

To investigate the response to an extended distant source such as the Sun with a background source of diffuse irradiance a special implementation was prepared to operate outdoors in a clear sky day, see Figure 9. For Terrestrial atmospheric measurements, we expect channels C and D to be especially sensitive to the non-filtered contribution because their filters are centred on the ozone absorption band and therefore on Earth the signal in these channels shall be negligible. Any extra signal must be due to an unfiltered contribution.

Fig. 9. Engineering model for field campaign measurements. The direct beam (Sun) angle of incidence can be varied by adjusting the platform inclination.

Three different photodiodes types were used per channel: a flight model unit, this one shall be here named *fm*, and two engineering photodiodes that are manipulated. In these last two photodiodes, because of the manipulation (loss of encapsulated nitrogen, mechanical distortion and misalignment after cutting and gluing the housing) we may expect a variation in the response with respect to the flight model units. One of these photodiodes, which is simply opened and glued again, is used as control reference to discard the influence of the manipulation alone (namely loss of the encapsulated gas), this one is from here on named *op*. The other one is opened, its filter is blocked, and then it is glued again, this one shall be named *ob*.

3.1 Angular characterization of the unfiltered diffuse radiation

The photodiode platform was placed horizontally, facing the sky. During this day the maximal Solar Zenithal Angle (SZA) was at 57.5°, and therefore the trajectory was very low with respect to the FOV of the photodiodes. The recorded signal was thus the response to the sky diffuse irradiance contained within the solid angle of view of the photodiode. This includes both the diffuse irradiance within the nominal FOV (filtered contribution) and the rest, up to the critical angle FOV (unfiltered contribution). The measured current varies smoothly as the SZA changes along the day. For photodiode C, in the vicinity of the maximal SZA, the diffuse unfiltered contribution (*ob*) is about 30% of the total signal in the control photodiode (*op*) whereas for big SZA the unfiltered signal is about 50% of the total one, see Figure 10. This is a smooth function and may be easily interpolated. The *fm* response is shown for reference.

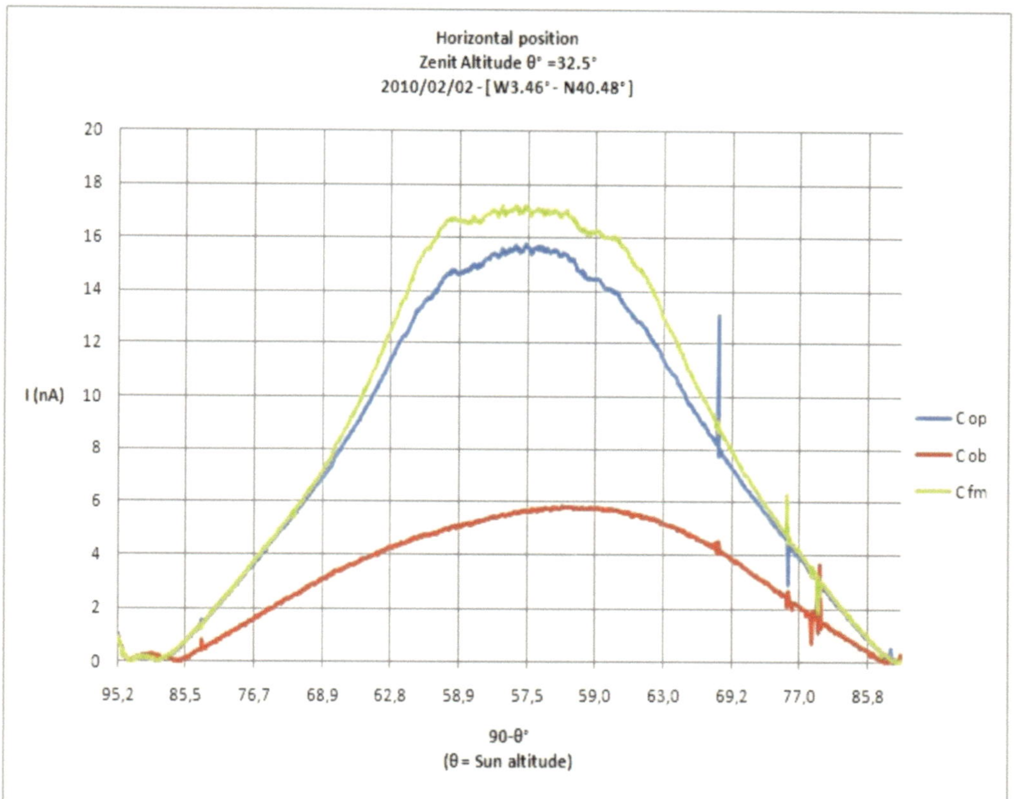

Fig. 10. Comparison of the evolution of the three C photodiodes: flight-model (*fm*), open-blocked (*ob*) and C open (*op*) measurement along the day, as the Sun traverses the sky. For this configuration the sun transit avoids both the nominal FOV and the critical angle FOV. Thus the measured current is produced in response to the diffuse irradiance alone.

In Figure 11 we compare this measurement for different *ob* channels. Since their filters are blocked their response is basically associated to the response of the SiC dice. When the filter is blocked, the order of magnitude of the signal of different blocked photodiodes is similar but there are differences of the order of up to 30% among them (which may be due to geometrical factors after manipulation or to differences in the coating of the lower part of filters). The averaged intensity of this current is about 50 times less the one of the unfiltered photodiode ABC, see Figure 11.

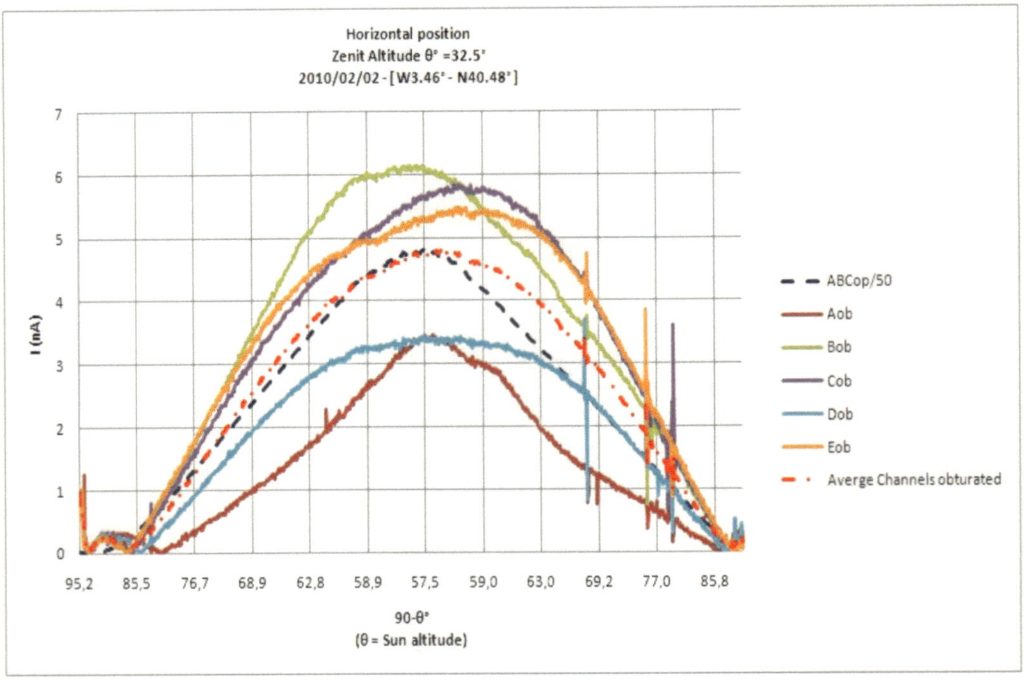

Fig. 11. Comparison of the unfiltered response of all photodiodes, averaged unfiltered response and rescaled (by 50) current of the total ABC photodiode in a configuration where only diffuse irradiance reaches the photodiode.

This suggests that, for this kind of photodiode, a fraction of the UV sky diffuse irradiance avoids the central filter of the photodiode and hits the bottom SiC diode directly producing a leakage signal of the order of about 2% of the *total UV (ABC channel)* induced current. If the channel of observation expects a very small signal for its nominal spectral range, this current leakage can be comparable or even greater than the one of interest. In particular, as shown above for the C channel, this represents almost 30% of the total measured current of the nominal channel, or even 50% for big SZA.

3.2 Angular characterization of the unfiltered direct beam contribution

For the next experimental setup the sensors are again facing the sky, pointing upwards. If the Sun during its trajectory is never within the nominal FOV of the photodiode but passes within the critical angle FOV there are two observable maxima at 45° that can be seen in the measurements of all photodiodes (both manipulated and original). See Figure 12 for an example of the response of the C photodiode when the solar transit never enters its nominal FOV. The two maxima at 45° correspond to direct beam wall reflections and thus appear also as unfiltered contribution in the *ob* photodiode. In this setup the photodiodes collect both the inner wall reflection of the direct beam rays (from the Sun) and the diffuse radiation rays (from the sky). The direct beam induced maxima are at about 45°. Beyond this point, when the Sun leaves the critical angle FOV, we should be able to obtain information about the envelope of the diffuse signal as in the example above. Indeed beyond 50° there is a change in the shape of the angle dependent response, at this moment the direct beam is fully reflected because of critical angle issues and the current is induced only by the background sky diffuse irradiance (the one studied above). Also in this case we observe that for big SZA the nonfiltered signal is about 50% of the total one.

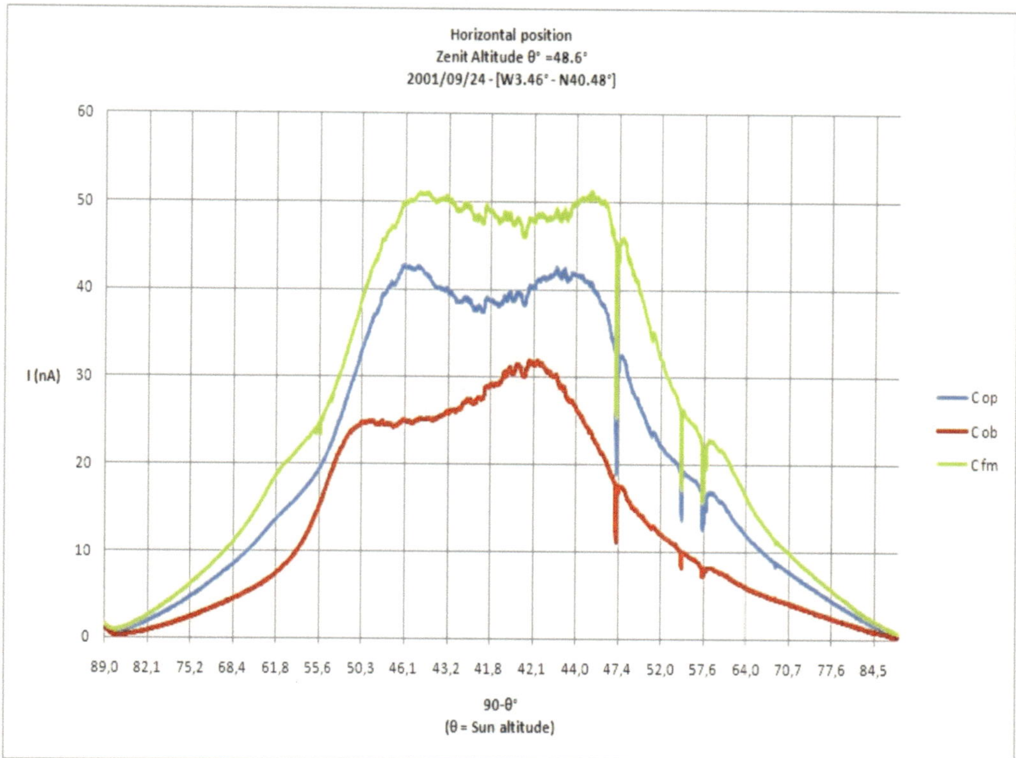

Fig. 12. Comparison of the evolution of the three C photodiodes: flight-model (fm), open-blocked (ob) and C open (op) measurement along the day, as the Sun traverses the sky. For this configuration the sun trajectory never enters the nominal FOV. However it is reflected in the inner photodiode walls and two maxima are observed at roughly 45°.

3.3 Angular characterization of the direct beam: filtered contribution and sunglint

On the next experiment the platform was inclined and pointing to the maximal solar altitude at 40.2°. Figure 13 shows the current evolution for an *fm* photodiode when the sensor platform is oriented towards the Sun maximal altitude position. When the angle of incidence falls between ± 30° with respect to the norm the direct beam is filtered. As was foreseen from laboratory measurements and from the field campaign measurements described in the previous sections, there are two additional maxima at about ± 40°-45° angle of incidence. This represents again secondary reflections of the direct Sun beam within the inner housing walls that manage to avoid the filter and produce a current leak. This unfiltered contribution is so big for the C photodiode that at this stage the total induced current is even bigger than at normal incidence. Beyond that point the signal decays rapidly due to critical angle reflection and the current is induced only by the background sky diffuse irradiance.

In Figure 14 we show for comparison the currents induced in two manipulated C photodiodes. This graph shows that the total irradiance contribution reaching the C sensing dice is partially filtered and partially non-filtered. We observe that for this channel the non-filtered contribution is about 50% of the total measured signal. Two maxima can be distinguished at ±45° with respect to the norm. These last two measurements, with the manipulated photodiodes, show an extra non-filtered maximum when the Sun in just above the norm to the sensors which happens when the solar altitude is at 40.2°. In this inclined configuration, this artefact is due to the Sun glint contribution. The direct beam is reflected on the ground and hits the photodiode with an incidence angle of 40.2°, see Figure 15. In the case of manipulated photodiodes (those that were opened, filled in by normal air, and glued again) this sunglint contribution passes giving a spurious signal. This sunglint contribution seems to be more efficiently rejected in the flight model unit probably due to critical angle issues. The sunglint is not important for REMS application, where the photodiodes will be facing the sky, but may be relevant for other applications with partial view of the ground.

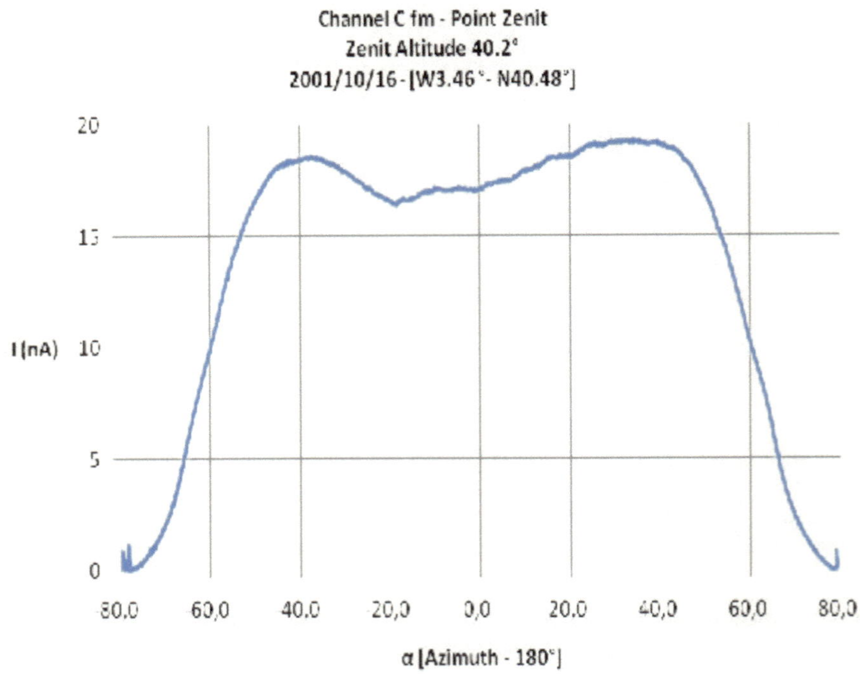

Fig. 13. Evolution of the C *fm* photodiode measurement along the day, as the Sun traverses the sky crossing the normal to the sensor. The platform was pointing to the maximal altitude position. There are to local maxima at zenithal angles of about 40°-45°.

4. Overall experimental verification and definition of mitigation strategy

Next we show a summary of this field site verification with all channels measuring simultaneously with an inclined platform configuration, pointing towards the Sun maximal altitude position at 47.9°. In this configuration when the opened photodiodes have been filled in by normal air, the signal shows again an enhanced unfiltered signal due to sunglint. This unfiltered sunglint contribution is clearly distinguished as a central extra peak in channels expecting weak signals (namely C and D).

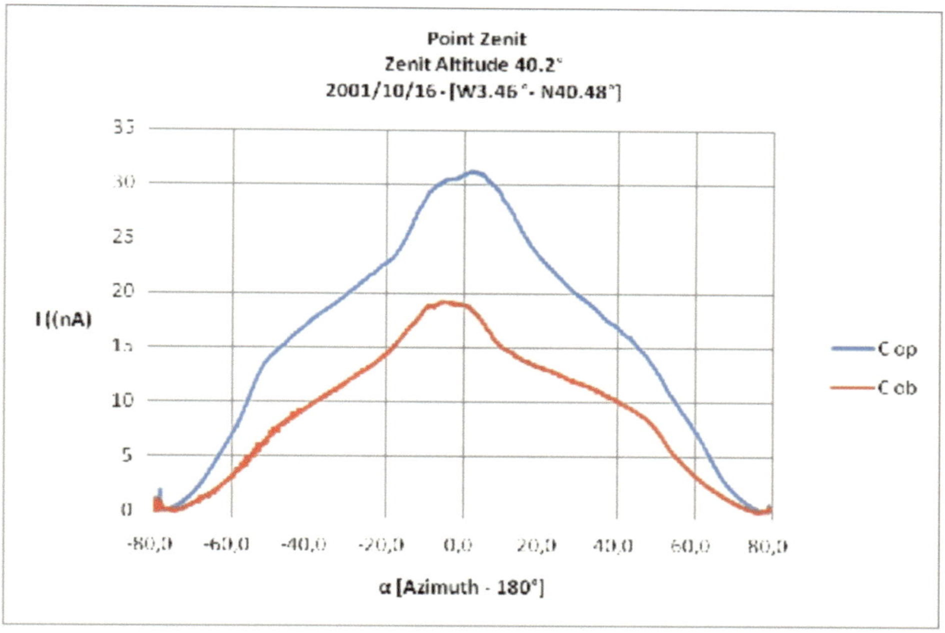

Fig. 14. Evolution of the response of manipulated C photodiodes. Comparison of the open-blocked (*ob*) and C open (*op*) measurement along the day, as the Sun traverses the sky. The open-blocked (*ob*) induced current is the signal induced by unfiltered reflected radiation only whereas the open (op) induced current is the sum of the currents induced on the SiC dice by the unfiltered and filtered radiation.

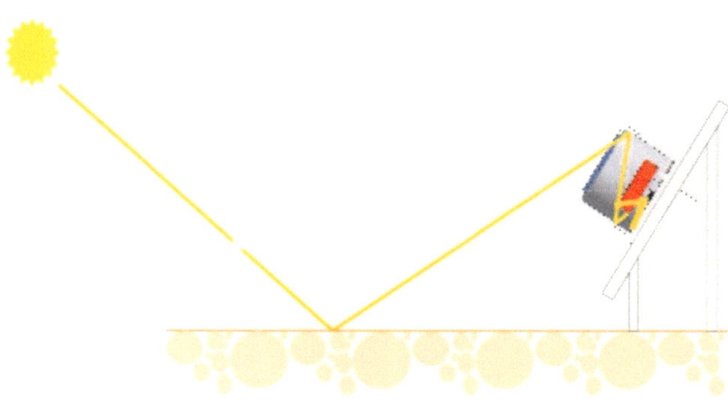

Fig. 15. Sunglint unfiltered contribution for inclined configurations.

For channels A and E the leakage is negligible compared to the nominal signal, see Figure 16. The induced current is smooth and follows the Sun trajectory. The red curve represents the unfiltered contribution (both diffuse and reflected direct irradiance) which in this case is almost negligible. The envelope of these graphs is the sum of the envelope of the background sky diffuse irradiance and, within $\pm 45^{o}$, the envelope of direct irradiance.
For channel B there is a significant, and almost constant, background of unfiltered contribution that creates a constant current leakage of about 10% of the maximal signal, see

Figure 17. Again, the envelope of these graphs shows one shape for the background sky diffuse irradiance and, within +/- 45°, another envelope for the diffuse plus direct response. The constant background of unfiltered contribution is due to the unfiltered diffuse irradiance of the sky.

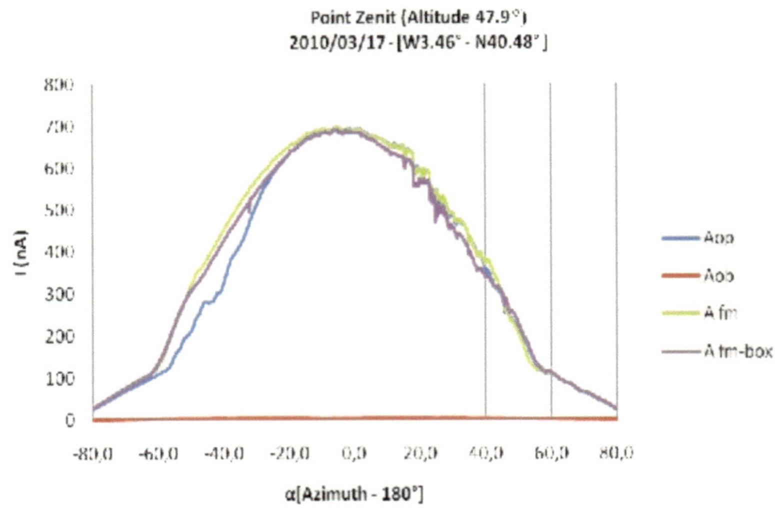

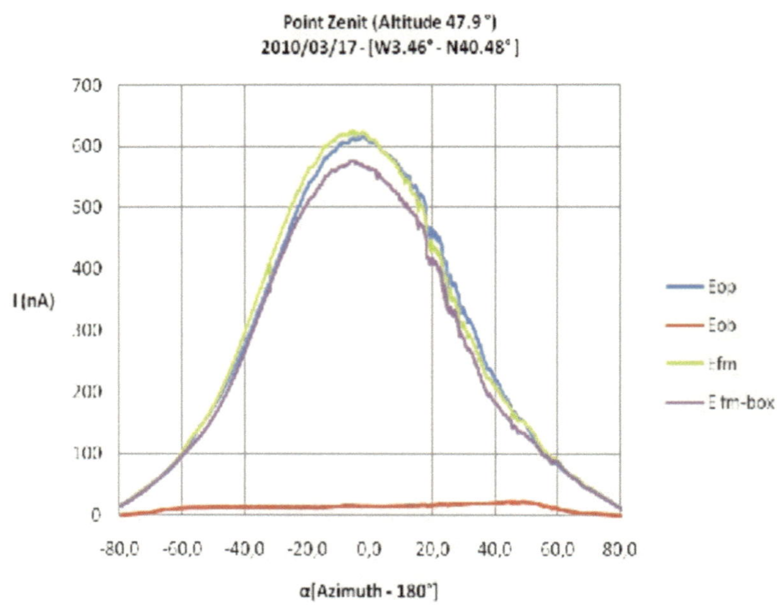

Fig. 16. Diurnal evolution of the measured current for an inclined configuration pointing to the maximal SZA. (Above) Channel A (Below) Channel E.

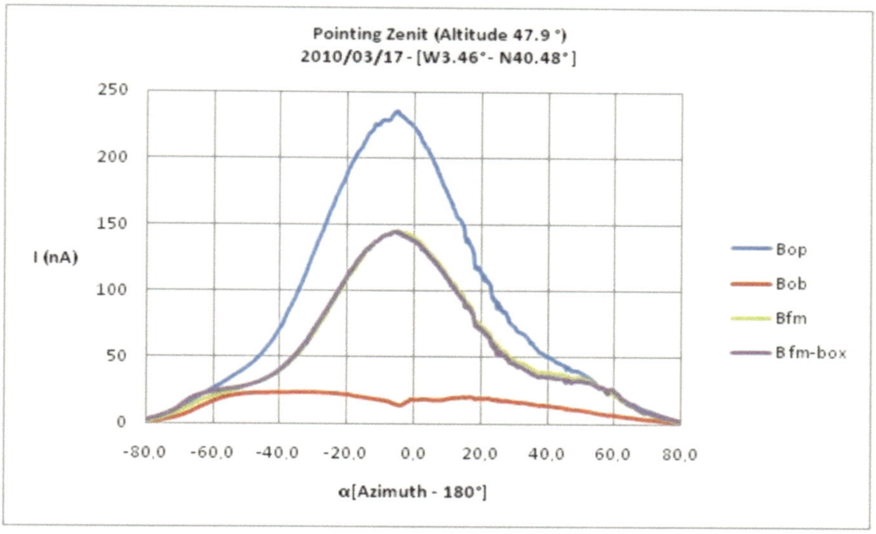

Fig. 17. Diurnal evolution of the B measured current for an inclined configuration pointing to the maximal SZA.

For the C and D manipulated photodiodes, the induced current shows an enhanced, unfiltered response, at SZA of 0 w.r.t to the norm because of sunglint unfiltered reflections, see Figure 17. The flight model units do not show this sun glint contribution. In these ones the two side peaks of direct beam reflections at about 45° are better seen. *Notice that beyond the FOV (beyond 45° roughly) all the photodiodes show the same response as the* ob *one, which means that beyond this point the induced current in both C and D photodiodes is pure unfiltered contribution of the background sky diffusive irradiance.*

The central enhanced peaks are seen in manipulated photodiodes with weak expected currents, namely C and D channels. And are not seen in the ABC opened, A ob, E ob and B ob. This confirms that is a current leakage of unfiltered contribution of the sunglint contribution of the ground.

Finally by comparing the average response of all photodiodes in the tails (namely beyond 45°) of the ABC signal, we conclude again that there is a continuous background unfiltered contribution which is of the order of 2-4% of the *total (ABC channel)* incident UV solar diffuse irradiance, see Figure 19. This 2-4% introduces a systematic error of current leakage due to diffuse irradiance in all filtered photodiode signals which should be treated as system background error to be subtracted.

In summary, with these observations we can adapt the REMS-UV operational scenario. Namely 1) measure the response of the ABC photodiode when the Sun is out of the FOV (roughly beyond ± 45°) which is the response to the sky diffuse irradiance alone, and fit its dependence with angle of incidence. 2) Discard those filtered channels whose measurement is comparable with the 2% of the current induced by diffuse radiation in the ABC channel. 3) Subtract this offset contribution to those channels with greater currents, namely, on Earth, A, B and E. 4) Discard measurements performed when the Sun is close to ± 45° with respect to the norm as they may produce unfiltered direct beam contributions due to inner wall reflections. These measurements should be neglected for absolute radiometric measurements (however they could be used for relative comparisons of day to day relative atmospheric changes).

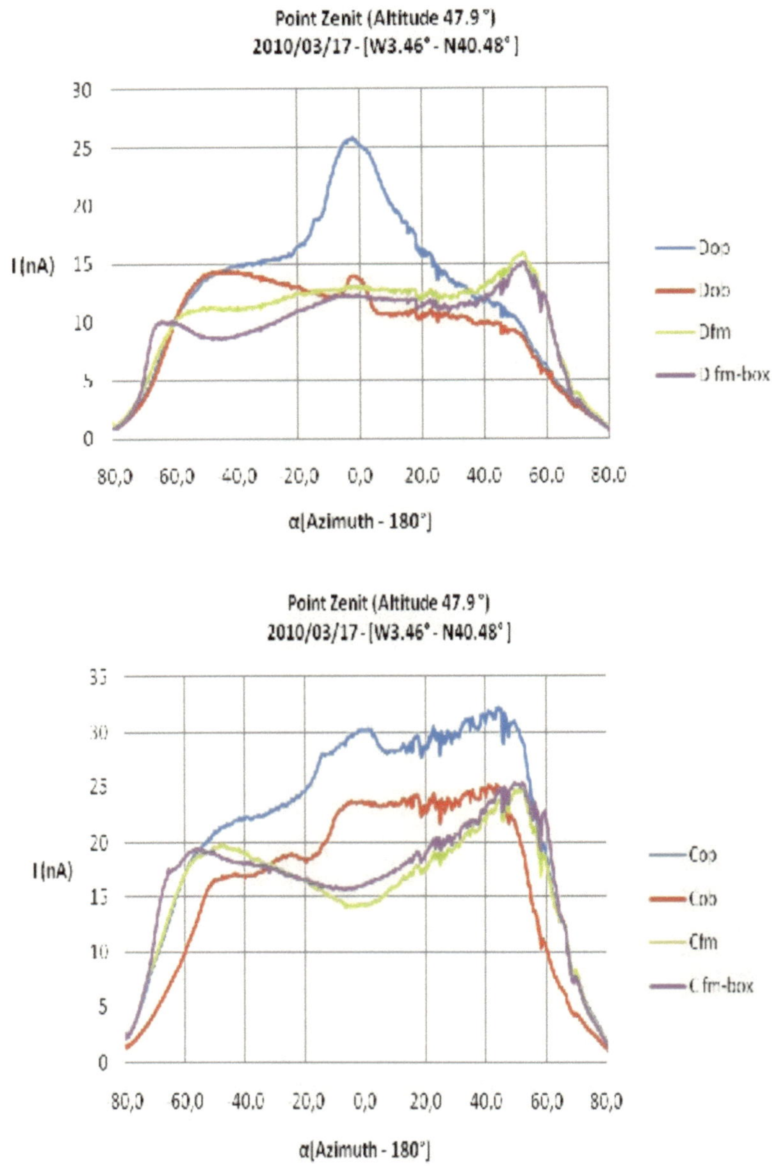

Fig. 18. Diurnal evolution of the measured current for an inclined configuration pointing to the maximal SZA. (Above) Channel D. (Below) Channel C.

Finally regarding REMS measuring strategy during operations on Mars we suggest to define a conservative measuring strategy using the recalibrated currents of photodiodes ABC, A, B and E; and use C and D corrected measurements as check points for interpolating algorithms [Zorzano et al. 2009] It must be clarified that on Mars channels C and D are expected to have much greater currents than on Earth since the UVC radiation is not blocked by the atmospheric ozone as on Earth. These measurements have not been performed to date on the Martian surface and thus if during the mission C and D currents are much greater than the 2% of the ABC diffuse signal these two measurements may be also included as inputs for the interpolating algorithms.

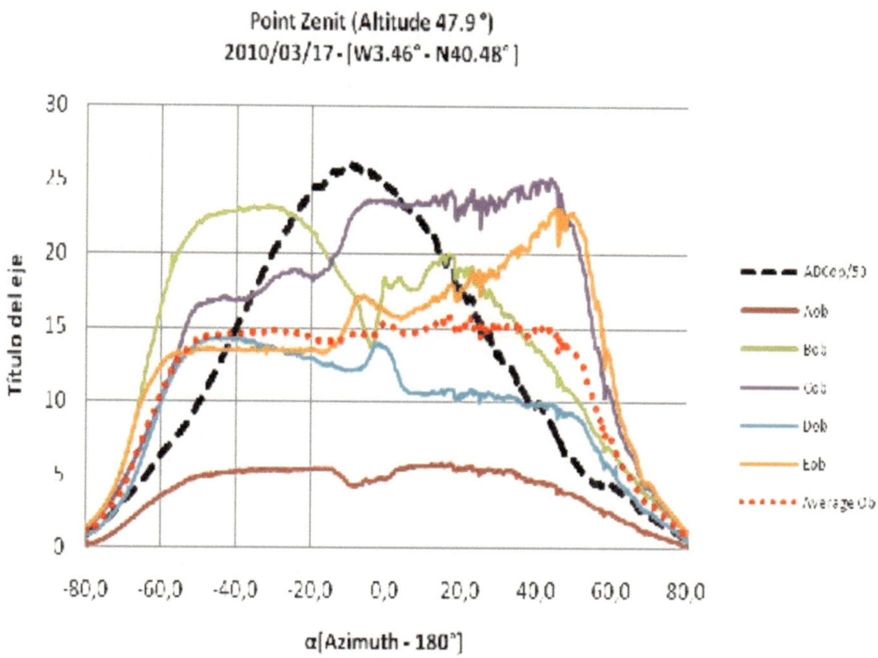

Fig. 19. Evolution along the day of the diffuse unfiltered contribution in filtered channels compared to the total diffuse signal of the total ABC channel.

5. Conclusions

Photodiode sensors are calibrated under controlled laboratory conditions, with collimated, plane light beams at normal incidence with respect to the sensing device. For any other implementation the photodiode needs to be calibrated under representative operation conditions. It is hard to simulate the UV diffuse radiation environment of an atmosphere under laboratory calibration setups. It is also hard to simulate sunglint or secondary reflections induced by reflecting surfaces with small lamps in calibration setups. Dedicated campaigns such as the ones described here should be implemented for this purpose.

In summary, through laboratory and field campaign measurements we have illustrated that:

- The response of the photodiode to a pure, direct, collimated light beam does not decay simply as a cosine law with a cut-off edge at the end of the FOV.
- The sensing dice is also excited by direct beams with angles beyond the nominal FOV that are reflected in the inner housing walls. A second maximal response may be expected for incident angles close to 45°, when the direct beam is reflected on the inner housing walls.
- For certain configurations, also sunglint contributions (or in general reflections of the direct beam source on nearby surfaces) may create unwanted reflected contributions that should taken into account.

- Equivalently, also the rays of diffuse irradiance with angles greater than the nominal FOV can hit the inner housing walls and reach the sensing dice.
- For the photodiodes used in this test, photon rays with incident angle beyond the nominal FOV, that are reflected in the inner walls of the housing (both the direct and diffuse beam contribution) may avoid the filter action.
- The fraction of radiance that gets to the dice avoiding the filter generates an unexpected unfiltered contribution to the photodiode current. This spurious current leak may be comparable to the nominal contribution, in narrow filtered channels or in spectral ranges where little incident radiation is expected. A mitigation strategy should be implemented to cope with this.

This problem may be common to other sensors used in radiometry or remote sensing applications. Under realistic operation conditions, the radiance comes from multiple paths and it is furthermore not isotropic and also non-normal. Some examples of this situations are:

- If part of the incident radiation hits a surface, a fraction of it may be reflected in an almost specular way and another fraction be scattered away. These other components of irradiance may reach the detector through secondary pathways (example satellite observation of ground reflectance contaminated with sun glint and diffused contribution from lambertian surfaces with different angle of incidence).
- If the medium between the source and detector include scattering agents (such as molecules or aerosols) the diffuse component of the radiation may become significant and therefore reach the sensor from other incident angles.
- Finally even in the case when the direct source beam points to the sensor from angles of incidence beyond its nominal FOV, secondary reflections within the detector housing may allow this radiation to reach the sensing device.

An alternative approach to the use of filters is the use of new tailor-made substrates where the band gap of the photo absorption layer is tuned to select the desired specific range of UV radiation. In particular, recently there is great interest in the development of the so-called solar-blind detectors for applications on Earth. As mentioned above, bellow 280 nm the UV radiation is absorbed by the terrestrial atmosphere and thus the highest-energy UV light photons from the sun cannot reach the surface. As a result, this region of the solar electromagnetic spectrum constitutes a 'black background' that can be used to detect and control artificial UVC-emitting sources such as flames (and thus detect for instance fires, or missiles). These measurements require detectors that are 'solar blind,' i.e., insensitive to light above 280nm, and that focus their spectral responsivity in the range of UVC. Current research on solar-blind photodetectors substrates include the development of new substrates such as nano-wires [Delaunay, 2011 and references herein] or $Al_xGa_{1-x}N$-based substrates [Razeghi, 2002]. This alternative is generally very costful -since it requires development and testing of new materials-, and is not always possible for any arbitrary spectral range. Furthermore the use of new sensing substrates may not be convenient for space applications where all the materials must be qualified for space and the maturity of the technology needs to be well proven.

6. Acknowledgements

We thank the support from the REMS team and MSL mission to Mars. This project is funded by the Industry Ministry (CDTI), Science and Innovation Ministry (project ESP2007-65862) and Defence Ministry of Spain. We also acknowledge the work of J. Barbero from ALTER Technology Group Spain, Roser Urquí from REMS project, and the National Physical Laboratory (UK) who have contributed to the calibration process.

7. References

Cockell, C.S., Catling, D.C., Davis, W.L., Snook, K., Kepner, R.L., Lee, P., McKay, C.P., 2000. The ultraviolet environment of Mars: biological implications. Past, present, and future. Icarus 146, 343.

Córdoba-Jabonero, C., Lara, L.M., Mancho, A. M., Marquez, A., Rodrigo, R., (2003). Solar Ultraviolet transfer in Martian atmosphere: biological and geological implications *Planet. Space Sci.* 51, 399-410.

Cordoba-Jabonero, C., Zorzano, M.-P., Selsis, F., Patel, M.R., Cockell, C.S., (2005). Radiative habitable zones in Martian polar environments. *Icarus* 175, 360-371.

Delaunay J-J, Li, Y., Tokizono, T., Liao, M., Koide, Y. (2011). Wide-bandgap nanowires for UV-light detection. Optoelectronics & Communications. SPIE. 10.1117/2.1201102.003466

Gómez-Elvira, J., and REMS Team, Environmental monitoring station for Mars Science Laboratory, Proceedings of *LPI Contributions*, 1447, p. 9052. (2008).

Holland, H. D., 1978. *The Chemistry of the Atmosphere and Oceans.* Wiley Interscience, New York.

Kinch, K.M., Merrison, J.P., Gunnlaugsson, H.P., Bertelsen, P., Madsen, M. B., Nørnberg, P. (2006). Preliminary analysis of the MER magnetic properties experiment using a computational fluid dynamics model. *Planet. Space. Sci.* 54, 28-44.

Mateshvili, N., Fussen, D., Vanhellemont, F., Bingen, C., Dodion, J., Muller, C., Depiesse, C., Perrier, S., Bertaux, J.L., Dimarellis, E. 2006. Martian clouds distribution obtained from SPICAM nadir UV measurements: preliminary results. *Second workshop on Mars atmosphere modelling and observations,* held February 27 - March 3, 2006 Granada, Spain. Edited by F. Forget, M.A. Lopez-Valverde, M.C. Desjean, J.P. Huot, F. Lefevre, S. Lebonnois, S.R. Lewis, E. Millour, P.L. Read and R.J. Wilson. Publisher : LMD, IAA, AOPP, CNES, ESA.

Montmessin, F., Quemerais,E., Bertaux,J. L. , Korablev,O. , Rannou,P. And Lebonnois,S. (2006). Stellar occultations at UV wavelengths by the SPICAM instrument: Retrieval and analysis of Martian haze profiles, J. *Geophys. Res.,* 111,No. E12, E12S06

Mukhin, L.M., Koscheev, A.P., Dikov, Y.P., Hurth, J., WRanke, H., (1996). Experimental simulations of the photodecomposition of carbonates and sulphates on Mars. *Nature* 379, 141.

Patel, M.R.; Zarnecki, J.C. and Catling, D.C. (2002). Ultraviolet radiation on the surface of Mars and the Beagle 2 UV sensor. *Planetary and Space Science,* 50(9), pp. 915–927.

Patel, M.R., Bérces,A., Kolb, C., Rettberg, P., Zarnecki, J. C. and Selsis, F., (2003). Seasonal and Diurnal Variations in Martian Surface UV Irradiation: Biological and Chemical Implications for the Martian Regolith. *International Journal of Astrobiology,* 2 (1), pp 21-34.Quin, R.C., Zent, A.P., McKay, C.P., (2001). Photodecomposition of carbonates on Mars. *Lunar Planet. Sci. Conf. XXXII,* 1463.

Patel, M.R., Bérces, A., Kerkgyrt, T., Ront, G., Lammer, H. and Zarnecki, J.C., 2004. Annual Solar UV Exposure and Biologically Effective Dose Rates on the Martian Surface. *Advances in Space Research*, 33 (8), pp 1247-1252.

Razeghi, M. *Short-wavelength solar-blind detectors-status, prospects, and markets. Proceedings of the IEEE*, Jun 2002, vol 90., 6, pp. 1006-1014.

Rodrigo, R., Garcia-Alvarez, E., Lopez-Gonzalez, M. J., Lopez-Moreno, J., (1990). A nonsteady one-dimensional theoretical model of Mars' neutral atmospheric composition between 30 and 200 KM. *Journal of Geophysical Research* (ISSN 0148-0227), vol. 95, Aug. 30 p. 14795-14810.

Vazquez L, Zorzano MP, Jimenez S. (2007). Spectral information retrieval from integrated broadband photodiode Martian ultraviolet measurements. *Optic Letters* 32(17) 2596-2598.

Zorzano, M.P., Vázquez, L., Jiménez, S. (2009). Retrieval of ultraviolet spectral irradiance from filtered photodiode measurements. *Inverse Problems*, vol. 25 pp.115023-115032 (2009).

Zorzano, M.-P., Mancho, A.-M., Vazquez, L. (2005). Numerical integration of the discrete-ordinate radiative transfer equation in strongly non homogeneous media. *Applied Mathematical and Computation*, 164, 263-274.

Zorzano M.-P., Cordoba-Jabonero C (2007). Influence of aerosol multiple scattering of ultraviolet radiation on Martian atmospheric sensing. *Icarus* 2007, vol. 190, no2, pp. 492-503.

11

Detection of VUV Light with Avalanche Photodiodes

Cristina M. B. Monteiro, Luís M. P. Fernandes and Joaquim M. F. dos Santos
Instrumentation Centre (CI), Physics Department, University of Coimbra Portugal

1. Introduction

Silicon avalanche photodiodes are alternative devices to photomultiplier tubes in photon detection applications, presenting advantages that include compact structure, capability to sustain high pressure, low power consumption, wide dynamic range and high quantum efficiency, covering a wider spectral range. Therefore, they provide a more efficient conversion of the scintillation light into charge carriers. Major drawbacks are lower gains, of few hundreds, higher detection limits and non-uniformities in the percent range.

Windowless APDs with spectral sensitivity extended downto the VUV region (~120 nm) have been developed by API [1], RMD [2] and Hamamatsu [3]. They have been used as photosensors for scintillation light produced in noble gases [4-6] and liquids [7-10] for X- and γ-ray spectroscopy applications. Up to now, the main application of APDs as VUV detectors is aimed for a neutrinoless double beta decay experiment using high pressure xenon [6].

Wide band-gap semiconductor photodiodes such as GaN and SiC are also alternative to photomultiplier tubes in UV detection. However, compared to Si-APDs, they present smaller active area of the order of the mm^2, with higher wafer non-uniformities, lower quantum efficiency and reduced spectral sensitivity in the VUV region (usually useful above 200 nm). On the other hand, they present some advantages, namely the lower biasing voltages, higher gains with lower leak currents, the solar blind capability. Recent reviews on these APDs can be found in [11-17] and references therein.

Through the last decade, we have investigated the response characteristics of a large area APD from API to the scintillation VUV light produced in gaseous argon and xenon at room temperature [4,5]. The emission spectra for argon and xenon electroluminescence is a narrow continuum peaking at about 128 and 172 nm, respectively, with 10 nm FWHM for both cases [18], and corresponds to the lower limit of the APD spectral response. For the 128 and 172 nm VUV light from argon and xenon scintillation, the effective quantum efficiency, here defined as the average number of free electrons produced in the APD per incident VUV photon is 0.5 and 1.1, respectively, corresponding to a spectral sensitivity of about 50 and 150 mA/W [4,19].

In this chapter, we review and summarize the results of our investigation, namely the gain non-linearity between the detection of X-rays and VUV light [20], the gain dependence on temperature [21,22], the behaviour under intense magnetic fields [23], the minimum detection limit, i.e. the minimum number of photons detectable above the noise level, and the statistical fluctuations in VUV photon detection [24].

2. APD operation principle

Figure 1 shows the structure of an APD, as well as the electric field profile inside its volume. When a high voltage is applied across the APD, only a small region of the p-layer, near the photodiode surface, remains undepleted - the drift region (1). This region has a residual electric field of some tens of V/cm [19]. In the depleted p-region (2), the electric field increases with depth until a maximum is reached, around 10^5 V/cm [19], near the p-n junction and decreases in the depleted n-region. VUV photons are absorbed in (1) and converted into electron-hole pairs. The resulting primary electrons are driven to the p-n junction by the electric field. Around the junction, they obtain a sufficient amount of energy to produce new electron-hole pairs by impact ionisation, leading to an avalanche process in the multiplication region (3). Charge gains of a few hundred are typical. Gain increases exponentially with the applied voltage, resulting in a significant improvement of the signal-to-noise ratio. Detailed operation principles of the APD have been presented in the literature [25-28].

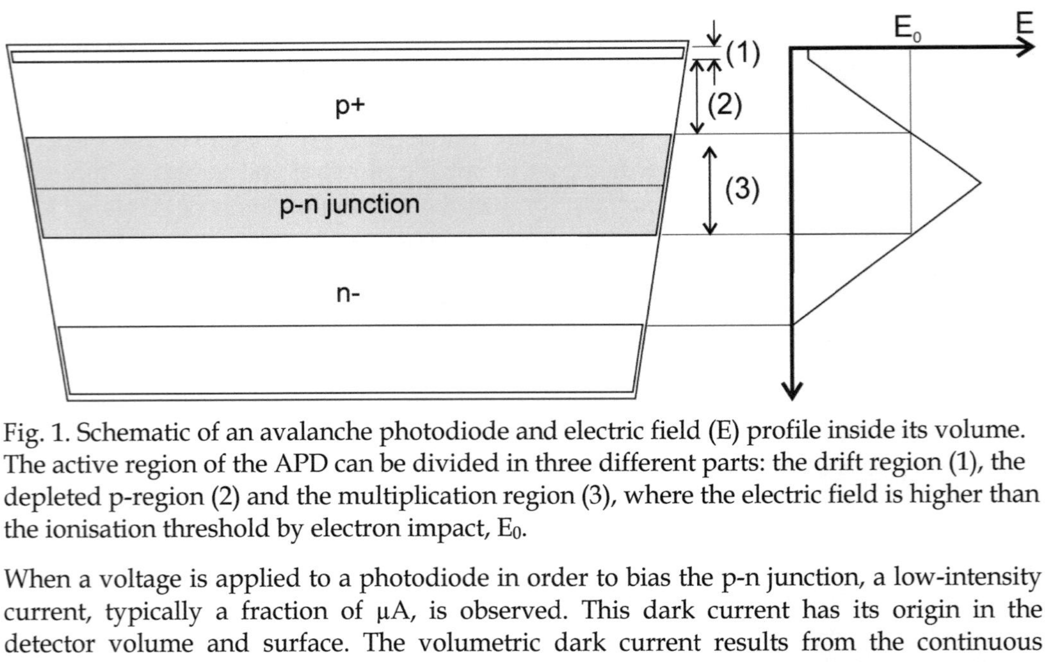

Fig. 1. Schematic of an avalanche photodiode and electric field (E) profile inside its volume. The active region of the APD can be divided in three different parts: the drift region (1), the depleted p-region (2) and the multiplication region (3), where the electric field is higher than the ionisation threshold by electron impact, E_0.

When a voltage is applied to a photodiode in order to bias the p-n junction, a low-intensity current, typically a fraction of μA, is observed. This dark current has its origin in the detector volume and surface. The volumetric dark current results from the continuous generation of charge carriers - minority carriers - on both sides of the junction, which are conducted through that junction, and from the thermal generation of electron-hole pairs in the depletion region, which increases with volume and decreases by cooling. The surface dark current is generated in the p-n junction edges due to high voltage gradients in its vicinity. Since dark current is a noise source and increases considerably with temperature, the electronic noise level can be reduced by cooling the APD, reducing the statistical fluctuations and the minimum detection limit of VUV photons [21,22].

Different studies have proven the APD response characteristics for VUV light to be different from those for visible light, which has been originally used to determine most of the characteristics of photodiodes [20-23]. This is due to the difference in the average interaction depth of the photons, which is approximately 1 µm for 520-nm photons and approximately 5 nm for 172-nm photons [29]. VUV photons interact mainly within the first atomic layers of the wafer, where the electric field is weaker. This results in higher diffusion of the charge carriers, which can be lost to the surface boundary and to impurities.

3. APD noise and statistical fluctuations

The energy resolution associated to radiation detection in avalanche photodiodes is determined by several factors, namely statistical fluctuations associated to the number of electron-hole pairs produced in silicon and the avalanche process (N); gain non-uniformity in the APD detection volume; detector noise, resulting from dark current, and electronic system noise. The total broadening in the energy distributions of APD pulses, ΔE, is the quadratic addition of those three contributions.

The output signal variance, in number of primary electrons, associated to the statistical fluctuations is given by [30]:

$$\sigma_N^2 = {\sigma_n}^2 + N\,(F-1) \qquad (1)$$

N being the number of primary electrons, ${\sigma_n}^2$ its variance and F the excess noise factor. F is related to the variance of the electron avalanche gain, ${\sigma_A}^2$, according to:

$$F = 1 + {\sigma_A}^2 / G^2 \qquad (2)$$

Due to the discrete nature of the multiplication process, as a result of electron avalanche fluctuations F is higher than 1 and varies with gain, G.

The relationship between F and G has been derived from the McIntyre model considering that photoelectrons are injected close to the p-zone surface [31]:

$$F \cong G\,k_{ef} + \left(2 - \frac{1}{G}\right)\left(1 - k_{ef}\right) \qquad (3)$$

k_{ef} being the effective ratio between the ionisation coefficients for holes and electrons. For lower gains, $k_{ef} << 1$ and $F \cong 2 - 1/G$.

For the useful gain range (G > 30), the variation of k_{ef} with voltage is very low and considering k_{ef} constant is a good assumption. As a result, the dependence of F on G should be approximately linear, e.g. see [32].

There is a clear difference between light and X-ray detection. In particular, the non-uniformity contribution is negligible in light detection if the whole APD area is irradiated, since the final pulse results from the average response to the entire number of photons interacting in the silicon.

For light pulse detection, the variance of the number of primary electrons is described by Poisson statistics, ${\sigma_n}^2 = N$. The statistical error, in number of primary electrons is, then:

$${\sigma_N}^2 = N\,F \qquad (4)$$

The noise contribution to the energy resolution results from two different sources, namely the detector dark current and the electronic system. Dark current presents two different components. One of them (IDS) does not depend on gain and corresponds to the surface current and to a small fraction of the volumetric current resulting from thermal generation of electron-hole pairs in the n-region, thus non-amplified. The other component (IDV) is amplified by the gain and corresponds to the volumetric current resulting from the production of electron-hole pairs in the p-region. The total current at the APD output is:

$$I = I_{DS} + G I_{DV} + G I_0 \tag{5}$$

where G is the APD gain and I_0 the non-amplified signal current, corresponding to electron-hole pairs produced by the absorbed radiation.

The noise associated to the electronic system is mainly originated in the FET (field effect transistor) at the preamplifier input. Fluctuations in the FET channel current are similar to the thermal noise and can be represented by a noise equivalent resistance (R_{eq}) in series with the preamplifier input [33].

A detailed noise analysis in avalanche photodiodes has been already presented in the literature [34,35]. If the preamplifier is connected to a linear amplifier with equal differentiation and integration constants, τ, the electronic noise contribution to the peak broadening in units of energy is:

$$\Delta E_N^{\,2} = \left(2.36\frac{e\varepsilon}{qG}\right)^2 \left[\frac{k_B T R_{eq}}{2\tau} C_T^{\,2} + \frac{\tau q}{4}\left(I_{DS} + I_{DV} G^2 F\right)\right] \tag{6}$$

q being the electron charge, $e \cong 2.718$ the number of Nepper, k_B the Boltzmann constant (1.38×10^{-23} J/K) and T the temperature (in Kelvin); C_T is the total capacitance at the preamplifier input, including detector and FET input capacitances.

The first term in (6) describes the electronic system noise associated to the detector and the second term corresponds to the dark current contribution. Both terms depend on the shaping time constants used in the linear amplifier. The noise contribution also depends on the gain and the excess noise factor.

4. VUV-light measurements

A Gas Proportional Scintillation Counter (GPSC) was used to provide VUV-light pulses, with a known number of photons, to the APD. The operation principle of the GPSC is described in detail in [36] and a schematic of the counter is depicted in Fig.2. X-rays interact in the absorption region producing a known number of primary electrons. The primary electron clouds are driven to the scintillation region where the electric field is kept below or around the gas ionisation threshold. Therefore, upon traversing the scintillation region, the primary electrons gain enough energy from the electric field to excite a large number of noble gas atoms, leading to a VUV-light pulse as a result of the de-excitation processes of the atoms. Following the incidence of VUV photons on the APD, charge multiplication takes place within the APD volume, originating the final charge pulse at its output. The operational characteristics of the GPSC with APD are described in [4] and [5] for xenon and argon fill gas, respectively.

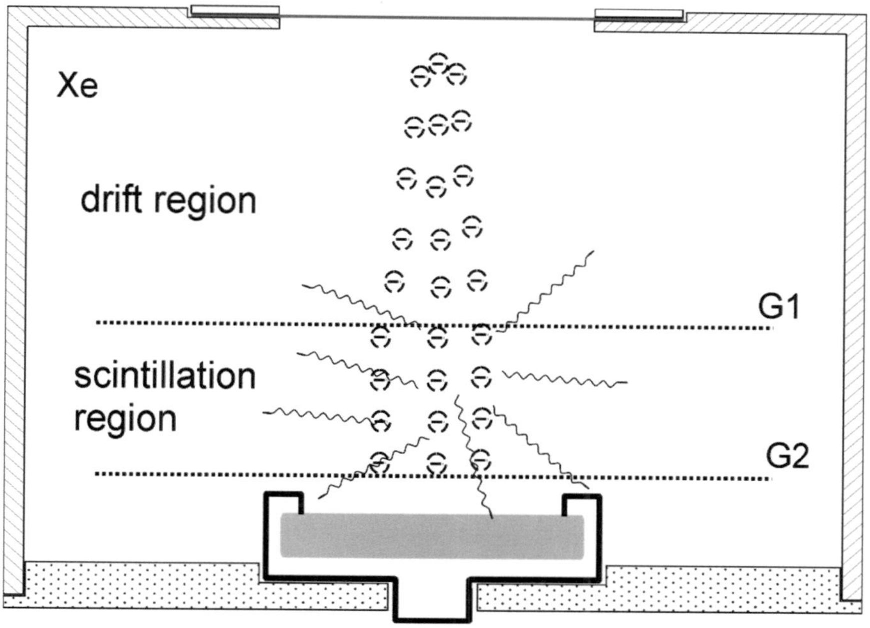

Fig. 2. Schematic of the Gas Proportional Scintillation Counter instrumented with an APD as the VUV photosensor.

The geometry of the GPSC was chosen to allow some of the X-ray photons to reach the APD without being absorbed in the gas, which allows direct X-ray interactions in the APD concomitant with X-ray interactions in the gas. A typical pulse-height distribution is presented in Fig.3 for a GPSC with argon filling, irradiated with 5.9-keV X-rays. The main features of the pulse-height distributions include the scintillation peaks resulting from the full absorption of X-rays in the gas and from events with subsequent argon fluorescence escape from the active volume, the so-called escape peaks, as well as the electronic noise tail. An additional peak resulting from direct absorption of the 5.9-keV X-rays in the APD is also present in the pulse-height distributions. This latter peak is easy to identify, since its amplitude depends only on the APD biasing and not on the GPSC biasing, being present even when the gas proportional scintillation counter biasing is switched off.

Knowing the *w*-value - i.e. the average energy to produce an electron/hole pair - of silicon for X-rays (w_{Si} = 3.6 eV), the peak resulting from the direct interaction of 5.9-keV X-rays in the APD can be used to determine the number of charge carriers produced by the VUV-light pulse. In the case of Fig.3, the amount of energy deposited in silicon by the argon scintillation pulse is similar to what would be deposited by 30-keV X-rays directly absorbed in the APD. This feature allowed the absolute determination of the argon and the xenon scintillation yields, given the quantum efficiency of the APD and the solid angle subtended by the APD relative to the region where the scintillation occurred [37-40].

The performance characteristics of the APD in VUV light detection has been investigated as a function of voltage applied to the APD, using the information of the successive pulse-height distributions obtained for each voltage. The relative positions of the VUV-light and the direct X-ray interaction peaks provide the information on the non-linear response of the APD to X-rays, important issue for the correct determination of the number of scintillation photons detected by the APD. Knowing the number of photons hitting the photodiode, the minimum number of photons above the noise that can be detected by the APD can be determined by the relative position of the noise tail. The width of the scintillation peak can be used to determine the statistical fluctuations resulting from the detection and signal amplification processes in the APD, provided that the statistical fluctuations associated with the X-ray interaction in the gas, i.e. in the number of primary electrons produced, and with the gas scintillation processes are known.

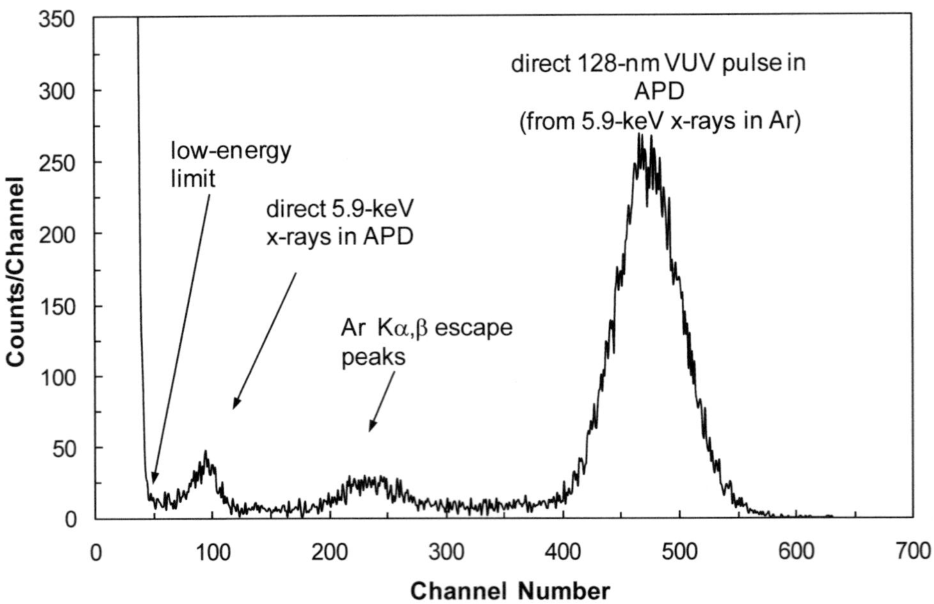

Fig. 3. Typical pulse-height distribution in APDs, resulting from both direct absorption of 5.9-keV X-rays in the APD and 128-nm scintillation absorption in the APD, resulting from the interaction 5.9-keV X-rays in argon.

The APD gain was obtained by normalizing the scintillation pulse amplitude to the manufacturer specification – a gain of 13.8 at 1577 V. The APD gain was also determined for the direct interaction of 5.9-keV X-rays. Typical APD gain variation with voltage applied to the photodiode is presented in Fig.4 for both 5.9-keV X-rays and 128-nm scintillation pulse interactions in the APD.

Figure 4 demonstrates the non-linear effects that are present in X-ray detection. While for light detection the VUV-photon interactions and, consequently, the charge carriers and subsequent electron avalanche are distributed through the whole APD, the point-like nature of the X-ray interaction results in the production of a charge carrier cloud and subsequent electron avalanche that is concentrated in a very small volume of the APD. Therefore, non-linear effects in X-ray detection are attributed to space-charge effects, reducing the local electric field, and to heating in the avalanche region [20,30]. This is confirmed by the non-linearity observed in the APD gain response between X-rays of different energies. When using higher energy X-rays, significantly higher gain reductions were measured, e.g. [41]. In addition, non-linear effects increase with increasing avalanche gain, i.e. with increasing voltage applied to the APD.

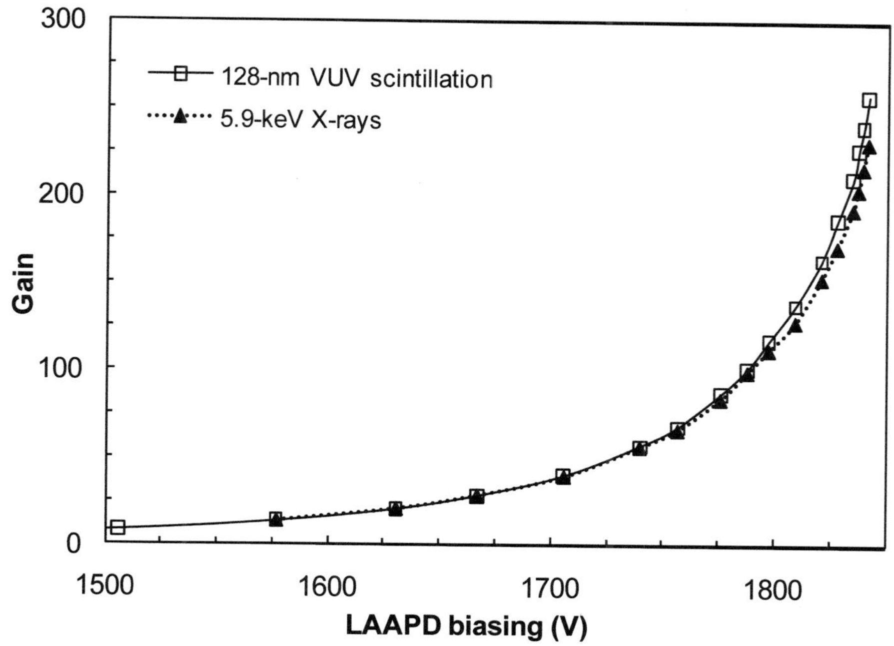

Fig. 4. APD gain for both direct 5.9-keV X-ray absorption in the APD and 128-nm scintillation absorption in the APD as a function of the APD biasing voltage.

5. APD characteristics for xenon scintillation detection (~172 nm)

For the present measurements the number of xenon VUV photons that irradiate the APD is about 2.4 x 10^4 photons per light pulse.

As mentioned above, significant non-linearity in APD gain response between X-rays and visible light was observed in different types of APDs, being the APDs from API those which present the lowest effects [28], reaching a reduction of 3% in X-ray gain response when compared to visible light. In Fig.5 we present the X-ray-to-xenon-scintillation amplitude ratio as a function of APD biasing. Non-linear effects are less than 3.5% and 7% for gains below 100 and 200, respectively, when considering 5.9-keV X-rays. These non-linearities are higher than those observed for visible-light detection [28,30] but are, nevertheless, smaller than those observed with other types of APDs.

Figure 6 presents the Minimum number of Detectable Photons (MDP) for xenon electroluminescence, defined as the number of photons that would deposit, in the APD, an amount of energy equivalent to the onset of the electronic noise tail. The MDP is approximately constant being, for the present conditions, about 600 photons for 172-nm VUV-light pulses and for gain values above 40, increasing significantly as the gain drops below that value and the signal approaches the noise level.

The obtained MDP can decrease if further efforts are made towards the reduction of the noise level achieved in the present setup. Nevertheless, the MDP can be reduced up to a factor of two by cooling the temperature of the photodiode to values below 0 °C, see section 7.

The results obtained with this APD for MDP at 172 nm are lower than those obtained with the peltier-cooled APD in [21] (~10^3 photons). The difference may be attributed to the differences in the APD dark currents, which limit the electronic noise and, thus, the MDP. It can also be attributed to the noise level present in both setups. Since the peltier-cooled APD has a different enclosure, with more wiring, it is more prone to electronic noise.

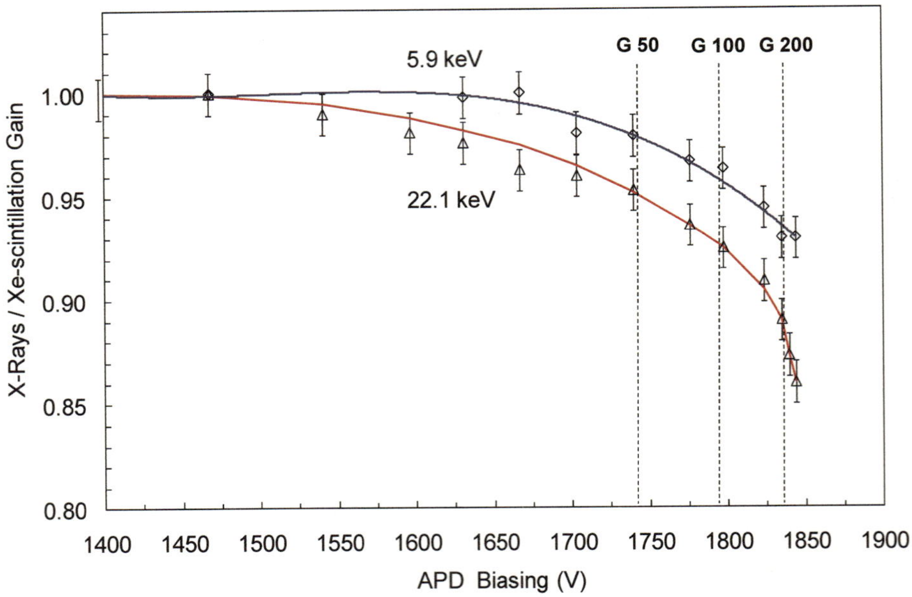

Fig. 5. X-ray to 172-nm scintillation pulse amplitude ratio as a function of APD biasing voltage, for 5.9- and 22.1-keV X-rays.

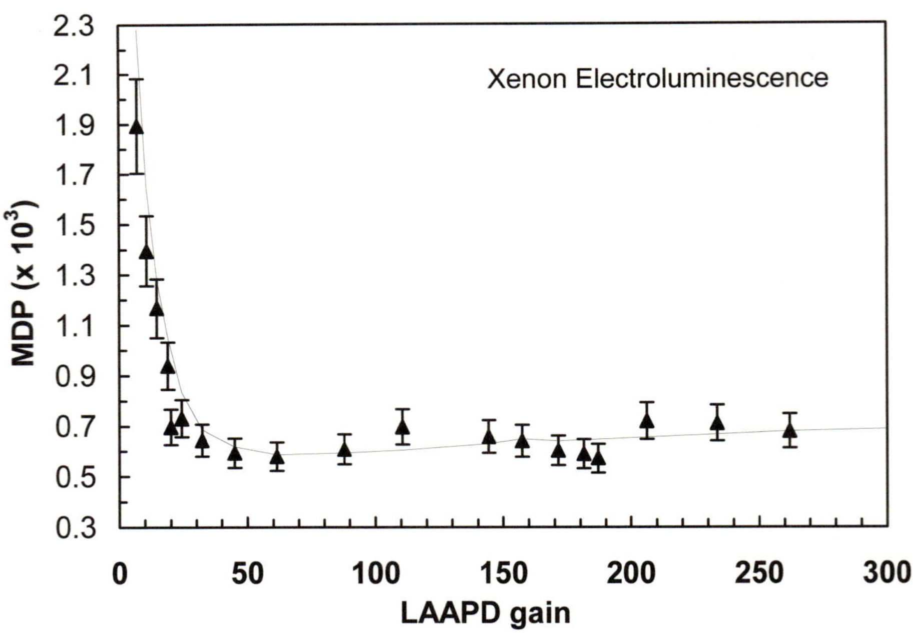

Fig. 6. Minimum number of detectable 172-nm VUV-photons as a function of APD gain. The line serves only to guide the eyes.

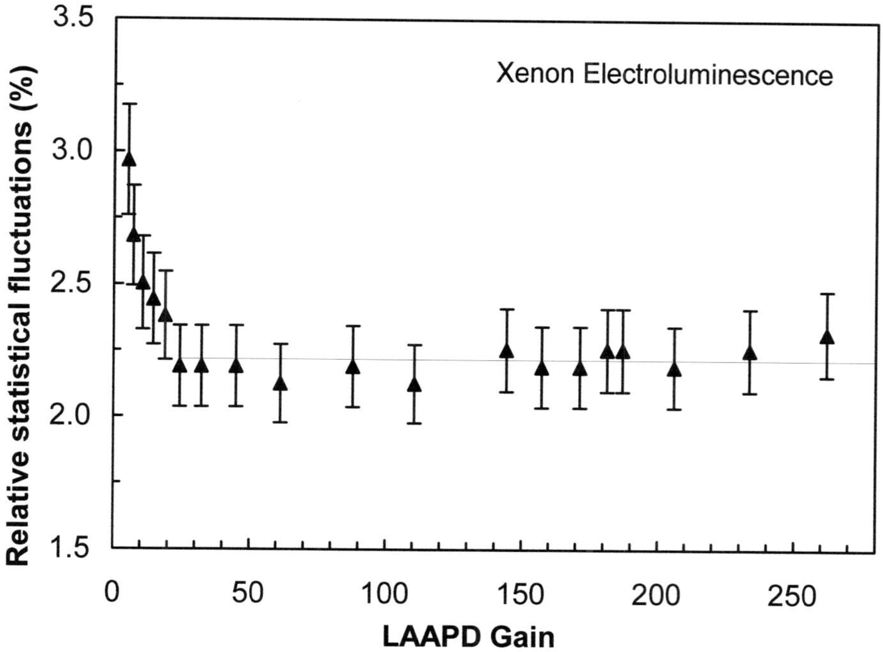

Fig. 7. Relative statistical fluctuations associated to the VUV detection of 2.4×10^4 photons of ~172 nm VUV-light pulses as a function of APD gain. The line serves only to guide the eyes.

The statistical fluctuations associated to the detection of VUV-light in the APD may be estimated from the energy resolution of the pulse-height distributions of 5.9-keV X-ray full absorption in the gas. The energy resolution of a conventional GPSC is determined by the statistical fluctuations occurring in the primary ionisation processes in the gas, in the production of the VUV scintillation photons and in the photosensor. Since the statistical fluctuations associated to the scintillation processes are negligible when compared to those associated to the primary electron cloud formation in the gas and to those associated to the scintillation detection in the photosensor, the energy resolution, R, of the GPSC, for an X-ray energy E_x, is given by [36]

$$R = 2.355\sqrt{\frac{F}{N_e} + \left(\frac{\Delta E}{E}\right)^2} = 2.355\sqrt{\frac{Fw}{E_x} + \left(\frac{\Delta E}{E}\right)^2} \tag{7}$$

where N_e is the average number of primary electrons produced in the gas by the X-rays, F is the Fano factor, w is the average energy to create a primary electron in the gas and E is the energy deposited by the VUV-radiation in the photosensor.
The statistical fluctuations associated to the VUV-photon detection can be, thus, obtained by

$$\frac{\Delta N_{UV}}{N_{UV}} = \frac{\Delta E}{E} = \sqrt{\left(\frac{R}{2.355}\right)^2 - \frac{Fw}{E_x}} \tag{8}$$

In the present case, E_x = 5.9-keV, w = 22.4 eV and F = 0.17 for xenon. The relative statistical fluctuations associated to the detection of 2.4×10^4 VUV photons for ~172 nm VUV-light pulses, as a function of gain, are depicted in Fig.7. The APD relative uncertainty decreases rapidly with the onset of gain, stabilizing for gains above approximately 30 and reaching values of 2.2%. This value can be reduced by cooling the photodiode operating temperature to values around 0 °C [22].

Figures 6 and 7 show that, for the detection of the light-levels of 172-nm photons presented in this study, best performance characteristics are achieved for gains around 40. However, gains as low as 20 are sufficient to achieve a nearly optimum performance, i.e. without presenting significant degradation of MDP and energy resolution. For lower light levels, higher gains may be needed to pull the signal of the light-pulse out of the noise and achieve the best possible performance.

6. APD characteristics for Argon scintillation detection (~128 nm)

For the present measurements the number of xenon VUV photons that irradiate the APD is about 1.4×10^4 photons per light pulse. As can be seen from Fig.4, the argon scintillation pulse deposits in the APD an amount of energy similar to what would be deposited by the interaction of ~30-keV X-rays in the photodiode.

In Fig.8 we present the X-ray-to-argon-scintillation amplitude ratio as a function of APD biasing. The non-linearity is higher than that found for xenon scintillation, being about 4.5% and 10% for gains about 100 and 200, respectively, when considering 5.9-keV X-rays.

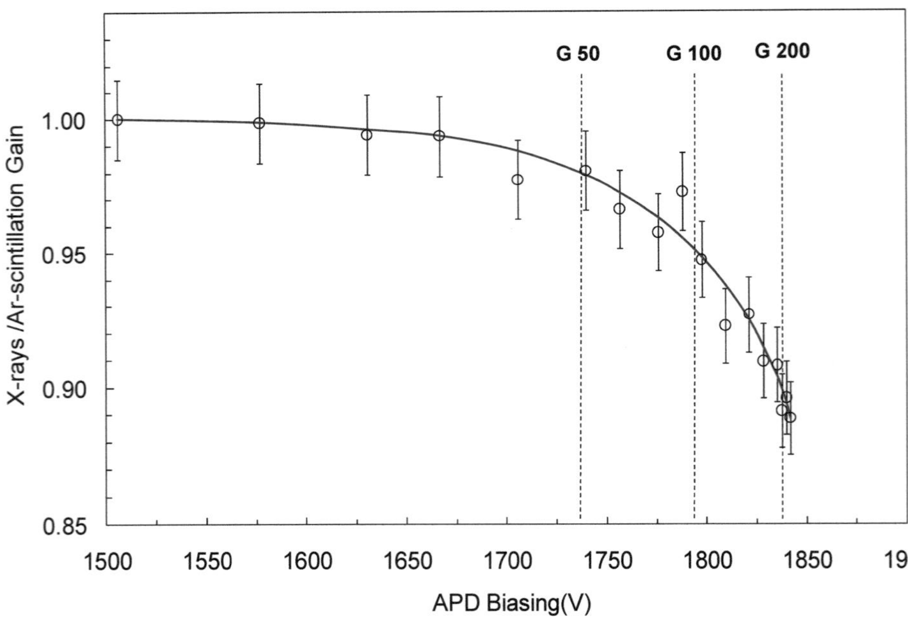

Fig. 8. X-ray to 128 nm pulse amplitude ratio as a function of APD biasing voltage, for 5.9-keV X-rays.

Figure 9 presents the minimum number of detectable photons (MDP) for argon electroluminescence, as defined for the xenon case. The MDP shows a similar trend as for xenon; it is approximately constant, being about 1300 photons for 128-nm VUV-light pulses for gains above 60, increasing significantly as the gain drops below this value.

As for the xenon case, the obtained MDP can decrease if further efforts are made to reduce the noise level of the present setup and can be reduced down to a factor of two by cooling the photodiode to temperatures below 0 °C, see section 7.
As for xenon, the statistical fluctuations associated to the detection of VUV light in the APD may be estimated from the measured energy resolution of the pulse-height distributions of 5.9-keV full absorption in the gas. The statistical fluctuations associated to the VUV-photon detection can, thus, be obtained from (8) where, for argon, w = 26.4 eV and F = 0.30. The relative statistical fluctuations associated to the VUV detection of 1.4×10^4 photons of ~128 nm photons VUV-light pulses, as a function of gain, are depicted in Fig.10. The APD relative uncertainty decreases rapidly with the onset of gain, stabilizing for gains above ~30 and reaching values of 3.9%.
Figures 9 and 10 show that, for the detection of the light-levels of 128-nm photons presented in this study, best performance characteristics are achieved for gains above 60. For gains lower than 60, the MDP increases significantly, while the statistical fluctuations remain constant down-to gains of 20.

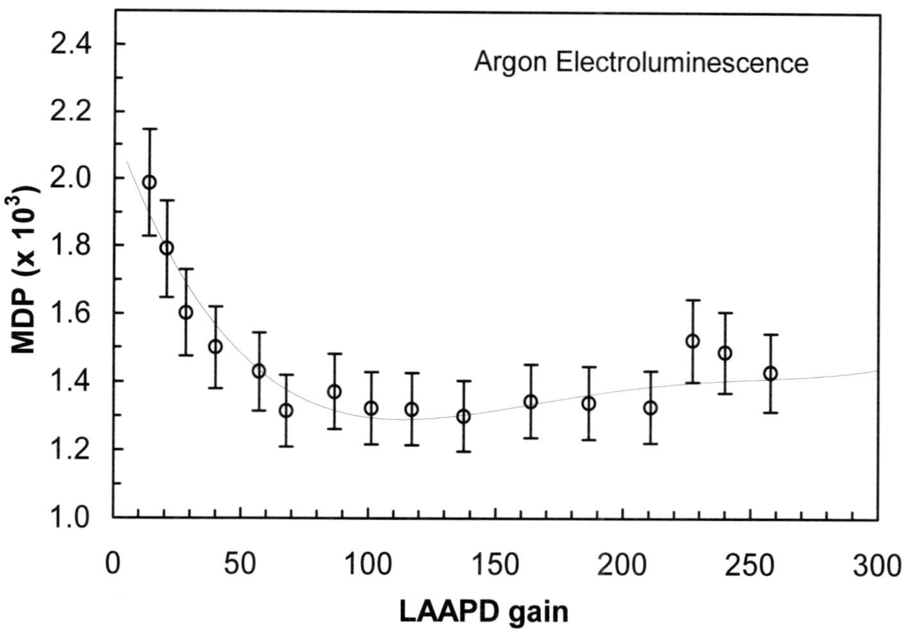

Fig. 9. Minimum number of detectable 128-nm VUV-photons as a function of APD gain. The line serves only to guide the eyes.

7. Temperature dependence

The APD response depends significantly on the operation temperature [22], in particular the gain and the dark current are two parameters that reflect significant dependence on temperature. Therefore, temperature control and stabilization during the measurements may be required, which is a drawback in many applications. In alternative, the knowledge of the gain variation with temperature may lead to corrections that take into account temperature variations during measurements.

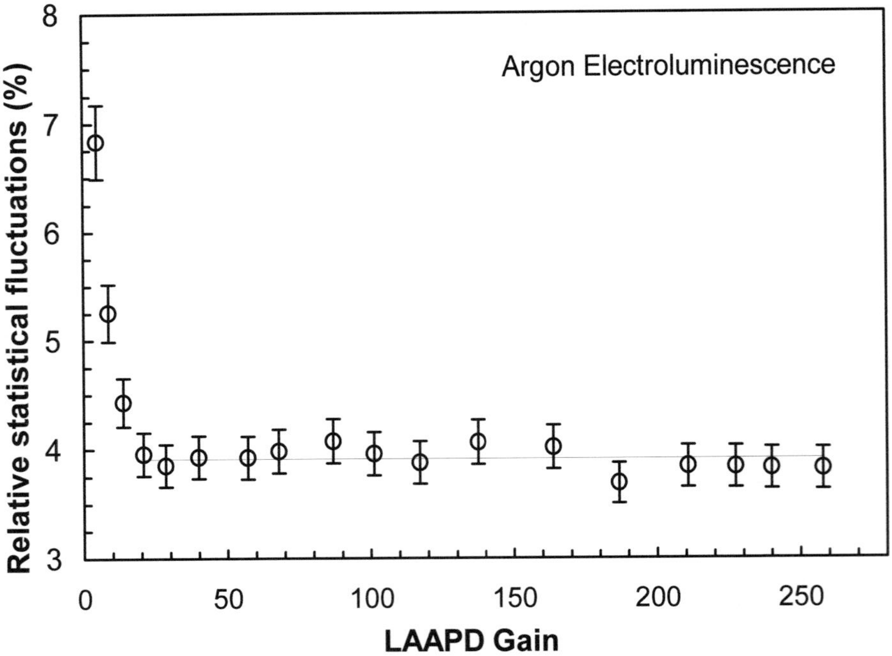

Fig. 10. Relative statistical fluctuations associated to VUV detection of 1.4×10^4 photons of ~128-nm VUV-light pulses as a function of APD gain. The line serves only to guide the eyes.

We have investigated the effect of temperature on gain, dark current, minimum number of detectable photons and statistical fluctuations using an APD with an integrated peltier cell capable of providing minimum operation temperatures of -5°C [21]. Figure 11 depicts the APD gain for VUV photons from xenon and for different operation temperatures. As expected, the gain increases with reducing temperature due to the increase of the silicon resistivity. The maximum gain increases from 300 to 700 as the temperature decreases from 25 to -5°C, respectively.

We can organize the data of Fig.11 and depict the gain as a function of temperature, for different biasing voltages, Fig.12. The gain decreases exponentially with increasing temperature. For each voltage, the relative gain variation is almost constant and increases from -2.7% to -5.6% per °C as the voltage increases from 1633 to 1826 V. The relative gain variation for high biasing voltages is almost a factor of 2 higher than the 3% reported by the manufacturer for visible light detection [19].

The increase in resistivity of the silicon wafer with decreasing temperature has impact on the APD dark current and, therefore, on the electronic noise. Figure 13 depicts the dark current as a function of gain for different operation temperatures. The dark current is reduced by about one order of magnitude as the temperature decreases from 25°C to -5°C. This reduction has a positive impact on the minimum number of detectable photons and on the statistical fluctuations in the APD, as shown in Figs. 14 and 15. The minimum detectable number of photons is reduced from 1300 to 500 as the temperature decreases from 25°C to -5°C.

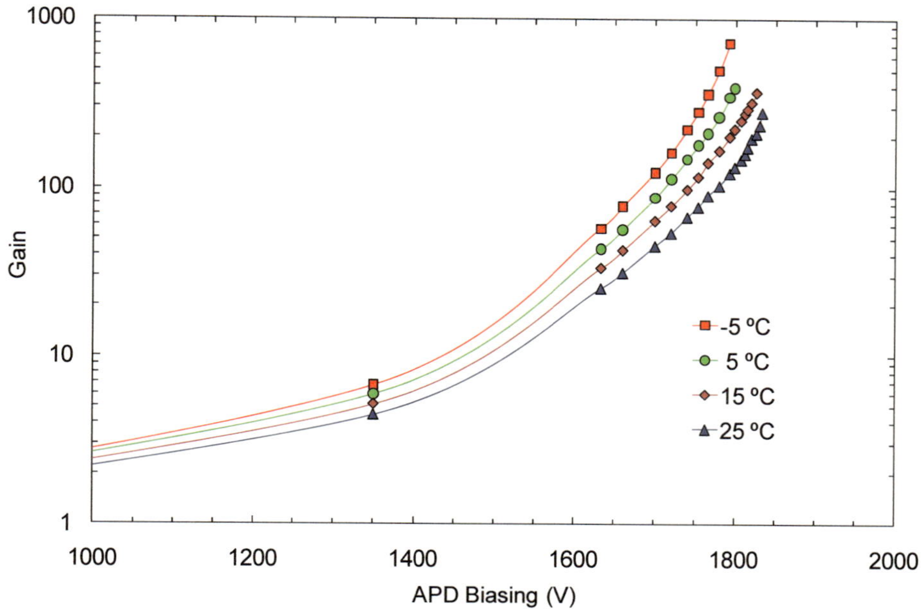

Fig. 11. APD gain for VUV scintillation as a function of APD biasing for different operation temperatures.

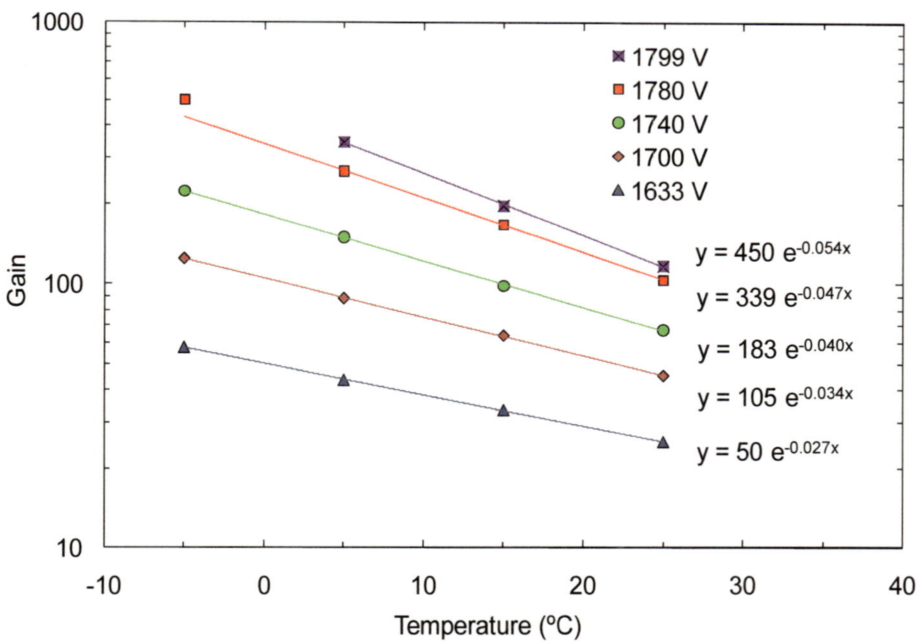

Fig. 12. APD gain for VUV scintillation as a function of APD temperature for different bias voltages.

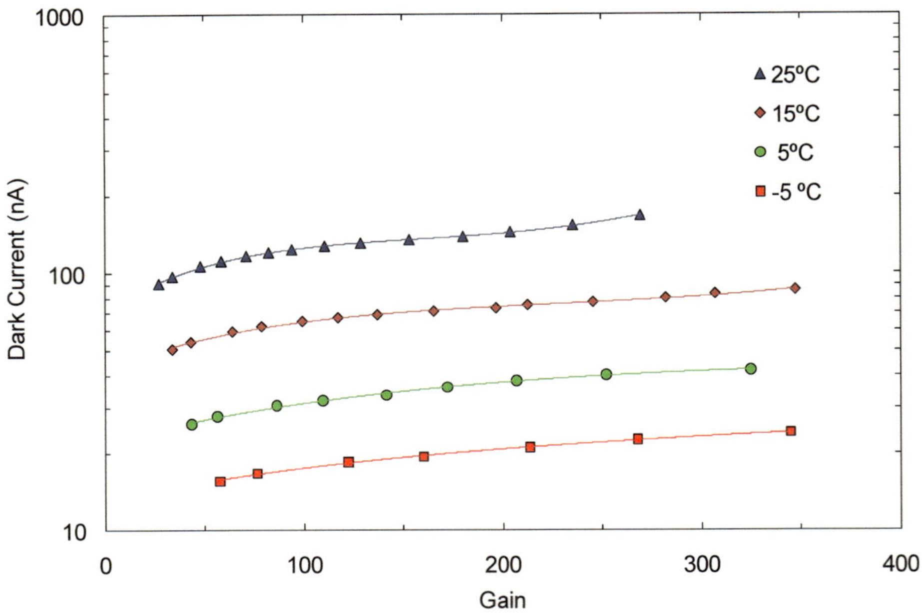

Fig. 13. APD dark current as a function of its gain for different operation temperatures.

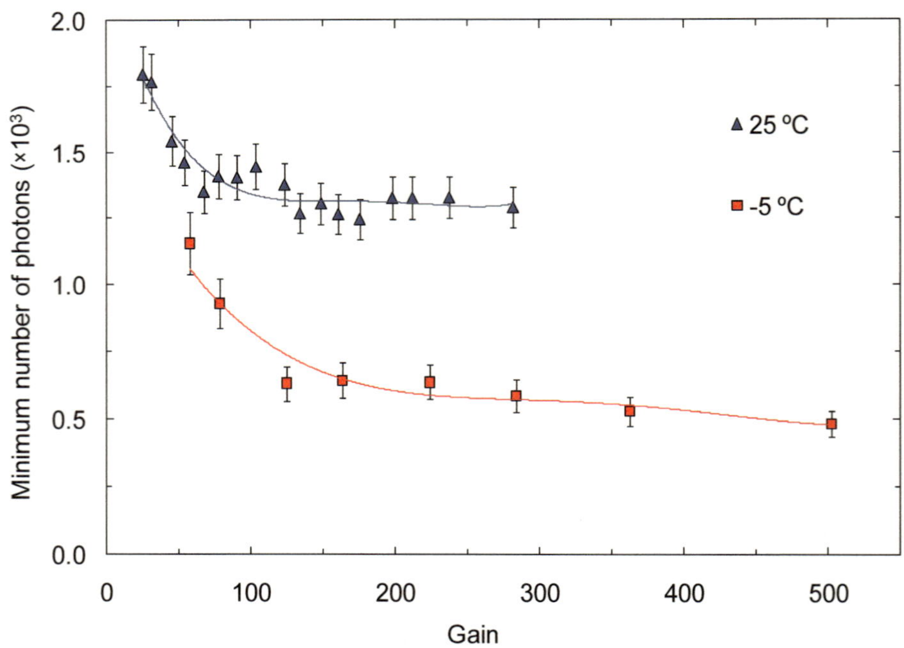

Fig. 14. Minimum number of detectable 172-nm VUV-photons as a function of APD gain for operation temperatures of 25ºC and -5ºC. The lines serve only to guide the eyes.

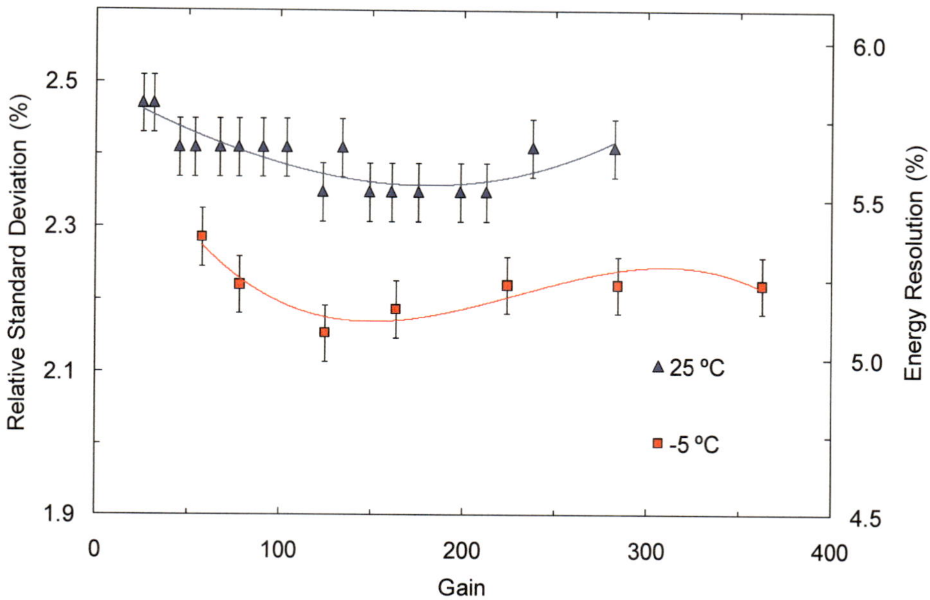

Fig. 15. Relative statistical fluctuations associated to the detection of 2.7×10^4 VUV photons of ~172 nm light pulses as a function of APD gain for operation temperatures of 25°C and -5°C. The lines serve only to guide the eyes.

8. Behaviour under intense magnetic fields

The insensitivity of the APD response to magnetic fields in X-ray and visible-light detection is well documented [23, 25, 42]. In opposition to visible light and X-ray detection, where the APD response is insensitive to magnetic fields, VUV detection with APDs is sensitive to magnetic fields. Figure 16 presents the pulse-height distribution obtained for VUV xenon-scintillation pulses for magnetic fields of 0 and 5 T, at room temperature. As can be seen, there is a significant reduction in the APD gain at high magnetic fields.

Figure 17 shows the APD relative pulse amplitude and energy resolution as a function of magnetic field for the xenon scintillation peak. As shown, for VUV-light the relative amplitude decreases gradually as the magnetic field is applied, reaching a 24% reduction at 5 T. The energy resolution increases from 13% to 15%. Since VUV photons interact within the first few atomic layers of silicon, where the electric field is weak, the magnetic field influences drift and diffusion of the produced photoelectrons, leading to partial charge loss to the dead layer at the APD entrance.

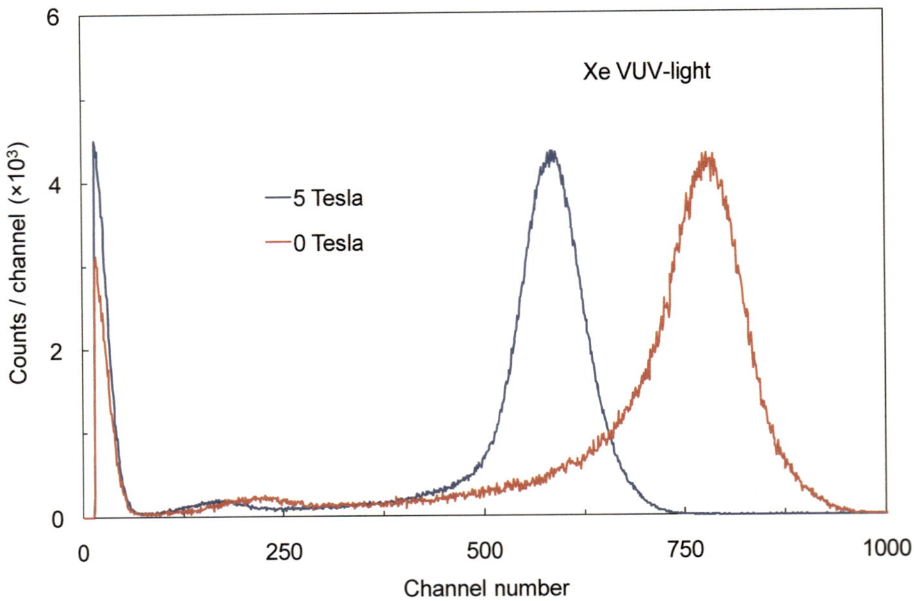

Fig. 16. APD pulse-height distributions for xenon scintillation light resulting from 5.9-keV X-rays absorbed in a xenon GPSC operating under magnetic field intensities of 0 T and 5 T.

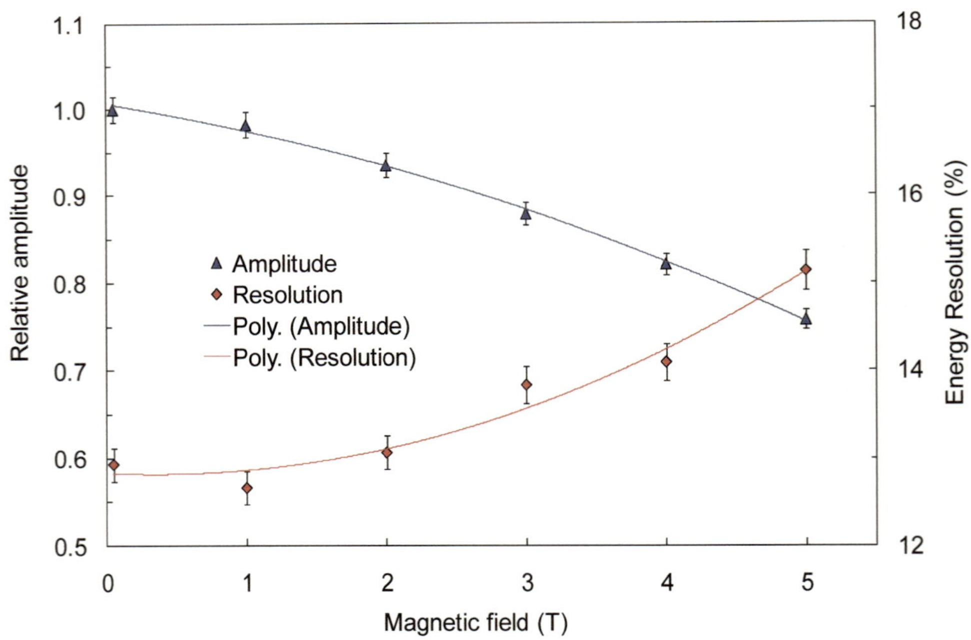

Fig. 17. Relative pulse amplitude and energy resolution as a function of magnetic field intensity for xenon VUV scintillation detection.

9. Conclusions

The APD is a suitable device for the detection of VUV light pulses of photons down to about 120 nm. PMTs present a sensitivity range down to 115 nm- with MgF_2 windows -, gains above 10^7, dark currents below a few nA instead of a few hundred nA for APDs, and are suitable for single photon detection. However, the photodiode compactness, reduced power consumption, simple operation and straightforward photon calibration are significant advantages over PMTs.

Figures 6 and 9 present the results for the Minimum number of Detectable Photons (MDP), defined as the number of VUV photons that would produce a signal in the APD with an amplitude equivalent to the onset of the electronic noise tail. The minimum number of photons that can be detected with the APD, for this experimental setup, are about 1300 and 600 for 128- and 172 nm, respectively, almost three orders of magnitude higher than is the case for PMTs. Therefore, the APD is not suitable for single photon detection and VUV-photon spectrometry. Nevertheless, it can be applied to synchrotron radiation in VUV-photon detection and to other areas of optics, where the light levels are adequate for its use.

The MDP for 172-nm photons is about half of that for 128 nm photons, achieving the lowest values for lower gains. This difference reflects the higher spectral sensitivity of these APDs for 172 nm, which is approximately 150 mA/W, corresponding to an average number of 1.1 free electrons produced in the APD per incident VUV-photon, when compared to 50 mA/W that corresponds to an average number of 0.55 free electrons produced in the photodiode per incident VUV-photon for 128 nm [19]. In fact, it is the number of primary charge carriers that defines the corresponding signal amplitude and the signal-to-noise ratio.

Figures 7 and 10 depict also the results of the relative statistical fluctuations associated to the VUV detection of 2.4×10^4 photons of 172-nm VUV-light pulses and $1.4\text{x}10^4$ photons of 128-nm, respectively. These values are 3.9% and 2.2%, respectively. This difference is consistent with the dependence of the APD resolution on the inverse of the square root of the number of the charge carries produced in the photodiode [30] and it reflects not only the difference in the number of photons involved in each case, but also the difference in the respective quantum efficiency. The numbers of charge carriers produced in the APD are

$2.4 \times 10^4 \times 1.1 = 2.64 \times 10^4$ free electrons for xenon

and

$1.4 \times 10^4 \times 0.55 = 7.7 \times 10^3$ free electrons for argon,

being

$$2.2 \times \sqrt{\frac{26400}{7700}} \cong 3.9\,.$$

This is also consistent with the results for visible (red) light from a LED obtained in [22], with an energy resolution of 7% for 2600 free electrons produced in the APD.

For the present charge carrier quantities, APD gains as low as 30 to 60 are enough to obtain best performances. However, gains as low as 20 and 30, respectively, are sufficient to achieve a nearly optimum performance, i.e. without presenting significant degradation of MDP and energy resolution. For lower light levels, higher gains will be needed to pull the signal of the light-pulse out of the noise and achieve the best possible performance.

The experimental results presented in this chapter show that both MDP and statistical fluctuations associated to light detection do not depend on photon wavelength, but rather on the number of charge carriers produced by the light-pulse in the APD. This is at odds with other effects like light-to-X-rays non-linearity and amplitude behaviour under intense magnetic fields, where the photon interaction in the first atomic layers of the wafer has a significant influence on these results.

In addition to space charge effects resulting from the point-like nature of X-ray interactions, our results suggest a dependence of the non-linearity on the light wavelength. The X-ray to light gain non-linearity for 128 nm VUV photons is higher than the one obtained for 172 nm VUV photons and both are higher than that reported for visible light [30]. The non-linearity must depend on the penetration depth in silicon of each type of light. For both VUV and visible light, photons are absorbed in the drift region of the APD, where the electric field is weak and the effect of capture of charges is more significant. Since the absorption is much more superficial for VUV light (~5 nm), capture is greater in this case but decreases with gain due to the electric field increase. Thus, the higher APD voltage results in a more efficient collection of the primary electrons produced near the entrance surface for argon, whereas this effect is smaller for xenon and most probably negligible for visible light since the penetration depth increases. The shallow penetration of the VUV photons may also explain the different behaviour of the APD response under intense magnetic fields for VUV and visible light detection.

10. Acknowledgements

"Project carried out under QREN, funding from UE/FEDER and FCT-Portugal, through program "COMPETE - Programa Operacional Factores de Competitividade", projects PTDC/FIS/102110/2008 and "Projecto Estrategico - Unidade 217/94".

11. References

[1] Windowless Series large area APD, Advanced Photonics Inc., 1240 Avenida Acaso Camarillo, CA 93012, USA.

[2] VUV-sensitive large area APD, Radiation Monitoring Devices, Inc., 44 Hunt Street, Watertown, MA 02472, USA.

[3] S-8664-55-SPL large area APD, Hamamatsu Photonics K.K., 325-6, Sunayama-cho, Naka-ku, Hamamatsu City, Shizuoka Pref., 430-8587, Japan.

[4] J.A.M. Lopes, J.M.F. dos Santos, R.E. Morgado, C.A.N. Conde, "A xenon gas proportional scintillation counter with a UV-sensitive large-area avalanche photodiode", IEEE Trans. Nucl. Sci. 48 (2001) 312-319.

[5] C.M.B. Monteiro, J.A.M. Lopes, P.C.P.S. Simões, J.M.F. dos Santos, C.A.N. Conde, "An argon gas proportional scintillation counter with UV avalanche photodiode scintillation readout", IEEE Trans. Nucl. Sci. 48 (2001) 1081-1086.

[6] R. Neilson, _, F.LePort, A.Pocar, K.S.Kumar, A.Odian et al., "Characterization of large area APDs for the EXO-200 detector", Nucl. Instrum. Meth. A 608 (2009) 68–75.

[7] V.N. Solovov, V. Chepel, M.I. Lopes, R. Ferreira Marques and A.J.P.L. Policarpo, "Study of large area avalanche photodiodes for detecting liquid xenon scintillation", IEEE Trans. Nucl. Sci. 47 (2000) 1307-1310.

[8] V.N. Solovov, A. Hitachi, V. Chepel, M.I. Lopes, R. Ferreira Marques, A.J.P.L. Policarpo, "Detection of scintillation light of liquid xenon with a LAAPD", Nucl. Instr. Meth. A 488 (2002) 572-578.

[9] K. Ni, E. Aprile, D. Day, K.L. Giboni, J.A.M. Lopes, P. Majewski and M. Yamashita "Performance of a large area avalanche photodiode in a liquid xenon ionization and scintillation chamber", Nucl. Instrum. Meth. A 551 (2005) 356.
[10] P. Shagin, R. Gomez, U. Oberlack, P. Cushman, B. Sherwood, M. McClish and R. Farrell, "Avalanche Photodiode for liquid xenon scintillation: Quantum Efficiency and gain", J. Inst. 4 (2009)P01005.
[11] J. C. Campbell, S. Demiguel, F. Ma, A. Beck, X. Guo, S. Wang, X. Zheng, X. Li, J. D. Beck, M. A. Kinch, A. Huntington, L. A. Coldren, J. Decobert, and N. Tscherptner, "Recent Advances in Avalanche Photodiodes," IEEE J. of Sel. Topics in Quantum Electronics 10 (2004) 777.
[12] Q. Wang, S. Savage, B. Noharet, I. Petermann, S. Persson, S. Almqvist, M. Bakowski, J.Y. Andersson, "Analysis and comparison of UV photodetectors based on wide bandgap semiconductors", Proceedings of SPIE-The International Society for Optical Engineering, Vol. 7602 (2010).
[13] R. McClintock, E. Cicek, Z. Vashaei, C. Bayram, M. Razeghi, M.P. Ulmer, "III-Nitride Based Avalanche Photo Detectors", Proceedings of SPIE-The International Society for Optical Engineering, Vol. 7780 (2010).
[14] E. Cicek, Z. Vashaei, C. Bayram, R. McClintock, M. Razeghi, M.P. Ulmer, "Comparison of ultraviolet APDs grown on free-standing GaN and sapphire substrates", Proceedings of SPIE-The International Society for Optical Engineering, Vol. 7780 (2010).
[15] S.M. Savage, A. Schoner, I. Petermann, M. Bakowski, "Sensitive and stable SiC APD for UV detection", Proceedings of SPIE-The International Society for Optical Engineering, Vol. 7726 (2010).
[16] G.A. Shaw, A.M. Siegel, J. Model, A. Geboff, S. Soloviev, A. Vert, P. Sandvik, "Deep UV Photon-Counting Detectors and Applications", Proceedings of SPIE-The International Society for Optical Engineering, Vol. 7320 (2009).
[17] S. Soloviev, A. Vert, A. Bolotnikov, P. Sandvik, "UV SiC avalanche photodetectors for photon counting", IEEE SENSORS, VOLS 1-3 (2009) 1819.
[18] T. Takahashi, S. Himi, M. Suzuki, J. Ruan, S. Kubota, "Emission spectra from Ar-Xe, Ar-Kr, Ar-N2, Ar-CH4, Ar-CO2 and Xe-N2 gas scintillation proportional counters", Nucl. Instr. Meth. A 205 (1983) 591-596.
[19] Advanced Photonix Inc. Application Notes: "Windowless large area APDs" (1999).
[20] L.M.P. Fernandes, J.A.M. Lopes, C.M.B. Monteiro, J.M.F. dos Santos, R.E. Morgado, "Non-linear behaviour of large-area avalanche photodiodes", Nucl. Instr. Meth. A 478 (2002) 395-399.
[21] J.A.M. Lopes, L.M.P. Fernandes, J.M.F. dos Santos, R.E. Morgado, C.A.N. Conde, "VUV detection in large-area avalanche photodiodes as a function of temperature", Nucl. Instr. Meth. A 504 (2003) 331-334.
[22] L.M.P. Fernandes, J.A.M. Lopes, J.M.F. dos Santos, P.E. Knowles, F. Mulhauser, L. Ludhova, F: Kottmann, R. Pohl, D. Taqqu, "LAAPD low temperature performance in x-ray and visible-light detection", IEEE Trans. Nucl. Sci. 51 (2004) 1575-1580.
[23] L.M.P. Fernandes, A. Antognini, M. Boucher, C.A.N. Conde, O. Huot, P.E. Knowles, F. Kottmann, L. Ludhova, F. Mulhauser, R. Pohl, L.A. Schaller, J.M.F. dos Santos, D. Taqqu, J.F.C.A. Veloso, "Behaviour of large-area avalanche photodiodes under intense magnetic fields for VUV, visible and X-ray photon detection", Nucl. Instr. Meth. A 498 (2003) 362-368.

[24] C.M.B. Monteiro, L.M.P. Fernandes, J.A.M. Lopes, J.F.C.A. Veloso, J.M.F. dos Santos, "Detection of VUV photons with large-area avalanche photodiodes", Appl. Phys. B 81 (2005) 531-535.

[25] A.Q.R. Baron and S.L. Ruby, "Time resolved detection of X-rays using large area avalanche photodiodes", Nucl. Instr. Meth. A 343 (1994) 517.

[26] E.M. Gullikson, E. Gramsch, M. Szawlowski, "Large-area avalanche photodiodes for the detection of soft X-rays", Appl. Optics 34 (1995) 4662.

[27] Pansart J P, Avalanche photodiodes for particle detection, Nucl. Instr. Meth. A 387 (1997) 186.

[28] M. Moszynski, M. Kapusta, M. Balcerzyk, M. Szawlowski, D. Wolski, I. Wegrzecka, M. Wegrzecki, "Comparative study of avalanche photodiodes with different structures in scintillation detection", IEEE Trans. Nucl. Sci. 48 (2001) 1205-1210.

[29] T.W. Barnard, M.I. Crockett, J.C. Ivaldi, P.L. Lundberg, D.A. Yates, P.A. Levine, D.J. Sauer, "Solid-state detector for ICP-OES", Anal. Chem. 65 (1993) 1231-1239.

[30] M. Moszynski, M. Szawlowski, M. Kapusta, M. Balcerzyk, "Large area avalanche photodiodes in scintillation and X-rays detection", Nucl. Instr. Meth. A 485 (2002) 504.

[31] R.J. McIntyre, "The distribution of gains in uniformly multiplying avalanche photodiodes: theory", IEEE Trans. Electron Devices 19 (1972) 703.

[32] L.M.P. Fernandes, J.A.M Lopes, J.M.F dos Santos, "Excess noise factor in large area avalanche photodiodes for different temperatures", Nucl. Instr. Meth. A 531 (2004) 566.

[33] P.W. Nicholson, *Nuclear Electronics*, Wiley-Interscience Publication (1974), John Wiley & Sons, London.

[34] R.J. McIntyre, "Multiplication noise in uniform avalanche diodes", IEEE Trans. Electron Devices 13 (1966) 164.

[35] P.P. Webb, R.J. McIntyre, "Large area reach-through avalanche diodes for X-ray spectroscopy", IEEE Trans. Nucl. Sci. 23 (1976) 138.

[36] J.M.F. dos Santos, J.A.M. Lopes, J.F.C.A. Veloso, P.C.P.S. Simões, T.H.V.T. Dias, F.P. Santos, P.J.B.M. Rachinhas, L.F.R. Ferreira, C.A.N. Conde, "Development of portable gas proportional scintillation counters for X-ray spectrometry", X-Ray Spectrom. 30 (2001) 373.

[37] C.M.B. Monteiro, L.M.P. Fernandes, J.A.M. Lopes, L.C.C. Coelho, J.F.C.A. Veloso, J.M.F. dos Santos, K. Giboni, E. Aprile, "Secondary scintillation yield in pure xenon", J. Inst. 2 (2007) P05001.

[38] C.M.B. Monteiro, J.F.C.A. Veloso, J.A.M. Lopes, J.M.F. dos Santos, "Secondary scintillation yield in pure argon", Phys. Lett. B 668 (2008) 167.

[39] C. M. B. Monteiro, A. S. Conceição, F. D. Amaro, J. M. Maia, A. C. S. S. M. Bento et al., "Secondary scintillation yield from gaseous micropattern electron multipliers in direct Dark Matter detection", Physics Letters B 677 (2009) 133.

[40] E.D.C. Freitas, C.M.B. Monteiro, M. Ball, J. J. Gómez-Cadenas, J.A.M. Lopes,, T. Lux, F. Sánchez, J.M.F. dos Santos, "Secondary Scintillation Yield in High-Pressure Xenon Gas for neutrinoless double beta decay search", Physics Letters B 684 (2010) 205.

[41] L.M.P. Fernandes, F.D. Amaro, A. Antognini, J.M.R. Cardoso, C.A.N. Conde, O. Huot, P.E. Knowles, F. Kottmann, J.A.M. Lopes, L. Ludhova, C.M.B. Monteiro, F. Mulhauser, R. Pohl, J.M.F. dos Santos, L.A. Schaller, D. Taqqu, J.F.C.A. Veloso, "Characterization of large area avalanche photodiodes in X-ray and VUV-light detection", J. Inst. 2 (2007) P08005.

[42] A. Karar, Y. Musienko, J. Ch. Vanel, "Characterization of avalanche photodiodes for calorimetry applications", Nucl. Instr. Meth. A 428 (1999) 413-431.

Part 5

Photodiodes for High-Energy Photon/Particle Detection

12

Quantitative Measurements of X-Ray Intensity

Michael J. Haugh[1] and Marilyn Schneider[2]
[1]*National Security Technologies, LLC,*
[2]*Lawrence Livermore National Laboratory*
USA

1. Introduction

This chapter describes the characterization of several X-ray sources and their use in calibrating different types of X-ray cameras at National Security Technologies, LLC (NSTec). The cameras are employed in experimental plasma studies at Lawrence Livermore National Laboratory (LLNL), including the National Ignition Facility (NIF). The sources provide X-rays in the energy range from several hundred eV to 110 keV. The key to this effort is measuring the X-ray beam intensity accurately and traceable to international standards. This is accomplished using photodiodes of several types that are calibrated using radioactive sources and a synchrotron source using methods and materials that are traceable to the U.S. National Institute of Standards and Technology (NIST). The accreditation procedures are described.
The chapter begins with an introduction to the fundamental concepts of X-ray physics. The types of X-ray sources that are used for device calibration are described. The next section describes the photodiode types that are used for measuring X-ray intensity: power measuring photodiodes, energy dispersive photodiodes, and cameras comprising photodiodes as pixel elements. Following their description, the methods used to calibrate the primary detectors, the power measuring photodiodes and the energy dispersive photodiodes, as well as the method used to get traceability to international standards are described. The X-ray source beams can then be measured using the primary detectors. The final section then describes the use of the calibrated X-ray beams to calibrate X-ray cameras.
Many of the references are web sites that provide databases, explanations of the data and how it was generated, and data calculations for specific cases. Several general reference books related to the major topics are included. Papers expanding some subjects are cited.

2. Brief introduction to X-rays: Characteristic spectral lines and bremsstrahlung

2.1 Characteristic X-ray spectral lines from atoms

The electronic structure of an atom, using Ag as an example and shown in Fig. 1, is:

$$^{47}\mathrm{Ag}:\ 1s^2\ 2s^2\ 2p^6\ 3s^2\ 3p^6\ 3d^{10}\ 4s^2\ 4p^6\ 4d^{10}\ 5s^1$$

Shown in Fig. 1 is the energy level diagram for the lowest four quantum numbers of the Ag ion, i.e., one of the 1s electrons, has been removed. How the electron is removed is covered in the next section.

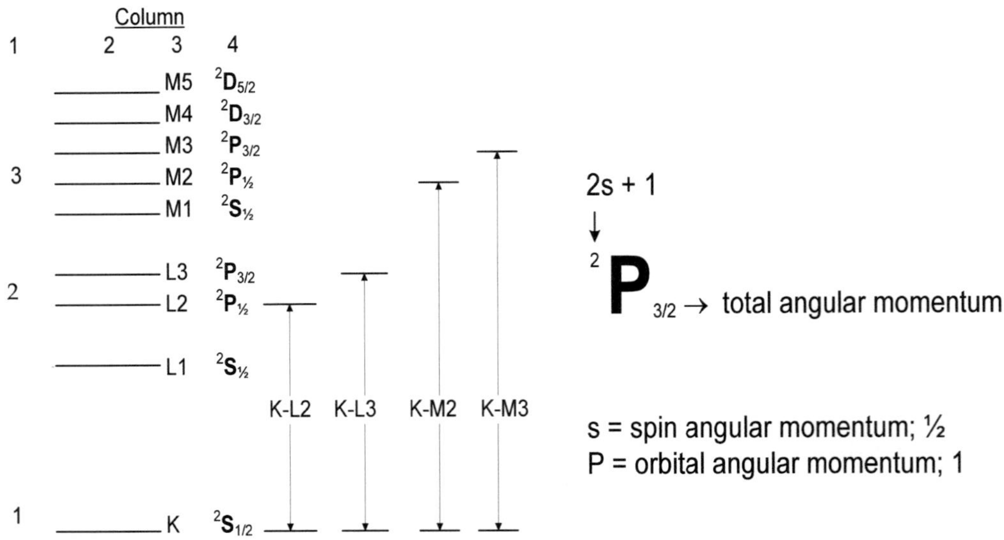

Fig. 1. The energy level diagram for the lowest levels of the singly ionized Ag atom.

Column 1 is the principle quantum number, column 2 indicates an energy level, column 3 is the IUPAC designation for the state, and column 4 is the state electronic structure.
We use the Russell-Saunders angular momentum coupling scheme (Herzberg, 1945) to describe the electronic structure of each state. The notation is illustrated in Fig. 1. The superscript "s" denotes 2s + 1 where "s" is the total spin of the state. In this case, it is a doublet state since there is an unpaired electron (one electron has been removed). The upper case letter indicates the orbital angular momentum (S means zero angular momentum; P means one unit angular momentum; D means two units, etc.) The subscript indicates the combination of orbital angular momentum and spin angular momentum. This description is somewhat simplified but it gives insight into transition probabilities and what X-ray transitions are expected along with their relative intensities.
When an electron has been removed from the lowest energy level, the $1^2S_{1/2}$ state, an electron can drop from a higher level with the simultaneous emission of a photon. The relative energies of these states can be obtained from the binding energies of the electrons in each state, and these are given in Table 1. A good source for this information is the Center for X-Ray Optics web site of the Lawrence Berkeley National Laboratory (CXRO reference).
The readily observed X-ray spectral lines to the electron deficient K level are shown in Fig. 1 column 4 and the energies and relative probabilities are given in Table 2.
Both the Siegbahn and the newer International Union of Pure and Applied Chemistry (IUPAC) notations for the transitions are given. The IUPAC notation is a bit less obscure but the Siegbahn notation is still more popular in current literature. In the IUPAC notation, the number refers to the order of the energies in the shell as shown in Fig. 1, so that L1 refers to the n=2 $s_{1/2}$, L2 refers to the $p_{1/2}$, and so on. Note that the spectral emission energy can be estimated by taking the difference between the corresponding binding energy given in Table 1. This estimate is reasonably accurate for K transitions, but care should be taken when using this estimate for higher energy level transitions. If the electron is removed from the n=2 shell, the spectral emission is referred to as an L line. The set of easily observed emission lines is given in Table 3.

Energy level	IUPAC	Energy, eV
K1s	K 1	25514
L2s	L1	3806
$L2p_{1/2}$	L2	3524
$L2p_{3/2}$	L3	3351
M3s	M1	719.0
$M3p_{1/2}$	M2	603.8
$M3p_{3/2}$	M3	573.0
$M3d_{3/2}$	M4	374.0
$M3d_{5/2}$	M5	368.3
N4s	N1	97.0
$N4P_{1/2}$	N2	63.7
$N4p_{3/2}$	N3	58.3

Table 1. Electron Binding Energies for the Ag Atom

Siegbahn Designation	IUPAC Designation	Spectral Line Energy, eV	Relative Intensity
Kα2	K-L2	21990.3	53
Kα1	K-L3	22162.92	100
Kβ3	K-M2	24911.5	9
Kβ1	K-M3	24942.4	16
Kβ2	K-N2,3	25456.4	4

Table 2. The K Type X-ray Spectral Lines for the Ag Ion

Siegbahn Designation	IUPAC Designation	Spectral Line Energy, eV	Relative Intensity
Lα2	L3-M4	2978.21	11
Lα1	L3-M5	2984.31	100
Lβ1	L2-M4	3150.94	56
Lβ2,15	L3-N5,6	3347.81	13
Lγ1	L2-N5	3519.6	6

Table 3. The L Type X-ray Spectral Lines for the Ag Ion

2.2 Spectral line widths, lifetimes, and competing processes

The X-ray spectral lines have a narrow width relative to the photon energy. The line widths for several fluorescence transitions in the Ag singly ionized atom are given in Table 4. One can estimate the fluorescence lifetime for the line width using the uncertainty relation given in Equation (1):

$$\Delta E * \Delta t \geq h \tag{1}$$

ΔE = fluorescence line width, eV

Δt = lifetime of the fluorescence state, sec

h = 4.135×10^{-15} eV•s, Planck's constant

The calculated lifetimes for the Ag transitions are given in Table 4. The fluorescence process can also be treated as a rate of decay from the higher energy state to the lower energy state with the rate constant given as the reciprocal of the lifetime. The rate constants for the two excited states Ag^{+} ion decay is given in Table 4.

Ag Transition	Line Width, eV	Lifetime, sec	Decay Rate Constant, sec^{-1}
K-L2	8.9	$4.65x10^{-16}$	$2.15x10^{15}$
K-L3	8.6	$4.81x10^{-16}$	$2.08x10^{15}$
L3-M4	2.2	$1.88x10^{-15}$	$5.32x10^{14}$
L3-M5	2.34	$1.76x10^{-15}$	$5.66x10^{14}$
L2-M4	2.4	$1.72x10^{-15}$	$5.80x10^{14}$

Table 4. Line widths and fluorescence lifetimes for several Ag transitions

There are other processes that compete with the fluorescence process. The process that most affects the X-ray fluorescence is called the Auger effect after Pierre Auger, although it was first discovered and published a year earlier by Lise Meitner. The Auger effect describes the transfer of energy that can occur when a vacant state is filled by an electron from the next higher state, but the energy for this transition is transferred to an electron in a higher state which is ejected from the ion, carrying the excess energy as kinetic energy. For example, consider an ion with a hole in the K shell that is filled by an electron from the L1 state. For the Auger process, the energy from the K-L1 transition is transferred to the L2 electron which is ejected from the ion with kinetic energy equal to the difference between the energy for the K-L1 transition and the binding energy of the L2 electron. The rate constant for this Auger process is about $1x10^{15}$ sec^{-1}. Comparing this rate to the fluorescence rate for the Ag K transitions, we note that the Auger rate is smaller so that the fluorescence yield for that condition will be about 85% of the of the total rate for filling the hole in the K shell. For K transition energies near 9keV (atomic number near 30), the fluorescence decay rate and the Auger rate are about equal and the yields will be about 50%. For lower atomic numbers, the fluorescence yield will be lower than 50% and conversely, the yield will be larger than 50 % for atomic numbers larger than 30. Recall that the line width of the transition is determined by the total rate of the excited state decay. For the Ag^{+} K transition the fluorescence rate is dominant. For L transitions, the Auger rate dominates up to the atomic number 100. It is only at this value of Z that the fluorescence yield is 50%. So the line widths for L transitions are determined by the Auger process and never drop below 2 or 3 eV. Refer to graphs of the relative yield as a function of atomic number (Podgorsak, 2010). There are several other internal conversion processes that compete with fluorescence but their rates are much lower and they will not be discussed here. The Auger effect is used for chemical analysis by measuring the kinetic energy of the Auger electron, a technique called Auger electron spectroscopy. The other competing processes also have a niche in analytical chemistry.

2.3 Electron impact to produce X-rays

The X-rays used in our measurements are primarily produced by the impact of electrons on solid materials. When an electron moving at a high velocity enters a solid material it deposits its energy in the solid in a variety of ways. Most of the energy ends up heating the anode but our interest is in that small percent of interactions that produce X-rays. Our strongest interest is in the collisions that remove an inner electron from the target material and produce the characteristic X-ray spectral lines from the atom. In fact, the X-radiation produced by the interactions of the electron with the solid material is a small fraction of the electron's energy loss processes. As can be seen in the NIST ESTAR tables (ESTAR), the radiation yield from Ti for electron impact in the energy range of 110 keV is less than 0.5%, and the majority of that radiation is bremmstrahlung. Bremmstrahlung is the spectrum of X-rays produced by the deceleration of electrons. Fig. 2 shows a typical emission produced by electron impact for a Ti anode target. The transition energies are given in Table 5.

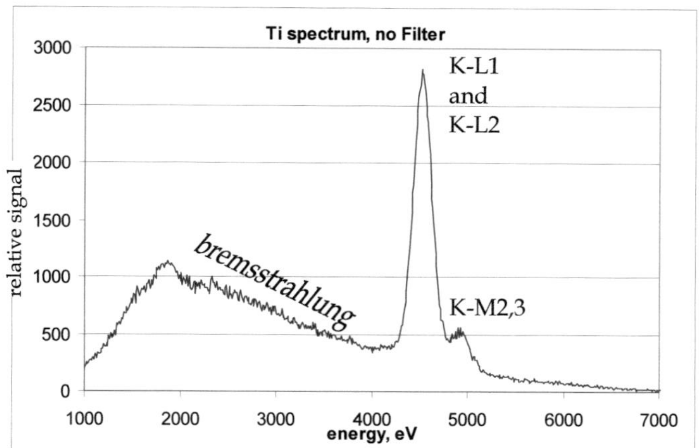

Fig. 2. This is a spectrum of the X-rays produced when an accelerated beam of electrons strikes a Ti anode.

Transition	K-L2	K-L3	K-M2,3	K-M4,5
Energy, eV	4504.92	4510.899	4931.83	4962.27

Transition energies are taken from the NIST X-ray database.

Table 5. Ti K X-ray Emission Lines

The anode was at 8,000 V and the heated filament electron source was near ground potential. The energy dispersive detector that was used to take this spectrum has a resolution near 240 eV. The tall, narrow band near 4500 eV comprises the K-L3 and the K-L2 Ti spectral lines. The spectral lines of Ti are approximately 2 eV wide. The bremsstrahlung is the broad band ranging from less than 1000 eV to 7000 eV and peaking near 2000 eV. The count of photons in the bremsstrahlung is 1000 times larger than the counts in the spectral lines.

The characteristic radiation depends on the anode material properties and the energy of the impacting electron. We have observed that the intensity of emission for characteristic lines follows this equation:

$$I_c = k\,(V_e - V_b)^n \qquad (2)$$

I_c = intensity of the characteristic X-ray line
k = proportionality constant
V_e= accelerating voltage of the electron
V_b= binding energy of the bound electron
n= a number somewhat greater than 2

Using this introduction to the basics of X-ray emission, the sources used to produce X-ray emission are presented in some detail in the following section.

3. X-ray sources

3.1 The diode source

NSTec laboratories have four X-ray sources that cover the X-ray spectral energy range from 50 eV to 110 keV. All the primary X-ray sources are the diode type; electrons are emitted from a heated tungsten filament, and then accelerated by an electric field to strike an anode. Two sources use a secondary beam that is generated when the primary beam strikes a sheet of material that fluoresces.

The diode sources produce spectral lines that are characteristic of the anode material and a broad spectrum of radiation known as bremsstrahlung, peaking near one-third of the accelerating voltage. A typical diode source is shown in Fig. 3. The filament is heated by an independent electrical circuit that is near ground potential. The anode is maintained at a high positive voltage so that the electrons emitted from the filament are accelerated and strike the anode at the energy determined by the voltage difference. The electric field is shaped using guide wires. X-rays are emitted in all directions and some exit the aperture, as shown in Fig. 3, and enter into the sample chamber. The anode is water-cooled so that a high beam current can be tolerated, thus giving a strong X-ray intensity. This intensity allows collimation of the X-ray beam with a pair of slits, as well as isolation of individual spectral lines using a diffraction crystal. The narrow band X-ray source can measure sample properties such as filter transmission, crystal reflectivity, and sensor efficiency. The source and sample chamber are in vacuum. The voltage supply is 20 kV, making the highest available spectral line nearly 17 keV (Zr K spectral lines).

The other diode source uses anodes that are cooled only by thermal conduction through the mechanical connections. This limits its operation to 10 W and 10 kV, with a usable spectral range from 700 eV to 8400 eV. It is often used to measure the absolute efficiency of X-ray cameras and the sensitivity variation across the sensor pixels.

The third source covers the spectral energy range from 8 to 111 keV. It uses X-rays from a diode source to produce fluorescent X-rays from a fluorescer material. This source is progressing toward NVLAP accreditation and will be described in detail in the next section.

The fourth source is currently being built and will cover the X-ray spectral region from 50 eV to several keV, also operating on the fluorescer principle.

The NSTec X-ray sources are used to calibrate and characterize components or complete systems that are used in the study of plasmas and similar efforts. A large component of our present calibration efforts is for diagnostics that are used on the NIF target diagnostics.

3.2 Reducing the band width of the source: filters, grazing incidence mirrors, and diffraction crystals

The emission from a diode source produced by the impact of the electrons on the anode has a broad band of bremsstrahlung and the characteristic spectral lines from the anode composition as was shown in Fig. 2. The large amount of bremsstrahlung X-rays does not allow one to use the raw emission from the diode source to accomplish calibrations such as measuring the energy dependence of a detector's sensitivity. There are several methods for reducing the spectral band width of the raw diode emission: (1) using thin sheets of solid materials that can act as high pass filters; (2) using a high pass filter combined with a grazing incidence mirror to make a band pass filter; (3) using fluorescers that produce only the spectral lines of the fluorescer sheet; and (4) using diffraction crystals to reflect only the X-rays that meet the Bragg angle requirements.

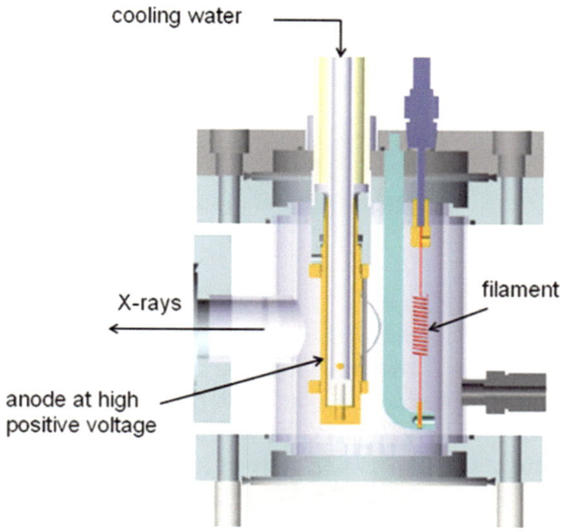

Fig. 3. Example of an X-ray diode

3.2.1 Filters

Thin sheets of solid material absorb X-rays and the transmission of the sheets depends upon the X-ray energy, the material thickness, and the atomic number Z of the material. Gases can also absorb X-rays but are not practical as filters for the applications described in this chapter. The transmission of a Ti sheet that is 25 μm thick is shown as Fig. 4(a).

The X-rays are absorbed in the Ti until the X-ray energy gets above 3000 eV. At the binding energy of the Ti 1s electron, 4966 eV, referred to as the K edge, the X-rays are again strongly absorbed. The sheet begins transmitting X-rays again when the X-ray energy rises above 6000 eV. This is the typical behavior of the X-ray transmission for solid materials. The transmission of materials for X-rays up to 30 keV is readily obtained using the CXRO web site. For higher energies, one can obtain absorption cross sections in the NIST tables. The transmission characteristics shown in Fig. 4(a) can be used to make a band pass filter for transmission of the Ti K lines when the electron accelerating voltage is at 8000 eV or lower and the Ti filter is sufficiently thick. This application will be discussed in more detail in the description of camera calibrations. High pass filters can be made from low Z materials and plastics are the most convenient. The transmission of 400 μm thick polyimide is shown in Fig. 4(b). The DuPont version of this material is called Kapton and the material is reasonably resistant to X-ray damage. The X-ray energy at 50% transmission is near 6 keV and the range of X-ray energy for the transmission range from 10% to 90% is 6 keV. This is a very broad cut off for the high pass filter.

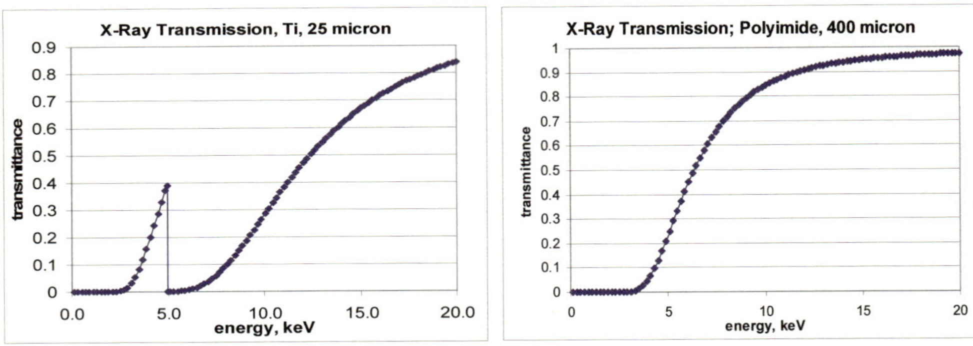

(a) (b)

Fig. 4. Graphs showing the X-ray transmission of (a) a 25 μm thick sheet of Ti, Z=22, and (b) a 400 μm thick sheet of polyimide.

3.2.2 Grazing incidence mirror

In materials, the index of refraction for X-rays is complex, with a magnitude slightly less than 1. The consequence of this is that an X-ray beam incident from vacuum onto a material is mainly absorbed, unless it is incident at a shallow (grazing) angle to the surface. Since the vacuum is the more optically dense region, the X-ray experiences "total internal reflection" and is specularly reflected. This forms the basis of grazing incidence X-ray mirrors. These mirrors reflect X-rays at the specular angle for angles less than a few degrees. As the mirror is rotated with respect to the direction of the X-ray beam, at some angle the reflected intensity will start to decrease and will eventually go to zero reflected intensity. The angle at which the X-ray intensity drops to 50% of the reflection at very low energies is referred to as the maximum reflection angle. The maximum reflection angle is a function of the X-ray energy, the mirror composition, and the mirror roughness. Calculated reflectivity curves for

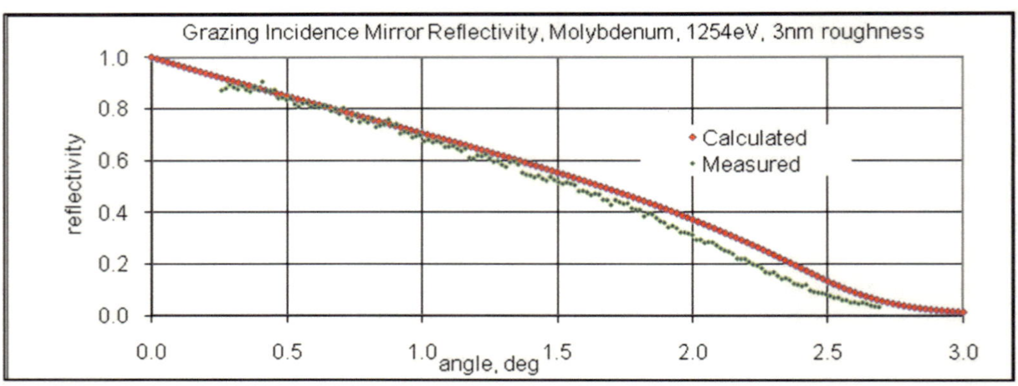

Fig. 5. This graph is a comparison of the measured reflectivity curve for the molybdenum grazing incidence mirror at an X-ray energy of 1254 eV with the calculated reflectivity with a surface roughness of 3 nm rms.

various materials and surface roughness can be obtained from the CXRO web site. A typical measured grazing incidence reflectivity curve is shown in Fig. 5 (green scatter). The corresponding calculated reflectivity curve is shown in red in Fig. 5. Given their angular dependence, grazing incidence mirrors are often used as low pass filters. The combination of a grazing incidence mirror with an appropriate thin sheet filter described previously forms a band pass filter.

The reflectivity curve for a grazing incidence mirror is affected by materials adsorbed on the surface. Water vapor and oxygen can significantly affect the reflectivity curve. For this reason, the grazing incidence mirror reflectivity curve is usually calibrated before it is used in experimental applications. This can be done using the NSTec sources. The synchrotron at Brookhaven is also used for these calibrations.

3.2.3 Diffraction crystal

Crystals are often used to isolate individual spectral lines from a diode source. They are used in plasma diagnostics as components of a spectrograph. The crystal reflectivity follows the Bragg law for the location of the maximum reflection as a function of X-ray energy:

$$n(12398.425/E) = 2d\sin\Theta \qquad (3)$$

n= an integer equal to the diffraction order
E= X-ray energy, eV
d= distance between the crystal planes, Å
Θ= angle between the X-ray beam and the crystal plane

For n=1 and a given Θ, only the X-rays having the energy E given by the Bragg law will be reflected. For a monochromatic plane wave the Bragg reflection curve has a finite width. Theoretical calculations of the reflection curves for many crystals can be obtained at the Argonne web site (Stepanov, 1997 & 2009). Real crystals can approach this theoretical width if properly made. Two of the NSTec sources have the ability to measure the reflectivity curve of flat and curved crystals such as those made of mica. (Haugh & Stewart, 2010) The use and calibration of crystals is not covered in this chapter.

3.3 The Manson type diode source: an X-ray system used for calibration

One of the NSTec diode type X-ray sources that is used for testing and calibrations generates X-rays in the energy range from 400 eV to 9 keV. We refer to this as the Manson source since this was the manufacturer. The source is not water cooled, and the power is limited to 10 W to avoid melting the anodes. The filament is shaped to a point near the anode. This produces a small spot, approximately 1 mm diameter, where the electrons impact the anode. This small X-ray emission spot acts as a point source providing a flat X-ray intensity in the sample region allowing us to do radiographic type measurements and to measure the sensitivity variation across the sensor array of a camera.

Fig. 6 shows a schematic diagram of the NSTec Manson system, looking down on it from above. The Manson comprises three compartments: the source chamber and two testing chambers which are the rectangular boxes in the figure. The two test chambers are connected to the main chamber by stainless steel vacuum components that include an isolation gate valve and a mechanical shutter. The diagnostic that is shown attached to the top arm in the figure is at vacuum. Components, such as filters, can be mounted inside the chamber

Each test chamber has its own vacuum pump and controls and can be isolated from the source chamber by a gate valve, then brought to atmosphere. Test chambers have photodiode and an energy dispersive detectors for measuring X-ray flux and the X-ray spectrum, mounted on push rods so that they can be moved into or out of the beam.

The X-ray beam paths that are used for testing are shown in red in Fig. 6. Filter 1, shown in the source chamber, is used to isolate a narrow wavelength band of X-rays. These filters are mounted in a vertical stalk that holds up to three filters. A light blocker prevents visible light emitted by the filament from entering the test chamber which would overwhelm the detectors and CCD.

The Manson system is a multi-anode device, holding up to six different anodes on a hexagonal mounting bracket. Two X-ray beams are isolated from the anode emission for use in the test chambers. A typical X-ray emission produced by the impact of electrons with a metal anode was shown in Fig. 2. The Ti spectrum that is observed when a 100 μm thick Ti filter is placed between the X-ray source and the detector as a band pass filter is shown in Fig. 7. See also Fig. 4(a) for the spectral characteristic of a thin sheet of Ti. Comparing the unfiltered Ti spectrum shown in Fig. 2 with the filtered spectrum shown in Fig. 7, we can see that the transmission is now limited to the spectral energy range between 4000 eV and 4966 eV, the latter being the K edge of Ti. The spectral content now includes the Ti K lines and the bremsstrahlung within the energy range given.

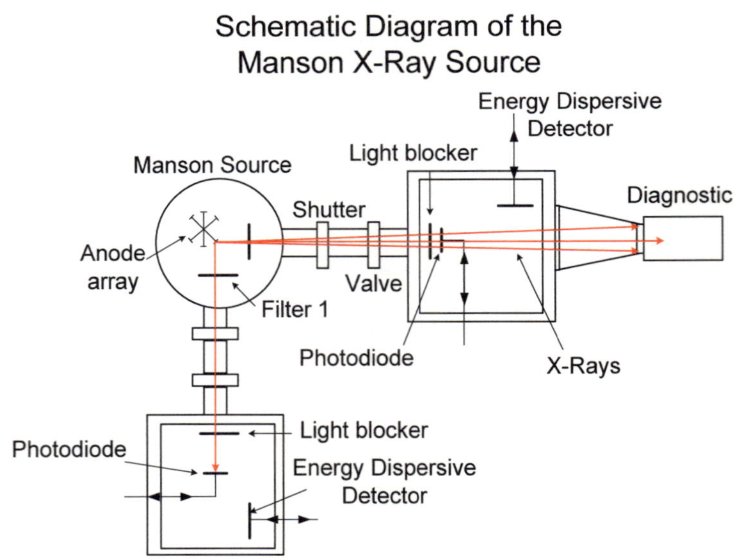

Fig. 6. Manson Schematic. The diagnostic being calibrated is shown directly attached to the chamber at the end of the upper arm. The red lines are the X-ray beam path.

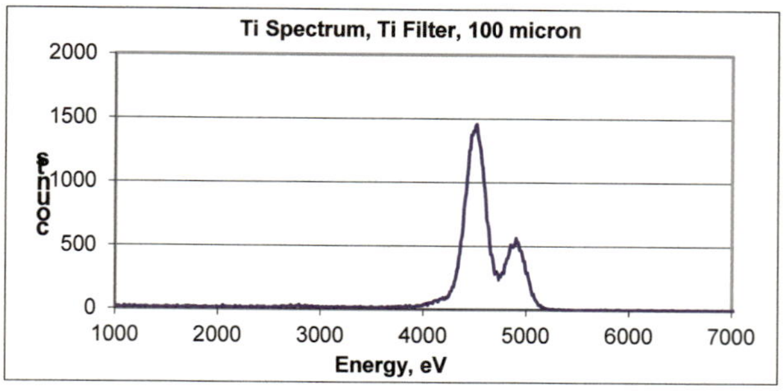

Fig. 7. The spectrum of Ti X-rays shown in Fig. 2 using a Ti filter 100 micron thick to limit the spectral bandwidth

3.4 Fluorescer source

The High Energy X-ray system (HEX) uses a diode type source to produce monochromatic X-rays. X-rays from the diode (a commercial 160 kV X-ray tube) excite characteristic X-ray lines in the fluorescer foil. The X-ray tube and the fluorescing targets are enclosed in a lead box. An exit collimator in the lead box shapes the X-rays into a beam. The fluorescer operation is illustrated by Fig. 8. For this example, the fluorescing material is a thin lead (Pb) sheet, with a thickness of approximately 250 µm, and the filter is a thin platinum (Pt) sheet. Table 6 gives the properties of the fluorescer and filter. The high energy X-ray lines are transmitted by the filter but the low energy lines are stopped by the filter.

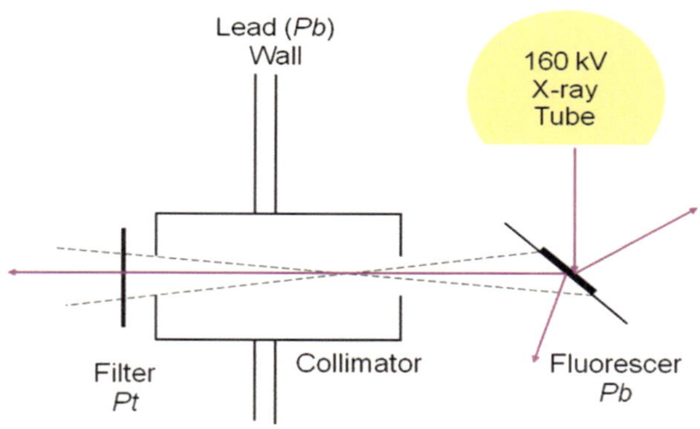

Fig. 8. Illustration of fluorescence principle

Filter	Fluorescer
Platinum (Pt), 50 µm thick	Lead (Pb)

Transmission	Spectral Lines, keV
0.72	73, 75
0.40	85
2×10^{-5}	10.4, 10.6, 12.6, 14.8
0	2.3

Table 6. X-ray Fluorescence

This method provides a reasonably narrow spectral energy that can be used to calibrate detectors at a range of well defined energies. The resulting spectrum from the arrangement is shown in Fig. 9.

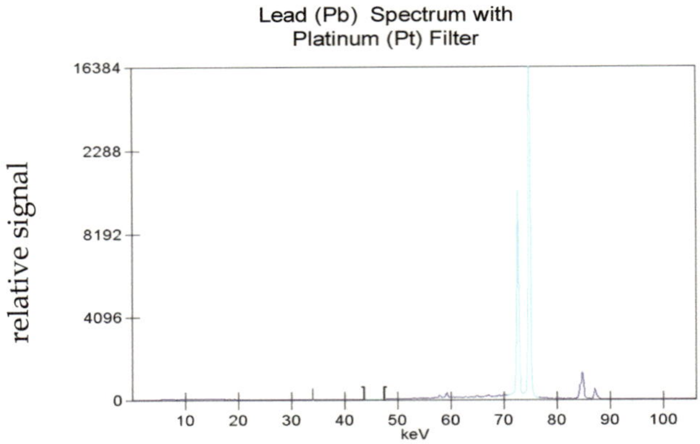

Fig. 9. Pb Spectrum, Pt Filter

The arrangement of the components is shown in Fig. 10(a) and (b).

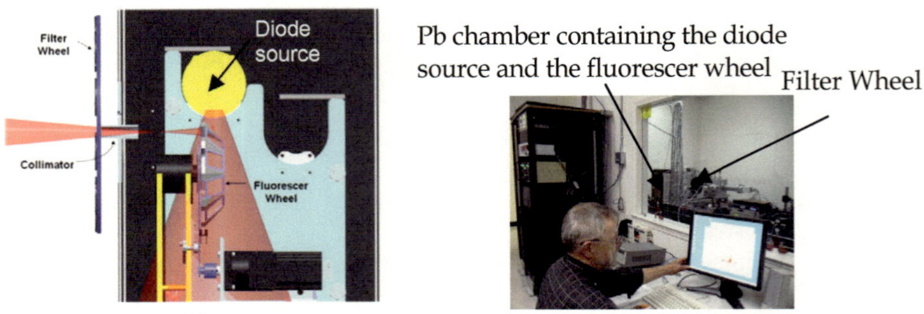

(a) (b)

Fig. 10. (a) HEX source component inside the Pb chamber and (b) a view of the control room looking through a window at the HEX optical table.

The end of the commercial X-ray tube is shown in yellow. The pink trapezoid that starts at the tube represents the primary X-ray beam. The fluorescers are mounted on the motorized wheel in the rectangles shown on the wheel. The fluorescer emits in all directions, but the X-ray beam is defined by the collimator inserted into the wall of the lead box, and the beam path is illustrated by the pink triangles. There is a filter wheel mounted downstream from the collimator, and it is also motorized. The fluorescer and the filter can be set from the computer in an adjacent control room, as shown in Fig. 10(b). The fluorescer is usually a thin sheet made of elemental metal, but metal compounds are sometimes used. The maximum intensities obtained when an 11.5 mm diameter collimator is used are on the order of $1x10^6$ photons per cm^2 per second, at one meter from the fluorescer, depending on the fluorescer material. The spectral lines used range from 8 keV to 115 keV.

Remote adjustment of the fluorescer wheel and the filter wheel is done through the control room computer. Data from the detectors and devices being calibrated are received in the control room.

The HEX source sits at the end of an optical table as shown in Fig. 11. The sample and the detectors are mounted on the optical table. The control room is separated from the HEX laboratory, as seen in Fig. 10(b).

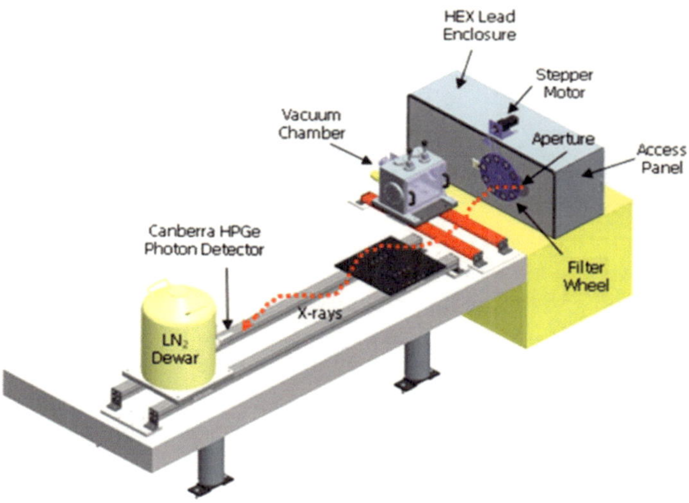

Fig. 11. HEX Layout

4. Solid state detectors

4.1 Introduction to semiconductor detectors

In the field of X-ray measurements, the term solid state detector usually refers to semiconductor type detectors. There are other X-ray detectors that are solid materials, such as scintillator -photomultiplier combinations, but for this work the term solid state detectors will include only semiconductor detectors. There are 3 types of semiconductor materials used for the work described in this chapter: silicon (Si), germanium (Ge), and cadmium telluride (CdTe). A semiconductor is defined as a material that has a small band gap (which can be manipulated) between the valence electrons and the conduction band, on the order of 1 eV to several eV. A metal has electrons populating the conduction band at any temperature, and an insulator has a large gap, on the order of 10 eV and higher. At normal temperatures, the semiconductor will have some population of electrons in the conduction band, and corresponding holes (positive charge where the electron vacated) in the valence band. The ratio of electrons in the conduction band to those in the valence band is given by the Boltzmann probability relation:

$$`N/N_0 = CT^{3/2}\ e^{-(E/kT)} \tag{4}$$

N= population of electrons in the conduction band
N_0= population of electrons in the valence band
E= band gap energy, eV
k= Boltzmann constant, 8.617343 x 10^{-2} ev/K
T= Absolute temperature, K
C= A material property

Consider an X-ray photon incident on the semiconductor. It has an energy that is many times that of the band gap. It interacts with the semiconductor material, primarily through the photoelectric effect, to produce energetic free electrons. These then produce electron-hole pairs. The number of electron-hole pairs produced is proportional to the energy of the X-ray photon. In general it takes several eV to produce an electron-hole pair, the exact energy depending on the semiconductor material. The number of electron-hole pairs produced by an X-ray photon depends upon the material and the X-ray photon energy as given by:

$$\mathbf{N_{e\text{-}h} = eV/\varepsilon} \tag{5}$$

$N_{e\text{-}h}$= number of e-h pairs produced
eV= X-ray photon energy, eV
ε= energy required to form an e-h pair in the semiconductor material
ε is usually referred to as the "ionization energy." Table 7 gives the ionization energy and the semiconductor band gap for the three materials we are discussing. The electron-hole pair formed will drift apart. But if a voltage is applied, the electrons will move in the direction opposite to the electric field, and the holes will move in the direction of the electric field, so the charges can be collected and measured.
The room temperature population of electron-hole pairs of the semiconductor (Eq. 4) will be noise in the measurements. Details of semiconductor behavior can be found in Knoll (2001), which describes the physics and behavior of semiconductor detectors.

Semiconductor material	ε, Energy required for one hole/pair production, eV	Semiconductor band gap, eV
Si	3.63	1.116
Ge	2.96	0.665
CdTe	4.43*	1.44

Table 7. Properties of semiconductors discussed in this section (*Quaranta, 1969)

There are several ways that the signal is lost and the solid state detector output fails to give the full measure of the X-ray intensity. Electrons produced by the X-ray interaction can recombine with another hole and electron lost from the charge collection process. Charge can be trapped by impurities or lattice defects and enhance the recombination process. For a properly designed semiconductor detector, this problem can be minimized. A semiconductor typically has a coating on the surface that protects the active semiconductor material from damage by the environment. This surface coating can absorb some of the X-ray beam and this will not generate a current out of the detector. This front surface absorption will have greater impact for the lower energy X-ray photons. For the higher energy photons, at some energy they will begin to pass through the detector's active material and will not be detected. Another signal loss mechanism occurs when the X-ray energy rises above the K edge of the semiconductor active material. The semiconductor can fluoresce and this energy is lost from the detector.

4.2 Photodiode used to measure X-ray power

Calibrated photodiodes are used to measure the X-ray power of X-ray sources. The sources are used to calibrate X-ray detectors. This description of photodiode operation will use the photodiode design that was used to measure the efficiency of X-ray CCD cameras and other imaging and non-imaging detectors. The photodiode is made from Si, 55 μm thick. If all of the X-ray photon energy is deposited in the active Si, and 100% of the e-h pairs formed are collected, the diode current **i** is given in Equation (6) as:

$$\mathbf{i = F \bullet (eV/\varepsilon) \bullet (1.6x10^{-19})\ amp} \tag{6}$$

$1.6x10^{-19}$ coulomb = Charge on the electron
F= Rate of X-ray photons hitting the photodiode, number/s
ε = Energy needed to form an electron/hole pair
eV = Photon energy

Applying Equation (6) to a Si diode with a 1 cm^2 area, with an incident beam of Ti Kα1 (4510.84 eV) X-rays, with a photon intensity of $2x10^5$ photons•cm^{-2}•s^{-1}, yields a current of $39.8x10^{-12}$ A. This can be readily measured using a commercial picoammeter.

The silicon photodiode in use was designed to measure X-ray intensity with nearly 100% efficiency for X-ray energies up to about 5 keV (IRD). Several of these photodiodes have been calibrated from 1000 eV to 60 keV at the Physikalisch-Technische Bundensanstalt (PTB Reference, Gottwald, 2006)). Within the measurement uncertainty, which is near 1%, the silicon photodiode is 100% effective up to the Si K edge, where it dropped several percent. The efficiency rises back to near 100% by 3 keV. The 55 μm thick Si photodiode begins to transmit near 5 keV X-ray energy, and at higher energies the diode efficiency follows the Si transmission curve.

4.3 Energy dispersive photodiodes

Some photodiodes are designed to measure the energy of the X-ray photons. The types of semiconductor materials that are commonly used include Si, Ge, and CdTe. A bias voltage is applied to the semiconductor and the electric fields generated require cooling of the detector. The voltage pulses produced by an individual X-rays are amplified and then counted according to pulse height. Pulses that have heights within a certain range are effectively assigned to channels according to the average pulse height by a processor referred to as a multi-channel analyzer. This produces an energy spectrum of the detected X-ray photons. The resolving power of the energy dispersive detector is generally limited to several hundred eV.

The detector sensitivity falls off at lower energies due to absorption at the front surface by a "dead" layer of the sensor and/or a window separating the vacuum chamber containing the detector from the environment. It falls off at higher energy when the photons begin to be transmitted by the sensor. The Ge and CdTe detectors are designed to operate optimally in the 10 keV to slightly over 100 keV. The X-ray photon interacts with the sensor material in ways other than forming electron-hole pairs, which can reduce its sensitivity. If a photon has sufficient energy, it can knock out a 1s electron of the sensor material, a higher state electron can transition down into the 1s vacancy. A second photon having an energy equal to the energy between these two states can then be emitted. The sensor then produces a corresponding pulse of electron-hole pairs that is smaller than the base peak by the energy difference between the incoming X-ray photon and the binding energy of the sensor material 1s electron. This is referred to as an "escape peak,"and effectively this means the incoming X-ray photon is not counted. There are other losses that are introduced by the detector fabrication details.

The Ge detector made by Canberra uses a high purity Ge disk, 8 mm diameter and 5 mm thick, that is cooled to liquid nitrogen temperature. The sensor is in a vacuum chamber that has a 4 mil thick beryllium (Be) window for X-ray beam entry. Ge has an escape peak near 11.1 keV. Fig. 12(a) shows a spectrum from the radioactive isotope of americium (Am) having an atomic mass of 241 (Am-241). This source emits gamma radiation (X-rays that are produced by nuclear transitions) at 59.5 keV and 26.4 keV, and X-rays that are produced by electronic transitions from the Am decay daughter neptunium (Np) ion are seen in the 13 keV to 22 keV range.

The CdTe sensor is a 5 mm square that is 1 mm thick. It is cooled sufficiently using a thermoelectric cooler so that a 400 V bias voltage can be applied without electrical breakdown. It also operates in vacuum with a 0.001 inch Be window. The energy required to form an electron-hole pair for this material is 4.43 eV. (Herzberg, 1945) Escape peaks occur near 27 keV and 23 keV. Electron-hole pairs formed near the back contact of the detector cause fluctuations in pulse height, and they are not seen as belonging to the true peak. This is a loss in sensitivity. The Am-241 spectrum for the CdTe detector is given in Fig. 12(b). Note that there are several peaks in the 30 keV to 40 keV energy range that are not in the spectrum when the Ge sensor is used. A small escape peak is seen near 32 keV, and a larger escape peak is seen near 36 keV.

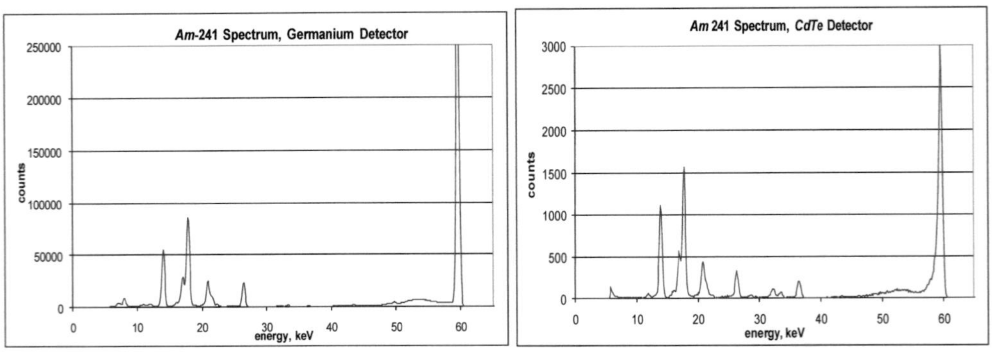

(a) (b)

Fig. 12. Am-241 spectrum. (a) Ge detector, showing γ-rays (gamma rays) at 59 keV and 26 keV, X-rays from Np from 22 keV to 14 keV, and escape peaks below 14 keV. (b) CdTe detector, showing X-rays and γ-rays as in (a), and escape peaks in the 30 - 40 keV range.

4.4 Imaging devices

X-ray imaging is used to obtain geometric and qualitative intensity information in many applications such as tooth damage, stellar luminosity, and temperature and density of laser-produced plasmas. In the past few decades, several types of solid state detectors have replaced film in most imaging applications. These include charge-coupled devices (CCDs), charge-injection devices (CIDs), image plates (photostimulable phosphor plates - PSP), and other devices and sensor materials. After a brief description of image plates, the work described in this chapter will be limited to the CCD and CID Si-based sensors.

4.4.1 Image plates

Image plates use principle of photostimulated luminescence (PSL) to to read the X-ray image after the image plate has been exposed. The exposed image is read and digitized with an image plate reader and stored in the computer as an image file. The image plate contains very small crystals (~5 μm) of barium fluorobromide phosphor with a trace amount of europium as a color center (Maddox, 2011). When exposed to X-rays, an atom of Eu^{2+} is converted to Eu^{3+} and the free electron is trapped in the barium fluorobromide lattice, creating a metastable state. The metastable electrons are freed by stimulation from the laser in the image plate reader, and they recombine with the Eu $^{3+}$ to give off a blue-violet fluorescence that is recorded by the image plate reader with a suitable detector. The digitized image can then be analyzed for geometric information and quantitative intensity information. We now use image plates extensively for characterizing the X-ray beam in a variety of arrangements. We have also calibrated image plates for quantitative intensity measurements that are then applied to plasma studies at LLNL (Maddox, 2011). LLNL and other groups have also evaluated their performance for medical imaging (AAPM, 2006). Image plates will not be considered further in this chapter since the subject of the book is photodiodes.

4.4.2 Silicon based cameras

The basic sensing function is the same for all Si-based sensors, and a camera operates in a manner similar to the principles described for the X-ray photon detectors described earlier. The photons interact with the Si semiconductor to produce electron hole pairs. The camera types vary in the method in that the charge is moved from the sensor pixel and eventually produces a digital output proportional to the charge formed. Janesick (2000) is a good source for the details of camera readout methods.

The charge coupled device (CCD) transfers the charge from a row of the sensor pixels to an adjacent row until it reaches the "read-out"row. The charge is read from each pixel sequentially in that "read-out" row as voltage. The voltage is then converted to digital counts. A major advantage of the CCD camera is low electronic noise because it is usually cooled to minimize the dark current from thermally-produced electron-hole pairs (Eq. 4). The major contribution to the noise is then from the read-out. The read-out noise is quite low, typically no more than a few counts. The dominant noise contribution for X-ray applications is then what was produced by the arrival time statistics of the photons themselves, a Poisson distribution. The standard deviation per pixel is then the square root of the number of X-ray photons the pixel absorbed during the exposure time.

An alternative electronic method for reading the charge stored in the individual pixels is the CID. Every pixel can be individually addressed using indexing row and column electrodes. A displacement current proportional to the stored signal charge is read when stored "packets" are shifted between capacitors within individually selected pixels. The displacement current is amplified and converted to a voltage, and fed to the outside world as a digitized signal. The CID technology offers certain advantages over CCD technology such as cost, reduced heat generation, and resistance to blooming. The disadvantage is the high readout noise. Improvements continue in this area but for X-ray photon intensity measurements, the readout noise far exceeds the photon statistical noise for a single image. When calibrating the CID, this can be mitigated by multiple imaging.

Most CCD cameras are front illuminated, the front side being where the gates are located. This region is not active to photon detection. This dead region will stop low energy X-rays and the front illuminated CCD cameras lose sensitivity as the X-ray energy drops below 2000 eV. Below 1000 eV they are not useful. For this reason many X-ray CCD cameras are back illuminated and back thinned and are sensitive even into the vacuum ultraviolet. The Si active region is typically 15 μm to 30 μm. The Si thickness thinning does reduce the sensitivity at the high energy side. Camera efficiency measurements for both front illuminated and back illuminated cameras are given in Section 6.

5. Calibrating primary detectors using radioactive sources and a synchrotron source

5.1 The calibration concept using radioactive sources

Radioactive sources provide a variety of spectral lines at well defined energies, as was described in the energy dispersive detector section (section 4.3). The photon output is directly proportional to the activity of the radioactive source, and the activity measurement is traceable to NIST. The uncertainty for the activity is provided by the vendor. The activity **R** at time t is given by:

$$\mathbf{R} = \mathbf{R_0} \exp((t \ln 2)/\tau_{1/2}) \tag{7}$$

$\mathbf{R_0}$ activity at time t=0 (disintegrations/sec)

$\tau_{1/2}$ radionuclide half life

The radioactive source is placed at some distance from the detector that is sufficiently large so that the source intensity is uniform over the detector area. The power **Ω** at the detector in photons per second for a selected spectral band is given by:

$$\Omega = \mathbf{R}\ \mathbf{B}\ (\mathbf{A_d}/4\pi r^2)\ \mathbf{T} \tag{8}$$

B branching ratio

$\mathbf{A_d}$ detector area

r distance between radioactive source and the detector

T X-ray transmittance through r cm of air

The branching ratio is the fraction of nuclear decays that produce the selected spectral band. It has an experimental uncertainty associated with it. If the activity was measured using a gamma emission, and the same gamma is used for the detector calibration, the uncertainty is that given is that given by the vendor in his activity certification. The detector efficiency η, is then given by:

$$\eta = \mathrm{S}/\Omega \tag{9}$$

S is the photon count per second as measured by the detector. The measurement arrangement is shown in Fig. 13. The instrument to be calibrated, the Ge detector, is seen on the right facing the radioactive source. An optical distance meter is located at the far left of the optical rail and is also at the same height as the source and detector and measures the distance from the radioactive source and to the detector window. The internal distances for source and detector are provided by the manufacturers to an accuracy of 0.5 mm or better. The distance sensor has an accuracy of ±1 mm and was calibrated at NIST within a month of the detector calibration measurements. Thus the source-to-detector distance accuracy is ±1 mm.

5.2 Measurement results

Measurement results are given in this section. The spectral lines that were used for these measurements are given in Table 8.

Fig. 13. Experimental arrangement for calibrating detectors using radionuclides

Spectral Energy, keV	Spectral Type	Radionuclide
5.97	Mn, K line	Fe55
22.16	Ag, K line	Cd109
26.34	γ	Am241
41.3	Eu, K line	Gd153
59.54	γ	Am241
88.00	γ	Cd109
97.43	γ	Gd153
103.18	γ	Gd153

Table 8. Spectral Lines Used for the Calibration of the Ge and CdTe Detectors

Quantum efficiency measurements for the Ge detector are shown in Fig. 14(a), and Fig. 14 (b) for the CdTe detector. These measurements show a precision near 3% at the 95% confidence level. The Ge detector shows a peak efficiency near 60 keV and falls off in efficiency at lower and higher spectral energies. The CdTe detector has a peak efficiency near 30 keV and also falls off in efficiency at lower and higher spectral energies.
It has been described earlier that escape peaks occur above the K edges (minimum energy needed to remove a 1s electron) of the detector materials, and this reduces the detector efficiency. This effect has been directly observed in the calibration of a Si photodiode in the vicinity of the Si K edge (1.39 keV). The K edges for Ge, Cd, and Te are 11.1 keV, 26.71 keV, and 31.81 keV respectively. (CXRO) In Fig. 14 (a) and (b), note the large gaps that exist between points. The situation regarding the large gaps can be improved by using more radionuclides but there are availability and economic restrictions that will limit this approach. We plan to fill in these energy gaps by calibrating an Si detector up to 60 keV using a synchrotron as described in the next sub-section. This calibration can then be transferred to the other detectors using the HEX source.

5.3 Synchrotron calibration of a silicon photodiode

The silicon photodiode used by NSTec for calibrations of camera efficiency that is described in Section 6 was manufactured by International Radiation Detectors (IRD) who claimed that its measurements are absolute (IRD Reference). This claim was substantiated by sending two photodiodes to the German synchrotron at the PTB. The photodiode efficiency was measured from 1 keV to 60 keV. The measured efficiency was 100% from 1000 eV up to the Si K edge at 1839 eV, then dropped about 3% to 97% due to the Si escape photon. The efficiency rose back to 100% by 3.5 keV. The efficiency begins to fall near 5 keV as the absorption of the Si falls below 100%.

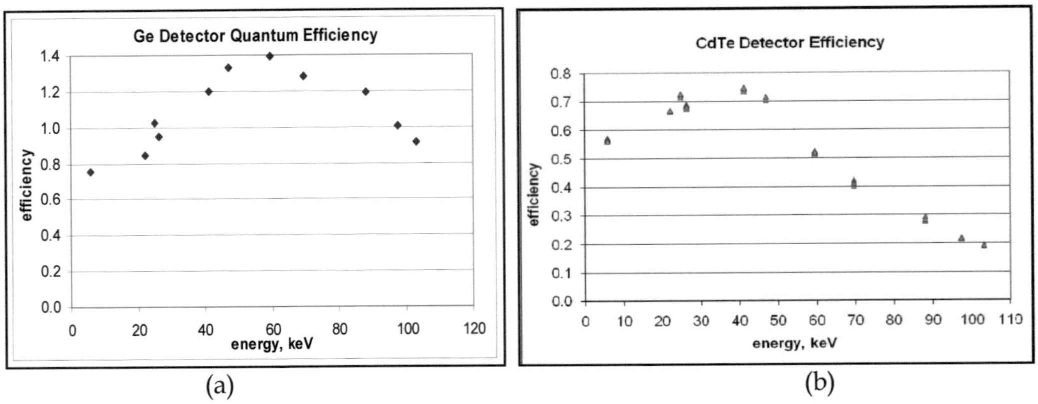

Fig. 14. Measurements of the quantum efficiency of the (a) Ge detector and (b) CdTe detector using the radionuclide sources.

NSTec has performed calibrations using the IRD Si photodiode, accepting the concept that it was an absolute photodiode and using the IRD measured thickness to correct for X-ray transmission at the higher energies. PTB calibration is an internationally accepted standard. The 2 calibrated photodiodes are then utilized as primary standards to cross calibrate other photodiodes which will be our working devices.

5.4 Accreditation procedure for calibration laboratories at NSTec

In order to provide substantiated traceability to international standards that would verify the quality of the NSTec calibration laboratories, we have been working with the National Voluntary Laboratory Accreditation Program (NVLAP). NVLAP is one of several organizations that can provide the certification that calibration procedures meet national and international standards. The NSTec laser timing laboratories have obtained this accreditation. The US National Institute for Standards and Technologies (NIST) does not have standards for X-ray intensity measurements; their concentration has been dosimetry. The calibration methods described previously are the starting point in the NSTec effort to achieve accreditation for the X-ray laboratories. Through the development of an accredited X-ray intensity calibration methodology, the diagnostics and system components for pulsed high energy physics experiments conducted at the national laboratories such as LLNL can now be calibrated traceable to national standards. Since the NSTec had already passed the NVLAP accreditation process on specific measurements in two laser laboratories, the groundwork for the X-ray laboratories was already done. The HEX laboratory would be the first one to be accredited. The required rigor of the quality assurance management system is already in place and approved. The required steps for accreditation are:

1. Determination of customer requirements
2. Development of calibration procedures and documentation of the procedures
3. Traceability

4. Evaluation of measurement reproducibility
5. Determination of uncertainty
6. Generation of calibration certificates.

The main customer requirement is to have the capability to calibrate customer detectors and components to an absolute value with known uncertainties.

The previous sections have described the detector calibration procedure used to calibrate our X-ray source. The radionuclides used are traceable to NIST, and the vendor of these radionuclides participates in a measurement assurance program with NIST. NIST regularly sends standard radionuclide sources to the vendor to conduct "blind" measurements; the activity of the sources is unknown to the vendor. The vendor then measures the activity and submits the results back to NIST. NIST then checks the measurements to verify that they meet the required accuracy. The distance measurement traceability was described earlier.

As with all traceable measurements, repeatability and reproducibility determination are requirements for accreditation. The variation in repeated measurements quantifies random errors. Only when calibrations using a common procedure and common equipment produce the results that vary within an acceptable range, even with different operators, is the technical process is ready for NVLAP review.

NSTec is in the process of evaluating the overall uncertainty in its calibration of X-ray detectors with radionuclide sources. A list of the factors being considered is given in Table 9. The procedures are being written at the time of this publication.

1.	Radionuclide signal measurement reproducibility.
2.	Radionuclide activity.
3.	Radionuclide branching ratio, when needed; daughter X-ray emission.
4.	Distance measurement, source to detector.
5.	Detector size.
6.	Measurement duration.
7.	Choice of region of interest (ROI) for counting photons belonging to a spectral line.
8.	Radionuclide decay rate.
9.	Energy accuracy.
10.	Air transmission of the X-ray.
11.	Curve fit to points between the measured points.

Table 9. Quantities in the Calibration Procedure Contributing to the Overall Uncertainty

6. Calibrating X-ray cameras using the primary detectors

6.1 Measuring the quantum efficiency and its spatial variation (flat field) for an X-ray CCD camera using the Manson source

The Static X-Ray Imager (SXI) is an X-ray camera used on the NIF target chamber to measure quantities such as laser beam pointing and the sizes of the laser entrance hole in ignition targets (Schneider, 2010). The SXI records a time integrated X-ray image. The sensor is a back thinned CCD chip with 2kx2k pixels, 24 μm square (approximately 50 mm x 50 mm). The use of this camera at NIF requires the knowledge of camera sensitivity S(E) as a function of X-ray energy E.

The X-ray photon interacts with the Si sensor to produce hole-electron pairs that the CCD electronics process to produce the digital signal count S. The number of electron-hole pairs produced by an X-ray photon that interacts with the Si sensor is a function of the photon

energy and is slightly dependent upon the temperature (Janesick, 2000). The sensor is cooled to 253K when operating.
A useful model relating the camera signal to fundamental quantities is found in Janesick (2000).

$$S(E) = P \bullet A_{pix} \bullet T \bullet \eta \bullet QE \bullet K^{-1} \text{ counts/pixel} \quad (10)$$

$P =$ Photon rate at the CCD, photons/cm²/sec
$A_{pix} =$ Pixel area
$T =$ Exposure time
$QE =$ Quantum efficiency, fraction of photons that interact with the pixel
$\eta =$ Quantum yield, number of electron-hole pairs produced by the photon
$K =$ Camera gain, electron-hole pairs per count
For the Si CCD, the quantum yield is given by (Knoll, 2001):

$$\eta = E/3.66 \text{ hole pairs/photon} \quad (11)$$

The camera manufacturer's measurement of the gain constant was used for the calculations of quantum efficiency. The quantity being determined by the calibration is S(E), but the model given by Equation 10 is a valuable check of the calibration procedure, the calibration implementation, and is useful for troubleshooting camera problems. For the Si based cameras, the QE is related to the properties of Si as we shall see in the following results.
The photon intensity P′ was measured using a photodiode. We use a photodiode manufactured by IRD, model AXUV100. These detectors are designed with no doped dead region and zero surface recombination, so that they have near theoretical quantum efficiencies for the soft X-ray spectral region. One unit of this model was sent by the photodiode manufacturer for calibration to the synchrotron at the PTB. The results from this calibration showed agreement within 1% from 1000 eV to 1839 eV, and the photodiode gave a 5% lower reading above this energy. The thickness of the Si photodiode is 54.5 μm. The current **i** from the photodiode is related to the photon intensity as given in Equation 12:

$$i = P' \bullet \eta_{PD} \bullet 1.6 \text{x} 10^{-19} \text{ amp} \quad (12)$$

Here $\eta_{PD} = E/3.62$ and $1.6 \text{ x } 10^{-19}$ is the charge on the elctron in coulombs. The area of the IRD detector is exactly 1 cm². The photodiode and the CCD are not at the same distance from the anode, as can be seen in Fig. 6. These distances were measured to an uncertainty of 2 mm. The camera sensitivity ξ, defined as counts per photon:

$$\xi = S(E)/(P \bullet A_{pix} \bullet T) \quad (13)$$

$$\xi = S(E)/((i/(\eta_{PD} \bullet 1.6 \text{x} 10^{-19})) \bullet (d_{IRD}/d_{CCD})^2 \bullet A_{pix} \bullet T) \quad (14)$$

$P =$ Photon intensity at CCD, photons•cm⁻²•s⁻¹
$d_{IRD} =$ Distance from the photodiode to the anode
$d_{CCD} =$ Distance from the CCD to the anode
Note that **P'**, the photon intensity at the photodiode in Eq. 12, is given by:

$$P' = P \bullet (d_{IRD}/d_{CCD})^2 \quad (15)$$

6.1.1 Methods for imaging

The SXI's CCD camera was mounted on the diagnostic arm is shown Fig. 6. There was an extension between the camera and the Manson chamber of sufficient length that the X-ray beam uniformly illuminated the CCD. The camera calibration proceeded by the following steps:

1. Locate the bad pixels so that they can be masked out for image analysis;
2. Determine the linear range of the camera;
3. Measure the camera sensitivity;
4. Measure the uniformity of the CCD chip response over the area of the camera.

The cameras had a large number of bad rows and hot pixels. The bad rows were associated with the readout and identified using closed shutter images with a 3 ms exposure time. The hot pixels were identified by taking an image using the Ti anode and no filter, and using the same exposure time that was used for the experiments on the NIF target chamber experiments. A map was made that identified the bad rows and bad pixels.

The photon intensity was measured with the photodiode in arm #1 as seen in Fig. 6. An exposure time was chosen to be as short as possible to give a reasonable signal. Photodiode readings were taken before and after acquiring each CCD image. During imaging, the X-ray beam intensity was monitored continuously for beam fluctuations using the photodiode in arm #2. If there were beam intensity fluctuations observed during imaging, that image was discarded.

Flat field images are images where the CCD is uniformly illuminated in order to measure the uniformity of the camera response over its area. They were taken using the same anode voltage that was used for the camera efficiency measurements and maximum anode current. The exposure time was chosen to produce a signal that was 50% to 60% of saturation. Ten flat field images and ten background images were taken at each photon energy.

6.1.2 Image analysis

The camera images for the efficiency analysis had the background subtracted and the bad pixels replaced by the average of adjacent pixels. The mean pixel count was determined by randomly selecting 1000 regions 20x20 pixels in size, calculating the mean counts/pixel for each region and calculating the average of the means for each region. This is the signal S for that image. Then, for the flat field images, average all images that have the same exposure time, average the background images, and subtract the average background from the average flat field image.

6.1.3 Camera sensitivity

The camera sensitivity for one of the SXI cameras is given in Fig. 15(a). The Quantum Efficiency (QE) calculated using Eq. 10 through 14 and camera gain K=7.62 electrons per count is plotted as a function of photon energy in Fig. 15 (b). The data scatter as measured by the standard deviation was 1% or less at each point. The dip near 1800 eV and the fall-off after 2000 eV are properties of Si. Si that is 15 μm thick transmits up to 35% as it approaches the K edge at 1839 eV. It begins transmitting again above 2500 eV and is transmitting 80% at 8 keV. These QE results are similar to that obtained by Poletto (1999). There are two possible causes why the QE does not approach 1 when the photons are completely absorbed: (1) There may be absorption at the surface coating of the Si; (2) the

Quantum Yield may be less than the photon energy divided by 3.66 eV per electron-hole pair. Analysis of a large number of single photon events could show the relative contribution of each effect.

6.2 Flat field

The flat field source is the 1 mm diameter spot on the anode. The anode is 1405 mm from the CCD. This arrangement would produce a flat field within 1% if there were nothing between the anode and the CCD. There is a light blocker that has an aluminum coating on a polyimide film (Al 1054 Å ±50 Å; polyimide 1081 Å ±100 Å). This item does not affect the flat field within the 1% cited above. The filter can cause a variation in the beam intensity across the CCD if there is sufficient variation in thickness, foreign material, or misalignment with the anode. A comparison of all the flat field images implies that the maximum variation is ±1% peak-to-peak.

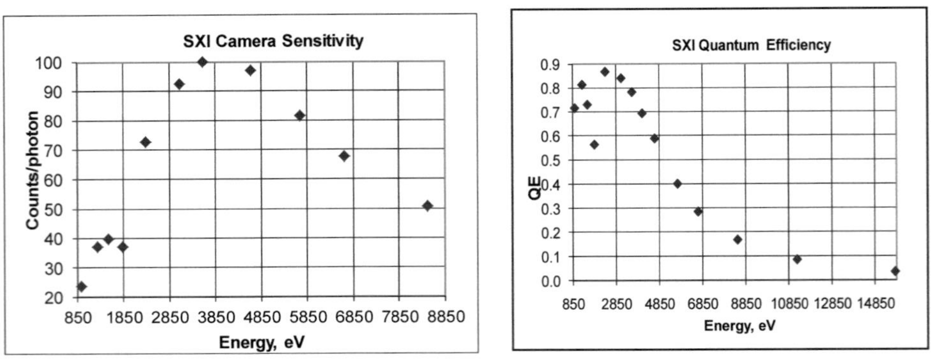

(a) (b)

Fig. 15. The SXI (a) camera sensitivity and (b) quantum efficiency as measured by the camera count per pixel for each photon of a given energy. The measurements made at X-ray energies below 8800 eV were done on the Manson. The higher energy measurements were done on the HEX.

Fig. 16(a) shows the flat field image for one of the SXI cameras at the Cu 8470 eV energy band. The image is set at high contrast so that the pixel signal variation shows clearly. A gross pattern is observed with the sensitivity at a maximum near the left center and decreasing slowly going away from the maximum. The image in Fig. 16 (b) is at Ti 4620 eV; it shows the same pattern but decreased magnitude. The pattern continues to decrease in magnitude until it is no longer visible at 3000 eV. Vertical lineouts averaged over a small horizontal width (see band in Fig. 16(b)) for three images at three different X-ray energies are shown in Fig. 17. The lineouts are normalized by dividing by the maximum counts in each image. The maximum sensitivity variation for each of the curves in Fig. 17 is 13% at 8470eV, 6% at 4620eV and 2% at 3580eV.

A flat field image of the Mg 1275 eV band is shown in Fig. 16(c) for comparison to the higher energy flat field images. There is no trace of the sensitivity variation pattern that is seen at higher energies. The 1275 eV lineout in Fig. 17 shows that the maximum variation is less than 1%, which is the measurement limit of our flat field procedure.

This sensitivity variation is a large scale effect; it includes groups of pixels and is probably related to the CCD manufacturing process. Any sensitivity variation of individual pixels is less than the photon noise associated with averaging 10 images.

A different phenomenon was seen at low energies. Small irregular patches having diminished sensitivity were observed that are readily seen in Fig. 18(a). This image shows a portion of the CCD. The effect on sensitivity in these regions also shows an energy dependence. Fig. 8b is a similar image taken at 3080 eV. The irregular patches have now become quite dim compared to what was observed at 1275 eV. At 4500 eV, these paths of low sensitivity have completely disappeared.

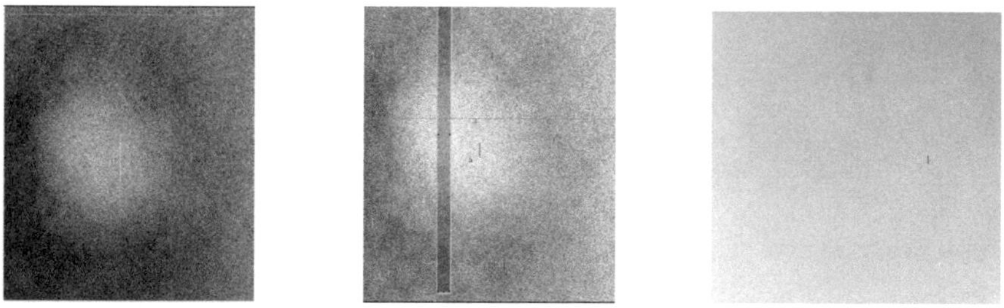

(a) (b) (c)

Fig. 16. Flat field image for the (a) Cu anode, 8470 eV and (b) Ti anode, 4620 eV, showing the pixel sensitivity variation (Signal range: 5200 to 7200 counts/pixel) The vertical band was the area used to calculate the cross section that is shown in Fig. 17. The same region was used for the cross section at the other energies. (Signal range: 5200 to 7200 counts/pixel) (c) Flat field image for the Mg anode, 1275 eV, showing the pattern observed at the higher energies shown in Fig. 16(a) and (b) has completely gone and the pixel sensitivity is flat.

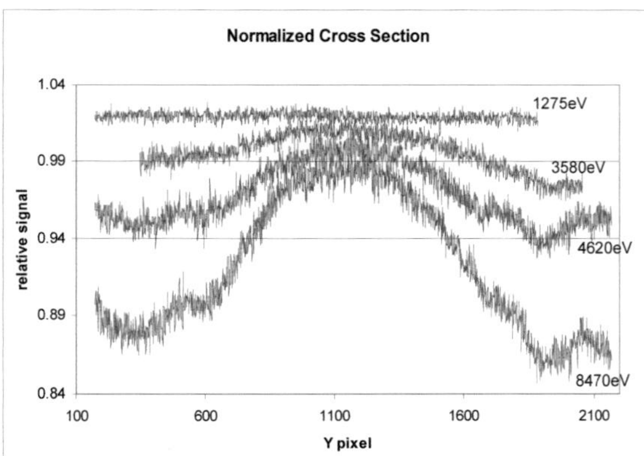

Fig. 17. Normalized vertical lineouts from flat field images at several X-ray energies. The lineouts were normalized to the maximum counts in each image. As the X-ray energy increases, the pixel sensitivity shows a greater vatiation.

There are several possible causes for these dark regions. Debris on the CCD surface could absorb X-rays and would be energy dependent, absorbing X-rays less as the energy increased. Damage to the CCD would likely cause an energy dependence that would increase the variance of the defective region from the surrounding pixels as the energy increased. Damage to the surface coating could produce this effect if the coating were thicker in that defective region. When we examined the CCD surface with a magnifying glass it did appear that the coating was deformed. It looked like a manufacturing defect.

It is difficult to correct these images using the normal method of flat field inversion. This could be done if you limit the energy range of the X-ray source. But the characterization always provides the information necessary for the effective use of the X-ray camera.

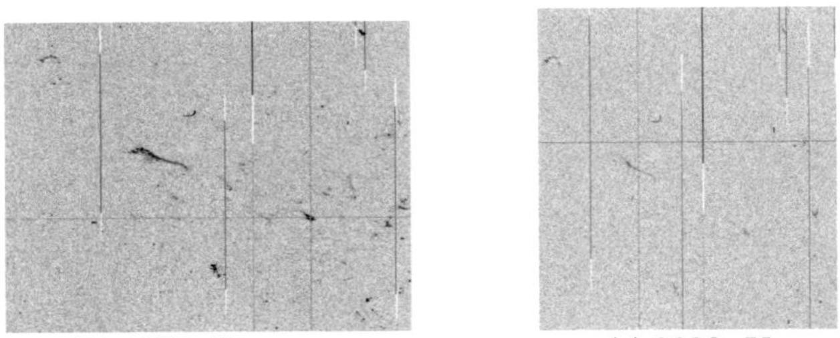

(b) 1275 eV (a) 3080 eV

Fig. 18. These are the same sections of a flat field image taken at two different energies, (a) 1275 eV and (b) 3080 eV. The sections cover about ¼ of the entire CCD. The dark regions are CCD surface defects causing diminished pixel sensitivity. For the 1275 eV section shown in (a) the blemishes are much darker than in the 3080 eV image shown in (b).

6.3 Calibrating a front illuminated CCD camera from 705eV to 22keV using the Manson and HEX sources

The SXI camera described above plays a critical role in the NIF operation, but this specific chip is no longer manufactured. There is another chip on the market with this large array, 2kx2k, 24 µm square, and we were requested to test the chip in a standard camera. The major concern regarding this chip was that it is front illuminated.

The QE measurements at X-ray energies below 10 keV were done using the Manson source following the procedures given in 6.1. These measurements are shown in the graph of Fig. 15. Compare this to the results shown in Fig. 19 for the QE of the back illuminated camera. The maximum QE for the front illuminated camera is QE=0.34 near 2300 eV. This is almost a factor of 3 lower than the QE measured for the back illuminated camera. The predominant difference begins to show below 1000 eV. At the Cu L lines, near 930 eV, the QE for the front illuminated camera is down by a factor of 10 from the front illuminated camera. At the Fe L lines near 705 eV, the QE is down by a factor of 100.

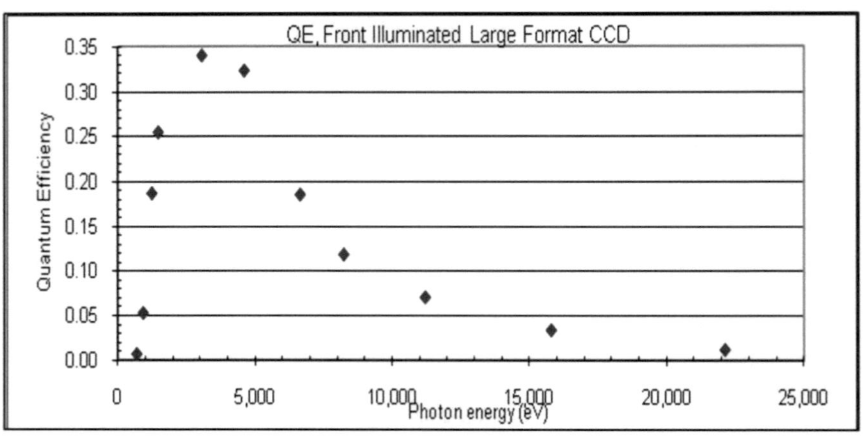

Fig. 19. The quantum efficiency measured for a front illuminated CCD sensor.

The measurements at 10 keV and lower energies were done on the Manson. The measurements at higher energies were made using the HEX. Compare this to the QE measurements shown in Fig. 15.

The Manson can only be used effectively up to the Cu K lines. The QE measurements at higher energies have to be done on the HEX. The CCD cameras must be kept in a vacuum since they are cooled and the HEX has a vacuum chamber on a rail as is seen in Fig.11. The chamber is very similar to that shown on the Manson. It differs in having a Be window on the side facing the Hex source. The camera is mounted on the opposite side from the Be window. The HEX fluorescer source is near 10mm diameter rather than the "point" source of the Manson. The X-ray beam is not flat across the entire CCD surface but is flat near the beam center. The camera is moved horizontally and vertically until the X-ray beam is centered on the CCD. The camera is then moved aside on the rail and the CdTe detector is placed at the same distance from the source as was the CCD. The beam center is then determined by moving the detector horizontally and vertically. These are the measurements used in Eq. 10 to determine the QE shown in Fig. 15 and Fig. 19 for the higher X-ray energies. The observation then is that the QE at these energies is the same for the front illuminated and the back illuminated cameras.

Measuring the sensitivity variation on the HEX requires that the X-ray intensity measurement be carefully measured over the entire area and an analytical representation be developed. This functionality is being developed now. We will use both the CdTe detector on a motorized X,Y positioner and image plates to measure the X-ray intensity distribution.

6.3 Single photon measurements using the Manson source

Images can be taken at sufficiently short exposure times so that most or all of the incidents recorded by the camera are caused by individual photons. These single photon images provide spectral information. This technique is used for astronomical measurements and laser plasma studies. The image shown in Fig. 20(a) was taken on the Manson source using a Ti anode and a Ti filter 100 μm thick. This is the same condition that was used to generate the spectrum shown in Fig. 7 using an energy dispersive detector. The camera used was a silicon CCD type having 1300 pixel x 1340 pixel array and the pixel size was 13 μm square. A background image using the same exposure time and no X-rays has been subtracted from the original X-ray image. The region shown in the figure is a 100 pixel square. There are approximately 95 single photon events in this 10000 pixel area, or about a 1% fill. This is the fill rate typically used in single photon measurements. Note that a significant fraction of the single photon events produce counts in more than one pixel, that is, the production of electron/hole pairs produces by the photon occurs in more than one pixel.

The graph shown in Fig. 20(b) is a histogram of the entire pixel array for the single photon image of the Ti X-rays. This plot shows the number of times a pixel has a given count as a function of counts. The histogram exhibits two peaks and they are above 400 camera counts. The two peaks are the Ti Kα photons occurring at 415 camera counts and the Ti Kβ photons occurring at 454 camera counts. These peaks represent single pixel events where the total number of electron/hole pairs produced by the photon is contained within that single pixel. As stated in the previous paragraph, there are many incidents in the image where the single photon produces counts in multiple pixels. These multi-pixel events produce the rising number of incidents in the graph going toward lower counts. There are no incidents at counts above the K-M band. Compare this spectrum to that shown in Fig. 7 where an energy dispersive Si detector was used. The spectral resolution is nearly the same for each detector. In general then, a camera is an energy dispersive detector when operated in the single photon mode.

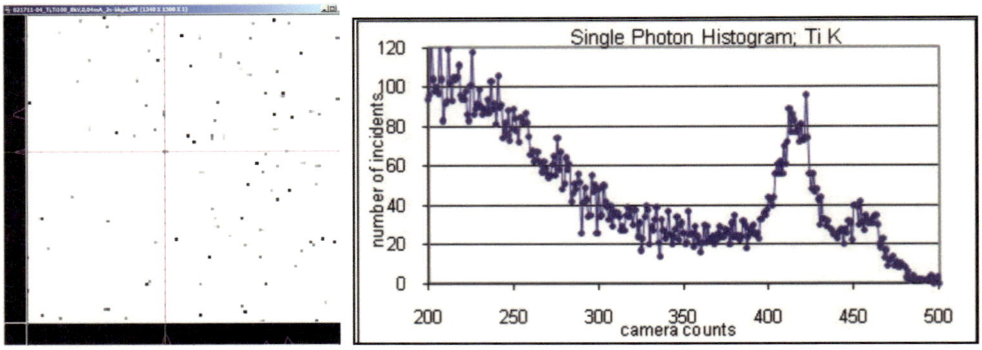

(a) (b)

Fig. 20. (a) This image shows single photon incidents on a CCD camera zoomed in to show the individual pixels in a small region of the camera active area. (b) This graph is a CCD active area showing the Ti K-L and K-M spectral bands. Compare this to the spectral scan of the Ti emission using the energy dispersive detector shown in Fig. 3.

The above description also describes a method for calibrating the camera count to spectral energy. As described earlier for the camera efficiency calibration, images are taken with several anode/filter combinations. The camera count for the peak center is then plotted against the literature value for the spectral energy (more precisely, a weighted average of the unresolved spectral lines).

More sophisticated software than a simple histogram can be devised that would capture a large portion of the multi-pixel incidents that are single photon events. This would reduce the noise that is seen in the histogram peaks. The method requires identifying significant pixels by a thresholding technique, then adding the counts of adjacent pixels to the central pixel. This represents a new image that generates a new histogram. The spectral peaks will be better defined because the noise is reduced.

6.4 Characterizing and calibrating an uncooled X-ray CID camera using the HEX source

This section describes the characterization of a CID camera that was planned as the detector in a spectrometer system that was to be used on the LLNL NIF target chamber. The initial interest was to measure the emission from highly ionized Ge so the camera was characterized in the 10 keV region using the HEX source (Carbone, 1998 and Marshall, 2001). The fluorescers chosen were Cu, Ge, and Rb giving weighted average for the K-L and K-M transitions of 8.13 keV, 10.01 keV, and 13.58 keV respectively.

The major use for this CID sensor is for dental X-rays. It is relatively cheap and therefore expendable, a desirable property for the NIF application. The camera operates at room temperature normally, which gave a challenging problem to the characterization on HEX. Since the CID operates at room temperature, the dark current can saturate the camera for exposure times less than 10 seconds. This not a problem on NIF since the exposure time can be less than 1 second with sufficient X-rays to provide a bright spectral image.

As indicated in the earlier description of CCD camera calibrations on the HEX, minutes of exposure time are needed to get a satisfactory signal. Preliminary experiments with the CID camera showed that we would be limited to three-second exposure times. It was determined that multiple exposures, on the order of 100 exposures, would be needed to obtain satisfactory photon statistics. The multiple exposures would also allow us to average the readout noise and get to the limit that photon statistics were dominant. A shutter control system was implemented for automatically taking the multiple images. We quickly found that drift in the dark current required us to take background images immediately after the X-ray exposure. The system was designed so that an image was taken with the shutter open to the X-rays, then the next image was taken with the shutter closed. In this way a pair of images were produced, one image exposed to X-rays and the other as a background, that were close enough in time that there was no observable dark current drift. A black Kapton sheet, 50 μm thick, was used to shield the camera from visible light. The same type shield is used for the camera on the NIF target chamber.

The X-ray beam was characterized geometrically using image plates to optimize collimator and distance choices. The intensity distribution was measured using the CdTe energy dispersive detector at multiple locations across the beam. Multiple images were taken with the CID, and then the detector was placed at the same location as the center of the CID had been located to verify that there was no drift in the X-ray source intensity. The multiple images were analyzed by subtracting each background from the previously taken X-ray image and summing the 100 resulting images. The final image then was effectively a 300 second exposure with the background removed. The measurements concentrated on the X-ray beam center for this initial effort. The CID camera efficiency, counts per pixel per photon, could then be calculated using the CdTe intensity measurements.

The results are shown in Fig. 21. The camera response was measured for two CID cameras at three spectral energies over the range of interest. The responses of the two cameras are the same within the experimental uncertainty. The expected response was modeled using the vendor's specification for camera gain and Si thickness and a typical surface coating. This is shown by the blue line in the figure. This did not fit the measurement data so a second model curve is shown using a thinner Si effective thickness.

The CID camera is now considered to be suitable for the spectrometer operation. The spectrometers will be incorporated as part of existing diagnostics at several locations on the NIF target chamber. All cameras will be calibrated using an extension of the procedure. It will extend to lower X-ray energies using the Manson source and measure the sensitivity variation of the CID over the full pixel array.

7. Conclusion

The chapter started with a presentation of basic X-ray physics needed to follow the description of X-ray detector calibration. The X-ray sources used at NSTec for calibrating detectors were described. The operation and characteristics of solid state semiconductor detectors was presented. Single sensor photodiodes, both current detectors and pulse counters, are used to measure the X-ray source beam intensities. The detectors are calibrated using either of 2 procedures: radioactive sources that are NIST traceable; a synchrotron beam that has an internationally accepted beam intensity accuracy. The chapter presented the methods used and the results obtained for calibrating several types of X-ray cameras.

The accreditation procedure for recognition of the X-ray calibration labs as certified to international standards is in process. This requires the full analysis of all uncertainties associated with the detector calibration. The calibrated photodiode has yet to be completed for the synchrotron calibration. It will then be used to better fill the efficiency curves of the energy dispersive photodiodes. There are several agencies around the world that oversee and certify the accreditation. NSTec will be working with one of them to achieve certification. The NSTec X-ray labs will continually improve existing procedures and develop new methods for calibrating X-ray detection systems and components.

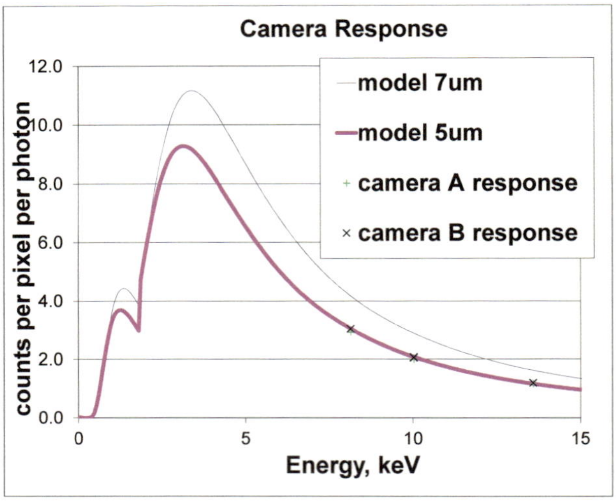

Fig. 21. The measurement results for the CID camera efficiency are shown as the crosses and the plus signs. The curves are model calculations for the CID camera response based on camera characteristics described in the text.

8. Acknowledgment

This manuscript has been authored by National Security Technologies, LLC, under Contract No. DE-AC52-06NA25946 with the U.S. Department of Energy. The United States Government and the publisher, by accepting the article for publication, acknowledges that the United States Government retains a non-exclusive, paid-up, irrevocable, world-wide license to publish or reproduce the published form of this manuscript, or allow others to do so, for United States Government purposes. This manuscript was done under the auspices of the U.S. Department of Energy by Lawrence Livermore National Laboratory under Contract DE-AC52-07NA27344.

There were many persons from both NSTec and LLNL involved in developing the X-ray laboratory calibration methods. I particularly thank Susan Cyr for special effort in putting this manuscript together.

9. References

American Association of Physicists in Medicine (AAPM) (2006). Report No. 93, Acceptance Testing and Quality Control of Photostimulable Storage Phosphor Imaging Systems, available from http://www.aapm.org/pubs/reports/rpt_93.pdf

Carbone, J., Zulfiquar, A., Borman, C., Czebiniak, S., & Ziegler, H. (1998). Large format CID x-ray image sensors, *Proceedings of SPIE 3301*, 90 doi:10.1117/12.304550, Solid State Sensor Arrays: Development and Applications II

Center for X-Ray Optics (CXRO) (n.d.). X-ray interactions with Matter, available from http://henke.lbl.gov/optical_constants/

ESTAR Program (n.d.). Available from http://physics.nist.gov/PhysRefData/Star/Text/ESTAR.html

Gottwald, A., Kroth, U., Krumrey, M., Richter, M., Scholze, F., & Ulm, G. (2006). The PTB high accuracy spectral responsivity scale in the VUV and x-ray range, *Metrologia* 43

Haugh, M. J. and Stewart, R. (2010). *Measuring Curved Crystal Performance for a High Resolution Imaging X-ray Spectrometer*, Hindawi Publishing

Haugh, M.J. & Stewart, R. (2010). X-Ray Optics and Instrumentation, Article ID 583620

Herzberg, G. (1945). *Atomic Spectra and Atomic Structure*, Dover

International Radiation Detectors (IRD) (n.d.). Available from http://www.ird-inc.com/axuvhighnrg.html

Janesick, J. (2000). *Scientific Charge-Coupled Devices*, SPIE Press, Bellingham, WA

Knoll, G. F. (2001). *Radiation Detection and Measurement*, 3rd edition, John Wiley & Sons

Maddox, B. et al (2011). High-energy backlighter spectrum measurements using calibrated image plates, RSI 82, 023111

Marshall, F. J., Ohki, T., McInnis, D., Ninkov, Z., Carbone, J. (2001). Imaging of laser–plasma x-ray emission with charge-injection devices, *Rev. Sci. Instru.* 72 713

Poletto, L., Boscolo, A., & Tondello, G. (1999), Characterization of a Charge-coupled Detector in the 1100-0.14 nm (1 eV to 9 keV) Spectral Range, *Applied Optics*, 38, 1 Jan 99

Physikalisch-Technische Bundensanstalt (PTB) (n.d.). available at http://www.ptb.de/index_en.html

Podgorsak, E. (2010). *Radiation Physics for Medical Physicists* 2nd edition, Springer

Quaranta, C., Canali, G., Ottavani, G. , & Zanio, K. (1969). Electron-hole Pair Ionization Energy in CdTe between 85K and 350K, *Lettere Al Nuovo Dimento*, 4, p. 908-910

Schneider, M.B., Jones, O.S., Meezan, N.B. et al (2010). Images of lthe laser entrance hole from the static X-ray imager at NIF, *Rev. Sci. Instru.* 81 10E538.

Stepanov, S. (1997). X-ray Server, available from http://sergey.gmca.aps.anl.gov/

Stepanov, S. (2009). X0h Program, avalable from http://sergey.gmca.aps.anl.gov/x0h.html

13

The New Photo-Detectors for High Energy Physics and Nuclear Medicine

Nicola D'Ascenzo and Valeri Saveliev
National Research Nuclear University
Russia

1. Introduction

One of the main methods for the detection of the energy of the elementary particles is the conversion of the particle energy into light photons due to the scintillation process and then the conversion of the light photons into the electronic signal due to the photoelectric process (Scintillator/Photo-detector Detection Systems). These detection systems are widely used in high energy physics and currently operated in running experiments, as for example in the *CDF* experiment at Tevatron (CDFII, 1996) and in the *ATLAS* experiment at Large Hadron Collider (ATLAS, 1999). The same method is implemented in Nuclear Medicine and is widely used in clinical practice. Detection systems for Positron Emission Tomography (PET), Single-PET and Gamma Camera are based on scintillators read out by photo-detectors.
The main requirement of such applications is the necessity of the detection of a low photon flux. A typical number of scintillation photons produced by modern scintillating crystals is about 25 photons/keV and about 1 photon/keV is generated by a plastic organic scintillator.
For a long time the main photo-detector for such detection systems was the Photomultiplier Tube (PMT), which was created more than 50 years ago and has many well known disadvantages (Toshikaza et al., 2006).
As alternatives to the PMTs, in the last decade, a new type of photo-detector was developed on the basis of the semiconductor technology, the Silicon Photomultiplier (SiPM) (Golovin & Saveliev, 2004; 2000). At the present time such devices are produced by few companies as Multi Pixel Photon Counters (MPPC), Hamamatsu, Japan (Yamamura, 2009), SensL, Ireland (SensL, 2010)
The SiPM consists of an array of space-distributed micro sensors. Each micro sensor is capable of detecting a single quantum of light - the array is detecting the photon flux. This innovative detecting structure is a great technological improvement in the efficient detection of low photon fluxes. The SiPM is rapidly being used and proposed in many experimental physics and nuclear medicine applications. Few examples give the impression of the extension of use of such detectors: read-out of the fiber/scintillator detectors in the neutrino T2K experiment (Yokoyama, 2009), calorimeter systems at the International Linear Collider (ILD, 2009), CMS hadron calorimeter at the Large Hadron Collider (Freeman, 2009) and others. Many projects are active on the design of PET and Gamma camera using SiPM-crystal detectors (Herbert, 2006).
The aim of this chapter is to show the advantages of using the SiPM for the low photon fluxes detection in scintillator-based high energy physics and medical applications. The examples

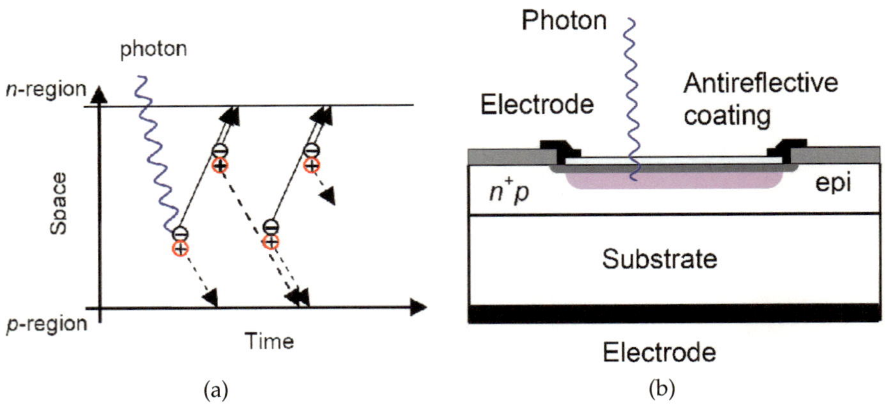

Fig. 1. Schematic of the avalanche process (a) and schematic of the avalanche breakdown micro-cell of a Silicon Photomultiplier n on p type with virtual guard ring (Saveliev, 2010).

of the hadron calorimetry at the International Linear Collider and the new generation PET are considered and the potential of the interplay between mathematical modelling and experimental study is analysed in the design and optimization of such applications.

2. The Silicon Photomultiplier

2.1 Photo-detector structure

The SiPM is a semiconductor-based photo-detector developed for the detection of low photon fluxes. It consists of an array of micro-sensors (microcells), which are designed to detect a single quantum of light with high efficiency. They are based on a special geometry *pn* juntion (Golovin & Saveliev, 2004). Under a reverse bias a depleted area with a high in-built electric field is formed inside the structure. The interaction of a visible photon in the depleted area is mediated through the photo-electric effect with the consequent creation of one electron-hole (e/h) pair. The detection of such small signal is a general problem due to the thermal noise of the detector itself and of the front-end electronics (Alvares-Gaume, 2008). The SiPM has the possibility of the detection of a single photon or single e/h pair through a very high intrinsic gain of the order of 10^6. The amplification is achieved by the avalanche breakdown process due to secondary impact ionization (Tsang, 1985). The schematics of the amplification process in one microcell of the SiPM is shown in Fig. 1a. When the e/h pair is created by a photon interaction, as in the figure, both the generated electron and hole are accelerated in the electric field and reach an energy higher than the ionization energy of the valence electrons and holes in the semiconductor. This initiates a self-sustaining avalanche process. The current rises exponentially with time and reaches the breakdown condition. The avalanche process is stopped via a quenching mechanism obtained by a serial resistor to every microcell. The rising current flows out from the microcell through a quenching resistor causing a voltage drop on the resistor and accordingly to the *pn* junction bias voltage. When the build-in electric field is lowered enough, the avalanche stops. After the quenching, a hold-off time is required to the microcell in order to restore the proper build-in electric field. The resulting intrinsic gain of the microcell is about 10^6 electrons per detected photon, which is well above the noise level of modern measurement electronics. A structure of the Silicon Avalanche Breakdown micro-cell is shown in Fig. 1b. The structure consists of a silicon substrate with a *p*-type epitaxial layer (epi). The avalanche breakdown structure

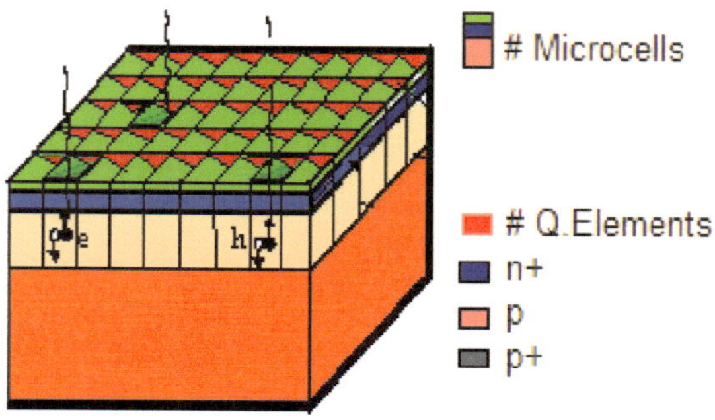

Fig. 2. Equivalent schematic of the structure of the Silicon Photomultiplier (Saveliev, 2010).

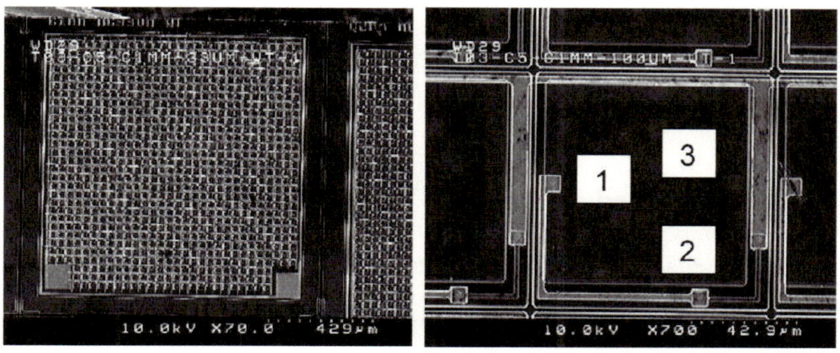

(a) (b)

Fig. 3. Micro image of a modern Silicon Photomultiplier. Overall view 1×1 mm^2 (a) and detailed view of the microcell area (b) (Saveliev, 2010).

is represented by the shallow *pn* junction (n^+p) in the silicon epitaxial layer with the so called virtual guard ring designed to prevent peripheral avalanche breakdown processes. The heavily doped n^+ region is connected to one electrode through a serial quenching resistor. The second electrode is formed on the back side of the substrate. The *pn* junction is designed to reach a very high in-build electric field of the order of 10^5 V/cm within the small thickness of the silicon layer of the order of few microns.
The schematic structure of the modern SiPM is shown in Fig. 2. It consists of an array of the above described *pn* junctions micro-cells (light grey squares) of typical size 30×30 μm^2 on a total sensitive area of few mm^2. Each microcell has the quenching element located close to the *pn* junction (grey and marked as Q element). The microcells are connected in parallel through a common electrode. The sum of signals from the array provides an output signal proportional to the number of detected photons.
The topology of the SiPM is shown on Fig. 3. On Fig. 3a is shown the top view of 1 mm^2 SiPM with micro-cells size about 30×30 μm^2. The total number of micro-cells is 1000 on 1 mm^2. The typical size of Silicon Photomultipliers is 1×1 mm^2 (up to 5×5 mm^2 without significant degradations in performances). In Fig. 3b is presented the microscopic view of a single avalanche breakdown micro-cell where the main elements of the structure are visible: the sensitive area (1), the quenching element (2), a part of the common electrode system (3). The microcells are also optically isolated in order to reduce the probability that optical photons

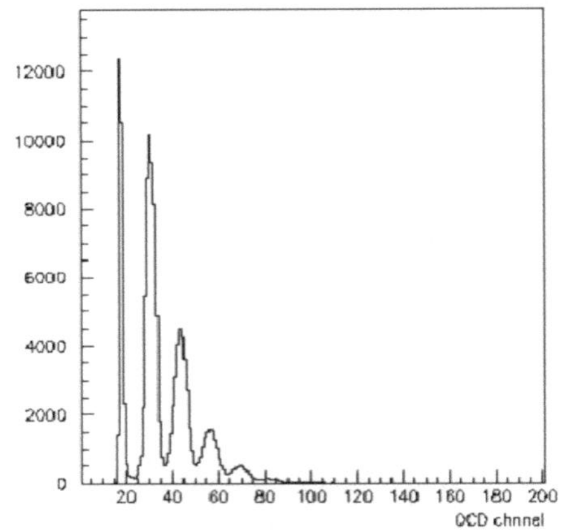

Fig. 4. Spectrum of a low photon flux signal in a SiPM.

produced in the avalanche process initiate an avalanche in neighbouring cells. The optical trenches around each microcell are also visible in the figure.

2.2 Silicon Photomultiplier performance

2.2.1 Single photon detection

The most challenging characteristic of the photo-detectors is the possibility of excellent single photon detection performance. The spectrum of a low photon flux detected with a SiPM is shown in Fig 4. The measurement is performed at room temperature. The resolution of the SiPM allows a precise analysis of the detected photon flux. The structure of the spectrum shows well defined peaks corresponding to the number of detected photons. The first peak corresponds to the noise of the measurement electronics (pedestal). The second peak corresponds to one photon detected, the third peak corresponds to two photons detected and so on. The typical Poisson distribution characterizing the photon statistic describes the spectrum. The SiPM introduces a significant improvement in the possibility of single photon detection in comparison with the traditional photomultiplier tubes (Toshikaza et al., 2006).

2.2.2 Photon detection efficiency of SiPM

The photon detection efficiency (PDE) of the Silicon Photomultiplier could be defined as:

$$PDE = \eta(\lambda) \cdot P_b(V) \cdot F \tag{1}$$

where $\eta(\lambda)$ is the quantum efficiency of the Silicon microcell structure, $P_b(V)$ is the probability of the avalanche breakdown in the silicon microcell structure, F is the filling factor of structure geometry (Saveliev, 2010).

The experimental determination of the photon detection efficiency of the SiPM is usually performed in two steps. First the photo detection probability of a single micro cell is measured relative to a calibrated photo detector with a monochromator light source. Then the result is rescaled to a full area SiPM multiplying by the filling factor, which in modern technologies is within the range 0.6–0.8.

The measured photon detection efficiency of the SiPM is shown in Fig. 5 as a function of the wavelength at 2V above the breakdown voltage. The PDE reaches a peak value of about 20% at around 500 nm. The quantum efficiency of photocathodes used in PMT shows a maximum

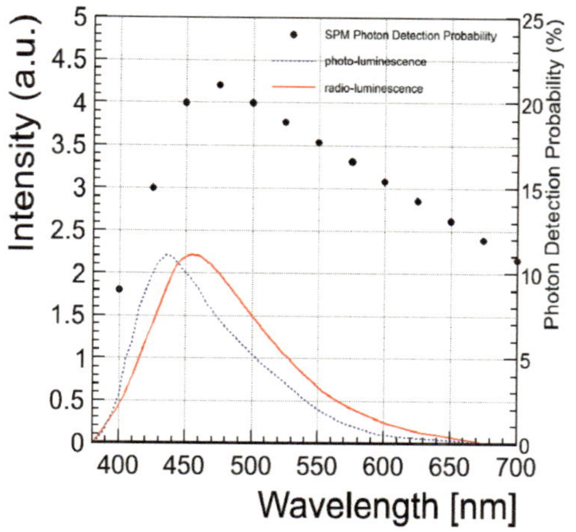

Fig. 5. Photon detection efficiency of the SiPM (black dots) Stewart (2008). Spectra of photo-luminescence (blue dotted line) and radio-luminescence (red continuous line) of a LSO crystal (Mao, 2008).

of 20-30 % within a spectral region between 350 nm and 500 nm (Toshikaza et al., 2006). Improvements are ongoing in the SiPM technology in order to achieve a high photon detection efficiency including the increasing the sensitivity in the blue spectral region. A realistic value of the photon detection efficiency of modern SiPM is $40 - 60\%$.

In the same Fig. 5 the photo-luminescence and radio-luminescence emission spectra of *LSO* are shown according to the reported experimental measurements (Mao, 2008). It is observed that the peak of the emission spectrum is 420 nm for the photo-luminescence and 450 nm for the radio-luminescence. The red shift of the radio-luminescence spectrum is probably due to a higher contribution of the irregular luminescence centre *Ce*2. The photon detection efficiency of the SiPM matches the requirements for the read-out of the scintillation light from *LSO*.

2.2.3 Time performance of SiPM

The time performance of the SiPM is defined by two factors: the rising time of the avalanche breakdown signal and the recovery time, which is defined by the process of reconstruction of the *pn* junction state after quenching the avalanche breakdown process and recharging through the quenching resistor. The rising time is defined by the generation time of the avalanche breakdown process and is characterized by the drift time of carriers under the high electric field. The drift velocity of the carriers under electric field of about 10^5V/cm is limited by the scattering process and in silicon structures it is approximately 10^7 cm$\times$ s^{-1}. As an example for the thickness of a depleted area of 4 microns the rising time is about 30 ps (Saveliev, 2010). The time resolution of the SiPM is measured as 27.54 ps, including the response of measurement system (Stewart, 2008).

2.2.4 Dynamic range and linearity of SiPM

The detection of photons by a silicon Photomultiplier is a statistical process based on the probability of detecting randomly space-distributed photons by the limited number of space-distributed sensitive elements. The photon detection efficiency and the total number of micro-cells determine the dynamic range of the Silicon Photomultiplier. The number of detected photons n_{dph} (number of micro-cells with signal) as function of the number of incident photons can be approximated by the following expression:

$$n_{dph} = N_{mic}\left(1 - e^{-\frac{PDE \cdot N_{ph}}{N_{mc}}}\right) \quad (2)$$

where N_{mc} is the total number of microcells, N_{ph} is the number of incident photons and PDE is the photon detection efficiency.
The Silicon Photomultiplier response is linear when the number of incident photons is much less than the total number of micro-cells. The Silicon Photomultiplier response begins to saturate when the number of fired pixels reaches approximately a quarter of the total number of micro-cells, but could be corrected by well known statistical functions (Saveliev, 2010).

2.2.5 Dark rate of SiPM

One of the main factors limiting the performance of the Silicon Photomultiplier is the dark rate. The dark rate is the frequency of a thermal e/h pair created in the sensitive area of the SiPM. Such e/h pair generates an output signal with amplitude equivalent of a single photon signal and could not distinguished from it. The typical dark rate value for the modern Silicon Photomultipliers is in the range 0.1-1 MHz per mm^2 (Saveliev, 2010).
The amplitude of the dark rate pulses is equivalent to the single photon signal amplitude and in applications dealing with tens-hundred of photons it could be neglected. For applications with very low photon flux the average dark rate can be measured and subtracted. However, the statistical variation in the dark rate cannot be subtracted and constitutes a noise source that determines the minimum detectable signal. As the dark rate of the Silicon Photomultiplier scales as its area and the acceptable dark rate is about 10^6 in low photon flux, the maximum designable area is limited to around few mm^2.

3. Recent advances of scintillator/SiPM detection systems in high energy physics

3.1 The scintillator/SiPM detection system for highly granular hadron calorimetry

A modern concept of high energy physics detection systems is the particle flow algorithm (PFA) (Thompson, 2007). This method is proposed for the experiment at International Linear Collider (ILD, 2009). Instead of performing a pure calorimetric measurement, as in traditional environments, the reconstruction of the four vectors of all the observable particles in the jet is proposed. The reconstructed jet energy is the sum of the energy of the individual particles. The momentum of the charged particles is measured in the tracker, while the energy of the neutral particles is measured in the calorimeters. The electromagnetic calorimeter is used for the measurement of the energy of photons and for the identification of photons and electrons. The hadronic calorimeter is used for the measurement of the energy of neutral hadrons and for the identification of hadrons. The muon chambers are used for the identification of muons. A detector optimized for the particle flow should have an excellent separation power of the components of the jets. The most important features in this respect are the spatial separation of the particles in the high energy jets, which is achieved with a high magnetic field, and high space resolution systems including the calorimeter systems. The key point of the hadron calorimetry designed for the particle flow technique is the granularity. It allows in fact to identify the single particles through the morphological properties of the shower. The calorimeter becomes an imaging device more than an energy measurement device.
Mathematical modelling studies were performed for the optimization of the performances of

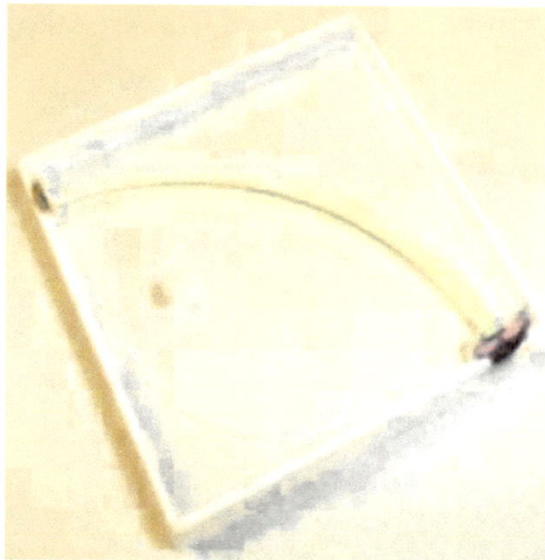

Fig. 6. Design of the scintillator/SiPM detection system for hadron calorimetry. The green sensitive SiPM is coupled to the scintillator through a wavelength shifter fiber (CALICE, 2010).

the hadron calorimeter for the International Linear Collider. It is shown that a transverse segmentation of 3×3 cm^2 and a longitudinal segmentation of about 1 cm satisfies the requirements in the experiment. Such modern performance calorimeter system, especially the hadron calorimeter, could be developed only on the basis of new technologies (ILD, 2009). One of the solutions is to use the modern scintillator/SiPM detection system.

The first proposed design of the scintillator/SiPM detection system consists of a $3 \times 3 \times 0.5$ cm^3 plastic organic scintillator tile read-out individually by a Silicon Photomultiplier (Fig. 6). The coupling between the scintillator and the photo-detector is performed via a wavelength shifter fiber. The scintillator/photo-detector system was optimized to yield about 15 photons on average in response to a minimum ionizing particle.

A simplification of the coupling between SiPM ad scintillator would be highly desirable, in order to extend the concept to a large scale detector. The new generation of SiPMs produced by Hamamatsu (MPPC) shows a better optical sensitivity in the 420 nm spectral region, making it possible to investigate the direct read-out of the scintillation tile (D'Ascenzo et al., 2007).

3.2 Mathematical model of a highly granular calorimeter based on scintillator/SiPM with individual read-out

In order to estimate and optimize the performance of the detection system with SiPM, a mathematical model of a scintillator/SiPM calorimeter is performed on the basis of the GEANT4 simulation framework. It allows to include the geometry and the physics processes in response to any final states resulting from the studied high energy particle collisions. The result of the simulation has maximal flexibility and can be studied with the same reconstruction and statistical analysis techniques developed for the application to real data.

The mathematical model includes a detailed geometrical description of the full detector system. The components of the mathematical model are shown in Fig. 7. In the barrel region the detector components are the vertex detectors (VTX, SIT), the tracker (TPC, SET), the electromagnetic and hadronic calorimeters (ECAL, HCAL) and the return yoke with muon system (YOKE). In the forward region the forward tracking detectors (FTD, ETD) the luminosity (LCAL, LHCAL) and veto detectors (BCAL) are included.

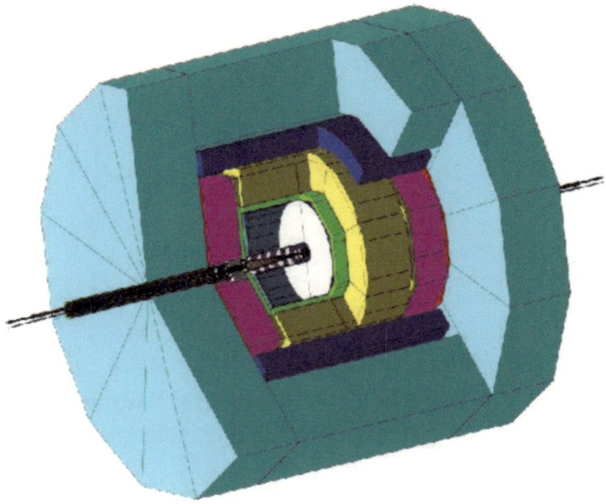

Fig. 7. Mathematical model of the detector for the International Linear Collider. The overview of the detector components is shown. From the inside to the outside, the detector components are the vertex detectors (VTX, SIT), the tracker (TPC, SET), the electromagnetic and hadronic calorimeters (ECAL, HCAL) and the return yoke with muon system (YOKE). In the forward region the forward tracking detectors (FTD, ETD) the luminosity (LCAL, LHCAL) and veto detectors (BCAL) are included (ILD, 2009).

The hadronic calorimeter consists of one barrel and two end-cap modules. The barrel module has octagonal shape with inner and outer radius respectively of 2.02 m and 3.33 m. The end cap modules has also octagonal shape and has a longitudinal thickness of 1.3 m, inner and outer radius of 329 cm and 3.33 m. Each module is composed of 40 layers in an alternating structure of 0.5 cm thick plastic scintillators as active material and 1 cm thick stainless steel as absorber.

The detailed geometry of the scintillator/SiPM detection system is introduced in the simulation. Each scintillator layer is segmented in 3×3 cm^2 scintillator tiles, individually read out by a SiPM. The total amount of calorimeter cells is about 10^6. The energy deposited in each cell is independently calculated and stored for further analysis.

The response of the scintillator/SiPM detection system is introduced with a parametric model based on the results of the experimental study of the hadron calorimeter system prototype (CALICE, 2010).

The detector is immersed in a magnetic field of 3.5 T.

The reconstruction of the particles in the final state from the simulated detector response is performed with a reconstruction software based on the Particle Flow Algorithm.

The dependence of the jet energy resolution on the size of the sensitive cells of the hadronic calorimeter is shown in Fig. 8. The energy resolution of jets with energy as low as 45 GeV is independent from the size of the HCAL cell, while higher energetic jets show a stronger dependence on the granularity of the hadronic calorimeter. This effect is due to the increased complication of the structures in the hadronic calorimeter in response to higher energy jets.

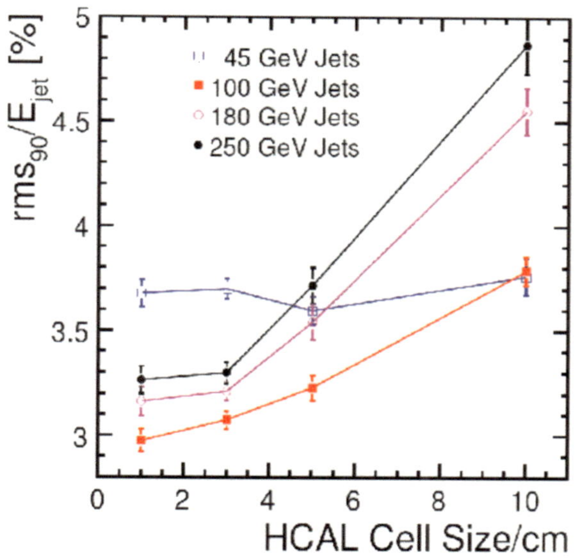

Fig. 8. Monte Carlo estimation of the dependence of the jet energy resolution on the sensitive element size of the highly granular hadronic calorimeter based on scintillator/SiPM detection system with individual read-out (ILD, 2009).

Consequently the best possible granularity is needed in order to identify each contribution. The result indicates that the ILC jet energy resolution goal is achieved with a 3×3 cm^2 scintillator tile segmentation (ILD, 2009).

3.3 Experimental study of the prototype of the highly granular hadron calorimeter based on scintillator/SiPM with individual read-out

The experimental study of the new detection system on the basis of scintillator/SiPM photo-detectors for application in highly granular hadron calorimeters was performed on a prototype of hadron calorimeter. The prototype consists of a sampling structure alternating 2 cm thick absorber steel plates with 0.5 cm thick sensitive layers. It has a total surface of 90×90 cm^2 and consists of 38 layers, for a total length of 5 λ_0.

Each sensitive layer is a array of 216 scintillators. The 30×30 cm^2 core has a granularity of 3×3 cm^2 and the outer region is equipped with tiles of increasing size — 6×6 cm^2 and 12×12 cm^2. Each scintillation tile is read-out individually by a SiPM (Fig. 6). The coupling between the scintillator and the photo-detector is performed via a wavelength shifter fibre arc (Kurakay WLS fiber Y11(200)). The fibre is inserted in a groove carved directly in the tile. A mirror is placed on one side of the tile in order to minimize the light losses along the fibre. The photo detector is installed directly on the tile coupled to a WLS fiber. A 3M reflector foil is applied on the surface of the tiles. The sensitive layer is housed in a steel cassette, with 2 cm thick rear and front plates. The calibration of the photo-detectors with light is done with a LED/clear fibre system. The fibres are embedded with the sensitive cassette itself and a proper electronic board controls the LED system (CALICE, 2010).

The electronic read-out of the signals of the Silicon Photomultipliers is made by the special electronics ILC-SiPM chip with 18 channels, each composed of a variable-gain,

Fig. 9. Set-up of the detectors in the CERN test beam. The AHCAL was tested in combination with a prototype of highly granular Silicon/Tungsten electromagnetic calorimeter (ECAL) and a strip-scintillator/steel Muon Tracker Tail Catcher (TCMT) (Behnke, 2007).

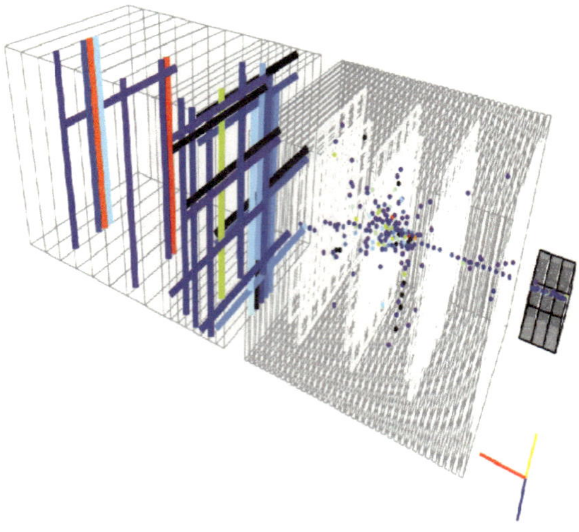

Fig. 10. Pion shower (12 GeV) identified in the data (CALICE, 2010).

charge-sensitive amplifier, a variable shaper, track and hold stage and a multiplexing unit (Blin, 2006).

The analog hadron calorimeter (AHCAL) was tested in combination with the prototypes of the highly granular silicon-tungsten electromagnetic calorimeter (ECAL) and of the scintillator strip tail catcher (TCMT) at the H6 beam line of the CERN SPS facility. The experimental

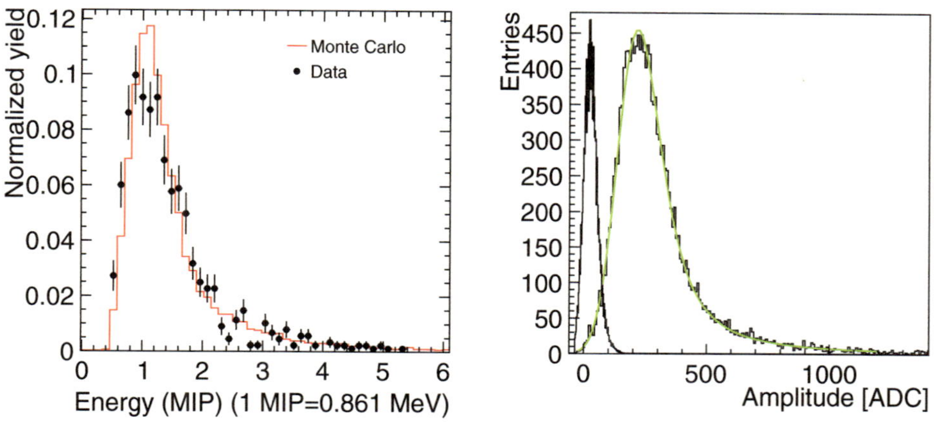

(a) (b)

Fig. 11. Response spectrum of a scintillator/SiPM detection system to muons in the hadronic calorimeter prototype (CALICE, 2010; D'Ascenzo, 2009).

setup is shown in Fig. 9.

The 450 GeV proton beam is used on a Beryllium target in order to generate a secondary beam of pions with a wide momentum spectrum in the range between 30 GeV/c and 205 GeV/c. In addition, a muon beam is also available due to contamination of the secondary pion beam. The average muon energy is hence approximately $0.8 \cdot E_{\pi}$.

A mathematical model of the test setup system based on the GEANT4 simulation framework is also implemented on the basis of the mathematical model of the full detection system. The simulation includes all the detailed components of the test beam experimental setup.

The first important goal of experimental study is the verification of the efficient detection of the high energetic particles, minimum ionizing particle (m.i.p.), as required by the PFA concept. As an example a 12 GeV pion shower identified in the data is shown in Fig. 10. Furthermore muons produced in the hadron shower are also identified as straight tracks which escape from the calorimeter and penetrate the tail catcher (D'Ascenzo, 2009). The study of the response to muons, which mainly deposit energy by the ionization process in the massive volume of matter, could give a good experimental evidence.

Fig. 11a shows the signal of a single calorimeter scintillator cell read-out by a SiPM produced by 120 GeV muons. On the same plot is presented the Monte Carlo result including the systematic effects of the detector. The experimental results are well described by the mathematical model.

The resolution of the m.i.p. signal in a scintillator cell of the hadronic calorimeter depends on the statistical effects of the photon detection. The poisson fluctuation of the number of photo-electrons ($N_{p.e.}$) generated in the SiPM is the main source of the smearing of the signal. Its effect on the resolution of the visible energy depends on the $\sqrt{N_{p.e.}}$; the most probable value of the m.i.p. signal is 861 keV and corresponds to 15 ± 3 photo-electrons, with a consequent relative statistical fluctuation of $\sqrt{15}/15 = 25\%$. Moreover, the poisson smearing doesn't affect the energy deposited in the single cell uniformly. According to a simulation of the energy response of the single AHCAL cell to muons, a Landau distribution with Most Probable Value at 861 keV and width 60 keV approximates the energy deposited in

the scintillator. The resolution is $60/861 \sim 5\%$.
The muon signal measured in the data can be fitted with a Landau distribution convoluted with a Gaussian distribution, which models the smearing of the detector read-out. The result of the fit of the response of a single cell to a 120 GeV muon is shown in Fig. 11b. The energy resolution of the m.i.p. signal is about 70% but the signal is well distinguished from the noise pedestal. In the full prototype an average S/N separation of about 9 is measured (CALICE, 2010).

4. Recent advances of scintillator/SiPM detection systems in nuclear medicine

4.1 The scintillator/SiPM detection system in Positron Emission Tomography

Positron Emission Tomography is a powerful functional imaging modality that provides dynamic, quantitative information on the biological characteristics of tumours and other tissues. While PET has mainly found clinical application in oncology, uses in cardiology, neurology and neuropsychiatry are expected to increase in the future. Recent studies showed the potential of PET for the measurement of tissue activation and perfusion in specific diseases, as brain neurological perfusion in Alzheimer and autism or hearth activation study in case of myocardial infarction (Boddaert & Zilbovicius, 2006; Buchsbaum, 2006).
It is required to develop various PET systems with significantly better performance than commercially available scanners, in particular concerning spatial resolution for earlier cancer detection and more accurate staging. Also the PET camera needs higher sensitivity to reduce scanning time, cost and patient exposure to radiation, good time resolution, operation at high magnetic fields for a combination with Magnetic Resonance Techniques and design flexibility. The detection system of PET is the key point which defines the main performance of the medical imaging systems and which is triggering the new clinical applications and new developments in molecular and cell biology. The modern advances in the SiPM development made it possible to develop a new type of scintillation crystals/SiPM detection system for application in Positron Emission Tomography.
The miniature size and the low material budget of SiPMs give the possibility to build flexible PET detection systems and include complementary methods for improving the performance. This feature is referred to as the depth of interaction (DOI) problem. The measurement of the DOI is realised quite simply with SiPMs and will improve imaging quality. The excellent time resolution of SiPMs and of the new scintillators gives the possibility of using the Time of Flight methods with a significant improvement of the signal to noise ratio of PET images. The effect on PET would be the ability to reduce the coincidence timing window by one order of magnitude. This would not only result in improvements in the noise equivalent counts (NEC) through the reduction in randoms, but also provides the ability to perform time-of-flight PET reconstruction. With a timing resolution of less than 0.5 ns, it becomes possible to define the site of positron annihilation within a line segment of less than 7.5 cm, and thereby to improve the reconstruction.

4.2 Mathematical model of a PET scanner based on LSO/SiPM detectors with individual read-out of crystals

In order to estimate the possibility to achieve the mentioned goals, a mathematical simulation study of a PET scanner with LSO crystals individually read-out by a SiPM is performed. The mathematical model for the LSO/SiPM detection system is developed on the basis of the GATE framework, which allows to include the geometry and the physics processes and also to perform the reconstruction by standard methods for the performance study (Strul, 2003).

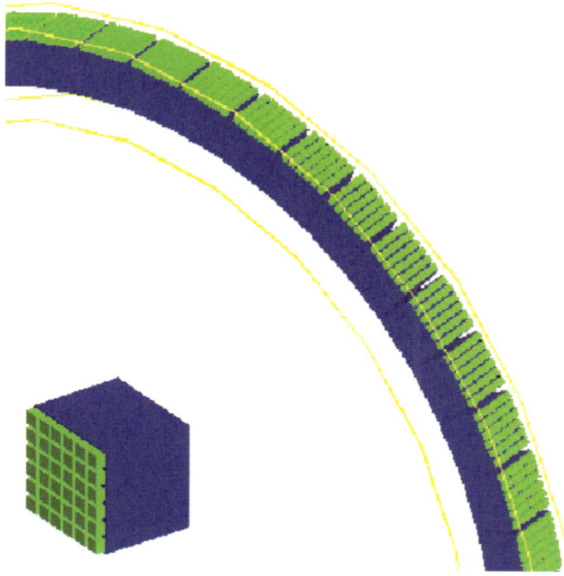

Fig. 12. Detailed geometry of the PET detection system on the basis of LSO scintillator crystal read-out individually by SiPM.

A detailed geometrical configuration of a detector ring for a PET scanner based on the LSO/SiPM detection system is shown in Fig. 12. One ring of 53.3 cm diameter is composed of detection modules placed around the axis in a cylindrical symmetry. The size of the system is typical of the state of the art high resolution brain PET scanners (Karp et al., 2003).
According to the NEMA NU2-2001 performance protocol (National Electrical Manufacturers Association, 2001) the source configuration used for the estimation of the space resolution is the β^+ emitter ^{18}F, arranged in a glass spherical capillary with internal and external radius respectively of 0.2 mm and 0.3 mm.. The initial activity is 10000 Bq.
Each detector module consists of a 6×6 array of LSO/SiPM cell. As an example, in case of $3 \times 3 \times 25$ mm^3 crystals, the crystals pitch is 3.1 mm and the size of one detector module is $18.6 \times 18.6 \times 2.5$ mm^3. The ring is composed of 85 modules with an angular pitch of 4.23°. LSO crystals are covered by a reflecting layer of Teflon, with the correct description of the physical and optical properties. The geometrical acceptance and the optic coupling of the crystals with the SiPM are included according to experimental estimations.
Light propagation and collection on the face of SiPMs are also included in the physics processes. The Photon Detection Efficiency of the SiPMs used in the simulation is shown in Fig. 5 and is reported from experimental measurements (Stewart, 2008).
The energy deposited in each crystal is calculated in the simulation and is converted into a photon flux via the scintillation processes. The scintillation photons are produced as gaussian distributed with a mean value (LY) of 27000 photons/MeV (Melcher, 1992) and a variance σ_{sc} equal to the expected Poisson statistic variance multiplied by a scale factor: $\sigma_{sc} = \alpha_s \sqrt{LY \times E_\gamma}$, where E_γ is the energy of the detected photon. The scale factor $\alpha_s = 4.41$ models the intrinsic not-linearity of LSO. The photon yield of each crystal is read-out independently by a SiPM and the detected light output of each SiPM is calculated.
The timing performance is included in the simulation as the scintillation process time dependence and the light propagation. The intrinsic time resolution of the SiPM is also

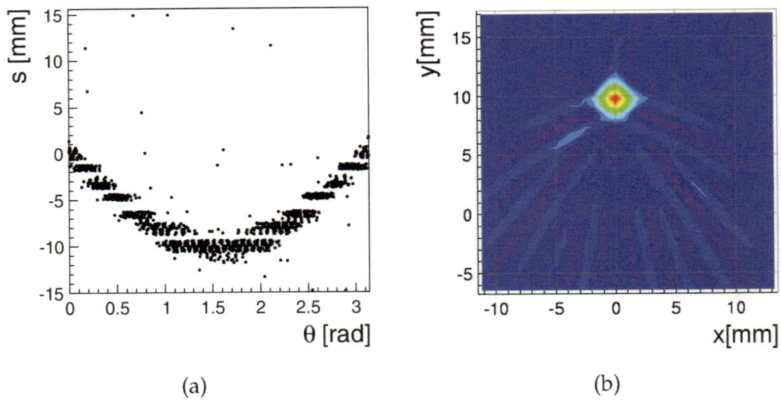

Fig. 13. Sinogram (a) and reconstructed image (b) resulting from the simulation of the response of the PET system based on LSO/SPM detectors to a ^{18}F source of 0.2 mm radius placed at a vertical distance of 1 cm from the centre of the tomograph. The detection module is composed of a 6×6 array of $3 \times 3 \times 25$ mm^3 LSO crystals.

considered in the simulation.

The coincidence condition is defined as two events in two opposite crystals with deposited energy within $\pm 3\sigma$ around the photo-peak and within a coincidence time window of 80 ns.

The reconstruction of the Lines Of Response (LOR) is performed by using the position of the centre of the two crystals found in coincidence. The sinogram is constructed from the LORs, without applying any rebinning or geometrical correction. A standard filtered backprojection algorithm FBP2 with Hammer filtering is applied to the sinogram for the reconstruction of the original image and for the study of the spatial resolution. The sinogram resulting from the simulation of the response of the PET system is shown in Fig.13a. As any rebinning is applied, the structure of the LSO array composing the detector block is visible. The reconstructed image is shown in Fig.13b. The transverse spatial resolution is estimated as $\sigma_x = (0.94 \pm 0.62)$ mm and $\sigma_y = (0.87 \pm 0.46)$ mm. The estimated average transverse resolution (FWHM) is (2.13 ± 1.26) mm. The axial resolution depends uniquely on the ring thickness. In this example case of a detecting module consisting of a 6×6 array of $3 \times 3 \times 25$ mm^3 LSO crystals, the ring thickness is 18.6 mm. The corresponding axial resolution is estimated as about $18.6/3.0 = 6.2$ mm.

The results of the study are shown on Fig. 14. The space resolution is studied of PET systems based on 6×6 arrays of $3 \times 3 \times 25$ mm^3, $4 \times 4 \times 25$ mm^3 and $5 \times 5 \times 25$ mm^3 individually read-out LSO crystals. The transverse space resolution (FWHM) ranges between about 2 mm and 4 mm. For a comparison with results reported in literature, a transverse spatial resolution of 4 mm was measured for a high resolution brain PET scanner based on an Anger-logic detector array with 4×4 mm^2 GSO crystals (Karp et al., 2003). The single crystal read-out introduces hence a sensitive improvement with respect to the traditional Anger-logic based PET systems. The axial resolution ranges between about 6 mm and 10 mm. These value refer to the ring thickness calculated using a 6×6 array of LSO crystals. The flexibility of the LSO/SPM detection system allows to optimize the ring thickness according to the specific clinical needs of the tomograph, resulting in lower or higher axial resolution.

The mathematical simulation shows a significant improvement of the performances and flexibility of the PET detection systems based on scintillator/SiPM detection systems.

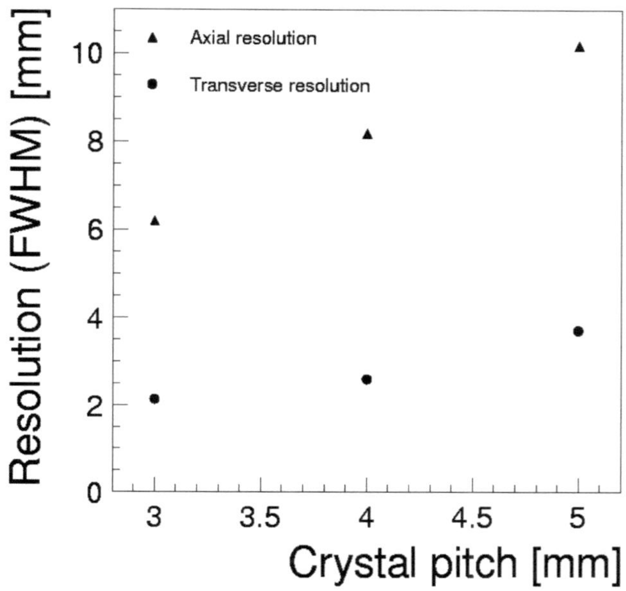

Fig. 14. Monte Carlo estimation of the transverse (dots) and axial (triangles) space resolution of PET systems based on 6×6 arrays of $3 \times 3 \times 25$ mm^3, $4 \times 4 \times 25$ mm^3 and $5 \times 5 \times 25$ mm^3 individually read-out LSO crystals. The space resolution is shown as a function of the crystal pitch.

4.3 Experimental study of the prototype of the PET detection system based on the LSO/SiPM detectors

The experimental study of the new detection system on the basis of LSO/SiPM photo-detectors for applications in medical imaging systems was performed on a prototype of PET detection system. The prototype consists of two LSO crystals coupled to a SiPM and positioned opposite to each other at 180°. The experimental setup is shown in Fig. 15a. The scintillator crystals used in this study are two $2.5 \times 2.5 \times 15$ mm^3 LSO crystals wrapped in two layers of 1.25 mm thick Teflon films. The crystals are fixed to two mechanical holders (plastic) and are positioned opposite to each other on an optic bench in a light tide environment. The distance between the LSO crystals is 1 cm in order to increase the acceptance angle for the efficient collection of the statistics. A SiPM is coupled to the surface of the LSO crystals without any optics coupling material. The SiPMs used in the test setup are 1 mm^2 Silicon Photomultiplier SPM, produced by SensL (Stewart, 2008).
The SiPM signals are read out on 50 Ω load resistors directly by 4 GHz Oscilloscope (Textronix TDS7404B) without any front end electronics. The signals is digitized with a sampling rate of 20 Gs/s, which corresponds to 100 ps time digitalising periods for two channels and 50 ps shift between the two signals.
A point-like positron source ^{22}Na is placed in the middle and aligned with the line of centers crystals connection. It is held by a thin plastic cylindrical support with 2 cm diameter and 2 mm thickness.
The digitized signal of the two SiPMs in coincidence correspondent to two 511 keV gamma quanta is shown in Fig. 16. The signal has typical amplitude of about 100 mV. The rise

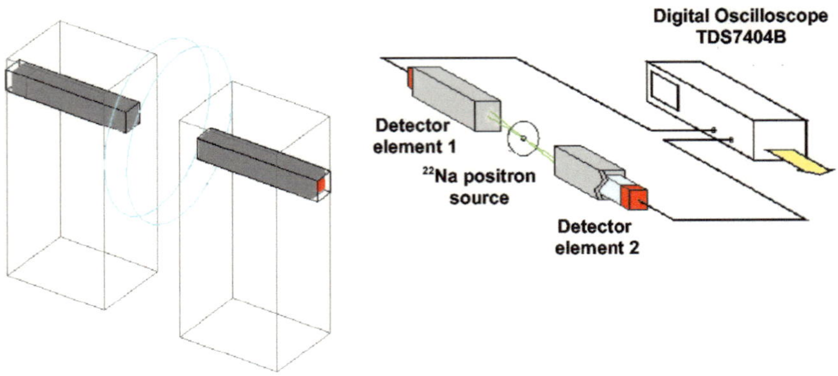

(a) (b)

Fig. 15. Mathematical model (a) and experimental setup (b) for the analysis of two LSO/SiPM (blue/red) system.

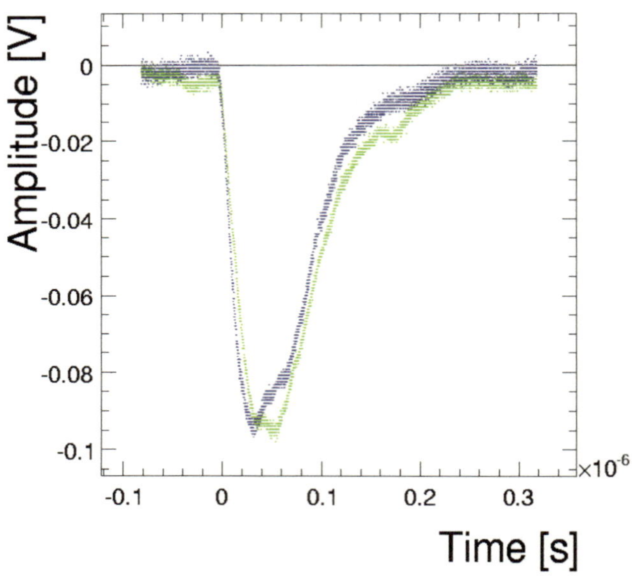

Fig. 16. Example of digitized signal of the two SiPMs (blue and green) when the annihilation photons from the ^{22}Na are detected in coincidence in the two opposite LSO crystals in the experimental setup.

time is 28 ns at the levels 10%-90%. The decaying component of the signal follows an exponential distribution with typical decay time of about 60 ns. The fully digitized signal gives a unique possibility to use powerful mathematical tools for the analysis of the main

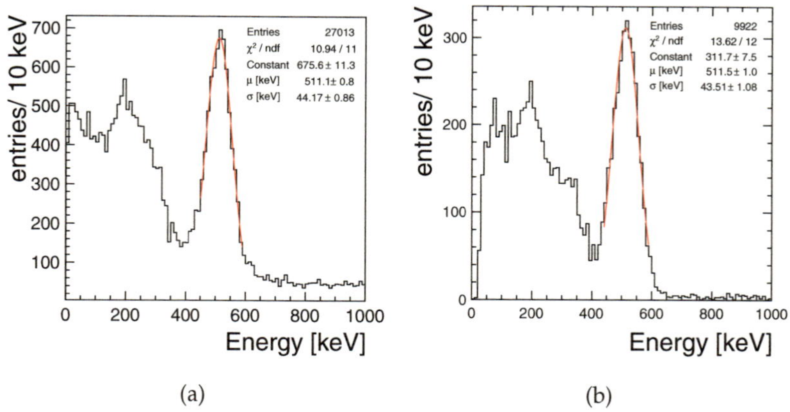

Fig. 17. Monte Carlo (a) and experimental data estimation (b) of the energy resolution in the experimental setup.

characteristics of the detection system based on the LSO/SiPM and for the precise verification of the mathematical model.

4.3.1 Energy resolution of the LSO/SiPM detection system

The energy spectrum measured in the test setup is shown in Fig. 17b. The energy deposited in the LSO crystal (number of photons detected in the SPM) are calculated as the integral of the output signal. The integration is performed in an off-line analysis of the stored digital waveforms of the two SPM signals. The typical features of a γ-ray spectrum can be individuated: the photoelectric-peak at the energy of incident photons (511 keV), the Compton continuum extending from the photo-electric peak down to the instrumentation threshold and the back-scatter peak at around 200 keV, due to the Compton interaction of the incident photon in the material around the crystal.

The energy resolution of the LSO/SiPM detection system for PET is defined in the region of the photoelectric peak as $R \equiv \left(\frac{\sigma}{511\ \mathrm{keV}}\right)$, where σ is the total variance and 511 keV is the mean value of the photo-electric peak.

The experimental energy resolution at the photo-electric peak is estimated with a gaussian fit as $R = (8.51 \pm 0.23)\,\%$.

The energy resolution of a LSO/SiPM system could be described by the total variance σ as the sum in quadrature of five independent contributions:

$$\sigma = \sigma_{LSO} \oplus \sigma_{stat} \oplus \sigma_{pdf} \oplus \sigma_{opt} \oplus \sigma_{el} \tag{3}$$

The intrinsic variance of the scintillation photons generated in the LSO is represented by σ_{LSO}. According to the experimental estimations reported in section 4.2, it corresponds to a resolution of $R_{LSO} = \frac{4.41\sqrt{LY \times 0.511}}{LY \times 0.511} = 3.76\%$.

The contribution σ_{pde} describes the broadening effect caused by the not uniform detection efficiency in the spectral range of the scintillation emission. It is estimated as $\sigma_{pde}\,[\mathrm{keV}]\,/511\,[\mathrm{keV}] = (3.77 \pm 0.54)\%$ for the combination of LSO/SiPM with the radio-luminescence spectrum and photon detection efficiency.

The impact of the reflection properties of the Teflon adds to the overall variance as an independent constant term σ_{opt}.

The optical transmission contribution of the experimental setup is estimated with the Monte Carlo as $\sigma_{opt}\,[\text{keV}]\,/511\,[\text{keV}] = (2.78 \pm 0.05)\,\%$. Dedicated experimental estimation of this contribution is also reported in the literature (Herbert, 2006).
The noise of the read-out electronics contributes to the total variance with a constant term σ_{el}. It is estimated from the experimental data as $\sigma_{el}\,[\text{keV}]\,/511\,[\text{keV}] = (1.68 \pm 0.11\%)_{el}$.
The binomial photo-statistics of the detection of the scintillation photons in the SiPM is included in the term σ_{stat}.
The detailed analysis of σ_{stat}, σ_{LSO} and σ_{pdf} is performed in analytical form with a statistical model, taking into account the photo-statistics of the generation and propagation of the optical photons in the crystal, the detection in the SiPM and the optical properties of the detection system.
The probability distribution $P(n)$, which describes the number of photons n detected in the SiPM if a γ-ray is detected in the LSO crystal, is expressed as:

$$P(n) = \int\int \frac{1}{\sqrt{2\pi\sigma_{sc}^2}} e^{-\frac{\left(N_{ph}-N_{LY}\cdot E_\gamma\right)^2}{\sigma_{sc}^2}} \times \\ \times \frac{1}{\sqrt{2\pi N_{ph}\alpha\cdot\epsilon(\lambda)(1-\alpha\cdot\epsilon(\lambda))}} e^{-\frac{\left(n-\alpha\cdot\epsilon(\lambda)N_{ph}\right)^2}{2N_{ph}\alpha\cdot\epsilon(\lambda)(1-\alpha\cdot\epsilon(\lambda))}} dN_{ph} P(\lambda)\, d\lambda \quad (4)$$

where:

- N_{LY} is the light yield of LSO (27000 photons/MeV).
- E_γ is the energy of the detected $\gamma-$ray. In this study $E_\gamma = 511$ keV.
- $\sigma_{sc} = 4.41 \cdot \sqrt{N_{LY} \cdot E_\gamma}$ is the intrinsic resolution of the LSO crystal for energy E_γ.
- $P(\lambda)$ is the radio-luminescence spectrum of LSO (Fig. 5).
- $\epsilon(\lambda)$ is the photo-detection efficiency of the SiPM (Fig. 5).
- α is the geometrical photon collection efficiency, which takes into account the photon losses due to the not perfect reflectivity of the crystal/Teflon surfaces. It depends on the geometry of the crystal and of the size of the SiPM.

The mean value of the detected photons $\bar{n}$ is from Eq. 4:

$$\bar{n} = \int n \cdot P(n)\, dn = \int \alpha \cdot \epsilon(\lambda) \cdot LY \cdot E_\gamma P(\lambda)\, d\lambda = \alpha \cdot \bar{\epsilon} \cdot LY \cdot E_\gamma \quad (5)$$

The second moment of the number of detected photons $\langle n^2 \rangle$ is:

$$\langle n^2 \rangle = \int n^2 \cdot P(n)\, dn = \\ = \alpha \cdot \bar{\epsilon} \cdot LY \cdot E_\gamma - \alpha^2 \cdot \langle \epsilon^2 \rangle \cdot LY \cdot E_\gamma + \alpha^2 \langle \epsilon^2 \rangle \sigma_{sc}^2 + \alpha^2 \langle \epsilon^2 \rangle \cdot LY^2 \cdot E^2 \quad (6)$$

where the quantities are defined:

$$\bar{\epsilon} = \int \epsilon(\lambda) P(\lambda)\, d\lambda \qquad \langle \epsilon^2 \rangle = \int \epsilon^2(\lambda) P(\lambda)\, d\lambda \qquad \sigma_\epsilon^2 = \langle \epsilon^2 \rangle - \bar{\epsilon}^2 \quad (7)$$

The quantities $\bar{\epsilon}$ and σ_ϵ represent the mean value and the total spread of the photon detection efficiency weighted over the radio luminescence spectrum of the LSO. The variance of the detected photons $\sigma^2 = \langle n^2 \rangle - \bar{n}^2$ is :

$$\alpha^2 \bar{\epsilon}^2 \sigma_{sc}^2 + LY \cdot E_\gamma \cdot \alpha \cdot \bar{\epsilon}\,(1 - \alpha \cdot \bar{\epsilon}) + \sigma_\epsilon^2 \left[\sigma_{sc}^2 + LY \cdot E_\gamma \cdot (LY \cdot E_\gamma - 1)\right] \quad (8)$$

The analytic formula for σ_{LSO}, σ_{stat} and σ_{pdf} is extracted from Eq. 8:

$$\begin{aligned} \sigma^2_{LSO} &= \alpha \cdot \bar{\epsilon}^2 \sigma^2_{sc} \\ \sigma^2_{stat} &= LY \cdot E_\gamma \cdot \alpha \cdot \bar{\epsilon} \, (1 - \alpha \cdot \bar{\epsilon}) \\ \sigma^2_{pde} &= \sigma^2_\epsilon \left[\sigma^2_{sc} + LY \cdot E_\gamma \cdot (LY \cdot E_\gamma - 1) \right] \end{aligned} \tag{9}$$

The performance of the LSO/SiPM is estimated with the mathematical model of the test setup. The best achievable energy resolution of a $2.5 \times 2.5 \times 15$ mm^3 LSO crystal is calculated in the case the crystal is read-out over the full area at one side by a perfect detector with photon detection efficiency equal to 1 over the whole LSO emission spectral range. The response of the crystal is simulated to a monochromatic 511 keV photons directed to the centre of the crystal. The energy resolution at the photo-electric peak is estimated as $R = (4.73 \pm 0.06)\,\%$, which corresponds to a total number of about 8100 photons. The result can be interpreted using the statistical model in Eq. 3, with the values $\bar{\epsilon} = 1$, $\sigma_\epsilon = 0$, $\alpha = 8100/(27000 \cdot 0.511) = 0.587$ and $\sigma_{el} = 0$:

$$(4.73 \pm 0.06)\,\% = (3.76\%)_{LSO} \oplus (0.71 \pm 0.01)_{stat} \oplus (2.78 \pm 0.05\%)_{opt} \tag{10}$$

The scintillator/SiPM detection system has the potential to reach the intrinsic energy resolution of the scintillator itself. This estimation is in fact in good agreement with reported experimental results, where an energy resolution of $(4.24 \pm 0.01)\,\%$ is obtained with a $3 \times 3 \times 15$ mm^3 LSO crystal read-out over the whole 3×3 mm^2 area by a SiPM (D'Ascenzo et al., 2007).

The energy spectrum calculated with the mathematical model corresponding to the conditions of the experimental measurements is shown in Fig. 17a. The typical features of a γ-ray spectrum can be individuated. An average number of 254 detected photons corresponding to the photoelectric peak is calculated in the mathematical model. The energy resolution is estimated with a gaussian fit around the photoelectric peak as $(8.64 \pm 0.18)\,\%$. The result of the mathematical model estimation is interpreted according to the analytic model in Eq. 3, with $\alpha\bar{\epsilon} = 254/(27000 \cdot 0.511)$:

$$\begin{aligned} (8.64 \pm 0.18)\,\% \approx\; & (3.76\%)_{LSO} \oplus (6.21 \pm 0.06\%)_{stat} \oplus \\ & \oplus (3.77 \pm 0.54\%)_{qpd} \oplus (2.78 \pm 0.05\%)_{opt} \end{aligned} \tag{11}$$

where σ_{opt} and σ_{LSO} are determined as described above, σ_{stat} is determined from direct calculation using Eq.9 and σ_{qpd} is determined from a subtraction in quadrature.

The measured energy resolution can be decomposed similarly in the independent components according to Eq. 3:

$$\begin{aligned} (8.51 \pm 0.23)\,\% \approx\; & (3.76\%)_{LSO} \oplus (5.79 \pm 0.74\%)_{stat} \oplus \\ & \oplus (3.77 \pm 0.54\%)_{qpd} \oplus (2.78 \pm 0.05\%)_{opt} \oplus \\ & \oplus (1.68 \pm 0.11\%)_{el} \end{aligned} \tag{12}$$

where σ_{el}, σ_{opt}, σ_{LSO} and σ_{qpd} are determined in Eq. 11 and σ_{stat} is determined subtracting in quadrature all the determined components from the overall measured resolution.

The experimental data are well described by the mathematical model and the results in Eq. 11 and 12 are in good agreement. This proves the accuracy of the mathematical model of the PET detection system on the basis of LSO/SiPM individual read-out.

The SiPMs used in the experimental set-up have a 1×1 mm^2 active area which is smaller than the crystal surface. Although their average photon detection efficiency in the LSO emission spectral region is around 20% (Fig. 5), the small active area limits the overall photon collection efficiency of the LSO/SiPM system.

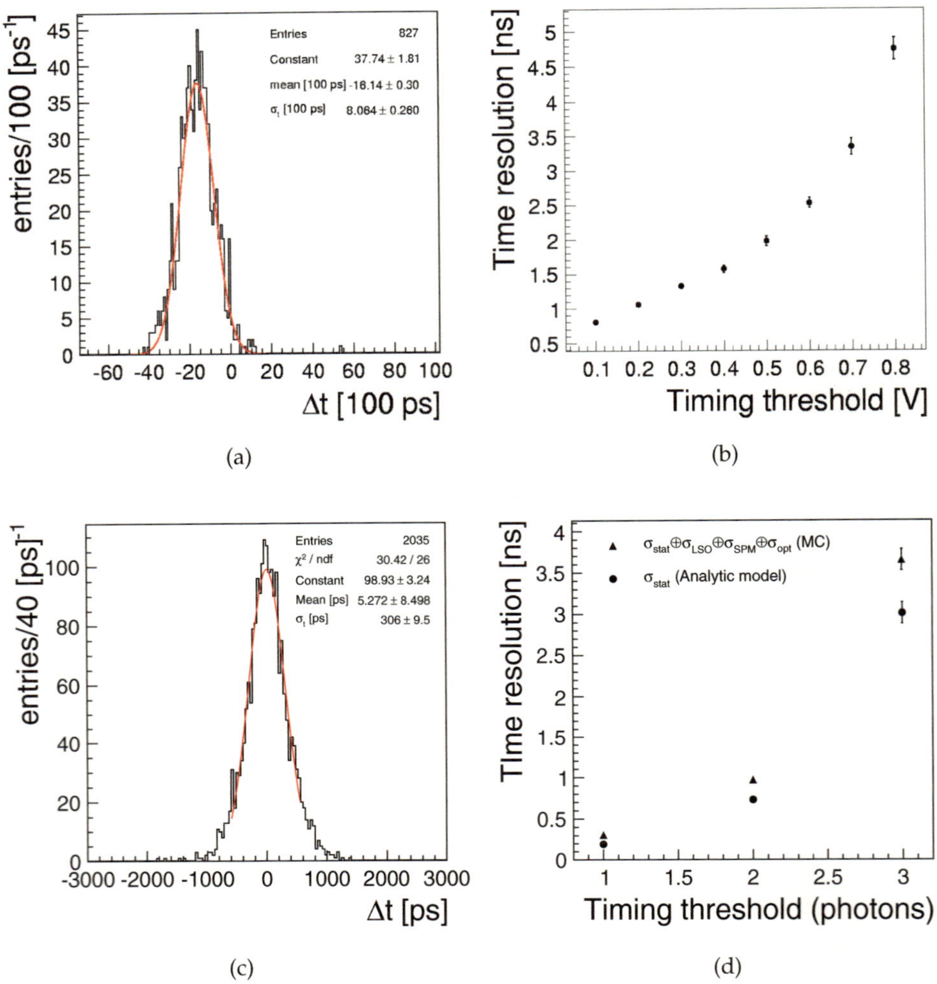

Fig. 18. Measured (a-b) and simulated (c-d) time difference distribution of two LSO/SPM detecting elements in coincidence of 511 keV signal from the β^+ emitter ^{22}Na. In (a,c) the timing threshold is set at $N_{ph} = 1$ photon. In (b,d) the time resolution is shown as a function of the timing threshold. The dots are the Monte Carlo estimation of the full energy resolution, the triangles are the analytical model described in Eq. 18.

4.3.2 Time resolution of the LSO/SiPM detection system

The measured time difference spectrum of the LSO/SiPM detection system in response to a Na^{22} source is shown in Fig. 18a. The events are selected in coincidence and if the integral of the corresponding signals is within a window of $\pm 3\sigma$ around the photo-electric peak. The threshold is selected to $N_{th} = -0.1$ V below the DC level of the signal. The time at which the signal crosses the threshold N_{th} is estimated with a linear fit around the negative edge of the detected signal. The mathematical simulation of the time response corresponding to the conditions of the experimental measurements is shown in Fig. 18c.
The time resolution $\sigma_t = 806 \pm 26$ ps is achieved from a gaussian fit to the coincidence time

spectrum.
The analysis of coincidence time resolution σ_t could be expressed as the sum in quadrature of five independent contributions:

$$\sigma_t^2 = \sqrt{2}\sigma_{LSO}^2 \oplus \sqrt{2}\sigma_{SPM}^2 \oplus \sigma_{stat}^2 \oplus \sigma_{opt}^2 \oplus \sigma_{el}^2 \tag{13}$$

where σ_{LSO} and σ_{SPM} are the time resolution respectively of LSO and SiPM.
The intrinsic time resolution of the SiPM σ_{SPM} is estimated from the rise time of the single photoelectron signal and is experimentally determined as $\sigma_{SPM} \sim 27$ ps (Saveliev, 2010).
The intrinsic timing response of the LSO crystal σ_{LSO} to 511 keV photons depends on two physics processes. The first is the time required for a 511 keV photon to undergo a photo-electric absorption. The second is the transit time of the photo-electron in the LSO crystal which determines the rise time of the scintillation photons emission. This means that if a 511 photon interacts through photoelectric effect in the LSO crystal within a mean free path λ, this process happens on average within $\sigma_{LSO} = \lambda/c \sim 29$ ps, where c is the speed of light.
The intrinsic overall coincidence time resolution of a LSO/SiPM detection system can be estimated as $\sqrt{2}\sigma_{LSO} \oplus \sqrt{2}\sigma_{SPM} = \sqrt{2} \cdot \sqrt{29^2 + 29^2} \sim 60$ ps.
The photo-statistics of the detection of scintillation optical photons, the optical properties of the Teflon/LSO surface and the read-out electronics affect significantly the overall coincidence time resolution of the LSO/SiPM detection system.
The contribution of the band-width and of the time-jitter of the read-out electronics introduces a constant finite resolution σ_{el}^2.
The reflection of the optical photons in the LSO/Teflon surface introduces a significant contribution to the coincidence time resolution at low photon fluxes. This effect is included in the overall coincidence time resolution with a constant term σ_{opt}. Its contribution to the coincidence time resolution of the LSO/SPM detection system is estimated with the Monte Carlo as $\sigma_{opt} \approx 250$ ps for the experimental setup.
The effect of the binomial photon detection efficiency to the detection of the timing threshold N_{th} photon is included in the term σ_{stat} and can be described analytically with a statistical model.
The contribution of the photo-statistics σ_{stat} is the variance of the probability distribution which describes the statistical process that after r intervals Δt, more than k photons are detected in the SiPM. The timing threshold is k detected photons. The distribution $P(t)$ of the detected scintillation photons is composed of two parts, according to the signal shape of the LSO/SPM system. It is approximately increasing linearly in the time range $(0, \tau_r)$ and is exponentially decreasing in the time range $(\tau_r, +\infty)$:

$$P(t) = \begin{cases} \frac{\bar{\epsilon}\alpha LY \cdot E_\gamma}{\tau\tau_r} t & 0 < t < \tau_r \\ \frac{\bar{\epsilon}\alpha LY \cdot E_\gamma}{\tau} e^{-\frac{t}{\tau}} & \tau_r < t < +\infty \end{cases} \tag{14}$$

where:

- LY is the light yield of the LSO crystal.
- E_γ is the detected energy in the crystal.
- τ_r is the rise time of the output signal of the detection system.
- τ is the average decay time of the LSO crystal.
- $\bar{\epsilon}$ and α are respectively the average efficiency of the SPM (Eq. 7) and the overall geometric photon detection efficiency.

The average value of the detected photons in the rth time interval is:

$$\mu\left(r\Delta t\right) = \int_{(r-1)\Delta t}^{r\Delta t} P(t)dt \tag{15}$$

The probability $P(n > k)$ of detecting more than k photons in the time interval Δt is described by the Poisson statistics:

$$P(n > k) = \sum_{n=k}^{+\infty} \frac{\mu^{n}\left(r\Delta t\right) e^{-\mu(r\Delta t)}}{n!} \tag{16}$$

The probability that more than k photons are detected after $r \times \Delta t$ ps is interpreted as the probability that more than k photons are detected at the rth trial:

$$P\left(r\right) = \left(\sum_{n=0}^{k-1} \frac{\mu^{n}\left(r\Delta t\right) e^{-\mu(r\Delta t)}}{n!}\right)^{r-1} \times \left(\sum_{n=k}^{+\infty} \frac{\mu^{n}\left(r\Delta t\right) e^{-\mu(r\Delta t)}}{n!}\right) \tag{17}$$

which is expressed as:

$$P\left(r\right) = \left(1 - \sum_{n=k}^{+\infty} \frac{\mu^{n}\left(r\Delta t\right) e^{-\mu(r\Delta t)}}{n!}\right)^{r-1} \times \left(\sum_{n=k}^{+\infty} \frac{\mu^{n}\left(r\Delta t\right) e^{-\mu(r\Delta t)}}{n!}\right) \tag{18}$$

The variance σ of this distribution can be estimated numerically. It is a non decreasing function of the threshold value k. This statistical effect makes the measurement very sensitive to the chosen timing threshold, as observed also in other experimental studies (D'Ascenzo et al., 2007). The contribution of the photon statistics to the total coincidence time distribution is $\sigma_{stat} = \sqrt{2} \times \sigma$.

The best possible time resolution is $\sigma_t = (54.07 \pm 1.30)$ ps at a timing threshold of 1 detected photon in case of a $2.5 \times 2.5 \times 15$ mm^3 LSO crystal wth perfect read out at one side. The statistically independent contributions to the time resolution of the system are according to Eq. 13:

$$\begin{aligned}(54.07 \pm 1.30 \text{ ps}) &\approx \sqrt{2}\,(29 \text{ ps})_{LSO} \oplus \sqrt{2}\,(29 \text{ ps})_{SPM} \oplus \\ &\oplus (9.00 \pm 0.02 \text{ ps})_{stat}\end{aligned} \tag{19}$$

where σ_{LSO} and σ_{SPM} are taken from the above estimation and σ_{stat} is estimated numerically from the probability distribution in Eq. 18 with parameters $\bar{\epsilon} = 1$, $\sigma_{\epsilon} = 0$ and $\alpha = 8100/(27000 \cdot 0.511) = 0.587$. The contribution of the optical reflections on the boundary LSO/Teflon surfaces σ_{opt} is negligible at this high detected optical photon flux. The Monte Carlo results agree with the estimation of best possible coincidence time resolution of the LSO/SPM detection system shown above on the basis of physical principles.

The time difference spectrum of the LSO/SiPM detection system estimated with the mathematical model of the test setup is shown in Fig. 18a. The expected time resolution is calculated with a gaussian fit as 306 ± 9 ps at a timing threshold $N_{ph} = 1$ photon. The independent contributions to the time resolution are estimated according to Eq. 13 as:

$$\begin{aligned}(306 \pm 9 \text{ ps}) &= \sqrt{2}\,(29 \text{ ps})_{LSO} \oplus \sqrt{2}\,(29 \text{ ps})_{SPM} \oplus \\ &\oplus (191 \pm 5 \text{ ps})_{stat} \oplus (239 \pm 8 \text{ ps})_{opt}\end{aligned} \tag{20}$$

where σ_{LSO} and σ_{SPM} are taken from the above estimation, σ_{stat} is estimated numerically from the distribution 18 with parameters $\alpha\bar{\epsilon} = 254/(27000 \cdot 0.511)$ and σ_{opt} is calculated as a

difference in quadrature from the other terms.
The dependence of the coincidence time resolution on the threshold is estimated by simulation (Fig. 18b). The time resolution degradates from 306 ps up to about 3.6 ns when the threshold is increased from 1 up to 3 photons. The prediction of the analytical model for the statistical term σ_{stat} is also shown in Fig. 18b. The statistical term is a non decreasing function of the threshold. This explains the degradation of the estimated coincidence time resolution with the increase of the threshold. The difference between the modelling estimation and the statistical model depends on the additional terms σ_{LSO}, σ_{SPM} and σ_{opt} which are not included in the analytic model but are considered in the simulation.
The worsening of the coincidence time resolution with the increase of the coincidence timing threshold is also observed in the experimental data (Fig. 18d).
From a comparison between the dependence of the coincidence time on the timing threshold in data and the Monte Carlo, the timing threshold applied in the experimental data can be estimated as about 2 detected photons.
Improvements in the experimental set up are needed in order to achieve the expected time resolution of the LSO/SPM detection system.

5. Conclusions

The Silicon Photomultiplier technology is mature for the efficient read-out of scintillators, with consequent improvements in High Energy Physics and Nuclear Medicine applications.
The direct read-out of plastic scintillators by SiPM is feasible and can be an elegant solution for a simplification of the design of highly granular hadronic calorimeters in new high energy physics experiments.
The read-out of inorganic scintillators by SiPM is also a promising solution for the design of highly granular Positron Emission Tomographs of new generation, with transverse space resolution down to 2 mm and excellent time resolution of few hundreds ps.
The measured performances of the scintillator/SiPM detection system are hence promising for the possible applications to calorimetry and Positron Emission Tomography. In the latter case, a benefit is found both for morphological and functional in vivo studies in which space and time resolution play a significant role.

6. References

Alvares-Gaume L. et al. (2008) Review of Particle Physics, Particle Detectors. *Physics Letters*, Vol. 667, No. 1-5, 2008 , 281-370.
ATLAS Collaboration (1999) ATLAS detector and physics performance, CERN/LHCC99-14.
Behnke,T.; Damerell,C.; Jaros,J. & Miyamoto,A. (2007) ILC Reference Design Report. Vol4: Detectors, www.linearcollider.org/about/Publications/Reference-Design-Report
Blin,S. (2006) Dedicated very front-end electronics for an ILC prototype hadronic calorimeter with SiPM read-out, *LC-DET-2006-007*.
Boddaert N. & Zilbovicius V. (2006) Functional neuroimaging and childhood autism, *Pediatr.Radiol.*, 32, 1-7.
Buchsbaum et al. (1992) Brief Report: Attention performance in Autism and regional brain metabolic rate assessed by Positron Emission Tomography, *Journal of Autism and Developmental disorders*, 22, 115-125.
CALICE Collaboration (2010) Construction and Commissioning of the CALICE Analog Hadron Calorimeter Prototype, JINST 5 P05007.

The CDF II Collaboration (1996) CDF Technical Design Report, FERMILAB-Pub-96/390-E.

D'Ascenzo, N.; Eggemann A.; Garutti E. & Tadday, A. (2007) Application of Micro Pixel Photon Counter to calorimetry and PET, *Il Nuovo Cimento C*, Vol. 30 N.5.

D'Ascenzo,N.; Eggemann, A. & Garutti,E. (2007) Study of Micro Pixel Photon Counters for a high granularity scintillator-based hadron calorimeter, *DESY 07-196, arXiv:0711.1287*.

D'Ascenzo,N. (2009) Study of the neutralino sector and analysis of the muon response of a highly granular hadron calorimeter at the International Linear Collider, *Phd Thesis*, DESY-THESIS-09-004.

Firestone,R.B. (1996) Table of isotopes, New York Wiley.

Freeman,J. (2009) Silicon photomultipliers for the CMS hadron calorimeter, *Nucl.Instr.Meth.A*, 617, 393-395.

Golovin,V. & Saveliev,V. (2004) Novel Type of Avalanche Photodetector with Geiger Mode Operation, *Nucl.Instr.Meth.A*, 518, 560-564.

Herbert,D.J.; Saveliev,V.; Belcari,N.; D'Ascenzo,N.; Del Guerra,A. & Golovin,A. (2006) First results of scintillator read-out with Silicon Photomultiplier detectors, *IEEE Trans. Nucl.Sci.* NS53(1), 389-394.

ILD Concept Group (2009) The International Large Detector: Letter of Intent, *FERMILAB-LOI-2010-03, FERMILAB-PUB-09-682-E, DESY-2009-87, KEK-REPORT-2009-6*.

Karp,J.; Suleman,S.; Daube-Witherspoon,M.; Freifelder,R.; Cardi,C.A.; Adam,L.; Bilger,M. & G. Muehllehner, Performance of a Brain PET Camera based on Anger-logic gadolinium oxyorthosilicate detectors *Journ.Nucl.Med.*, 44, 1340-1349.

Mao,R.; Zhang,L. & Ramsden,D. (2002) Emission spectra of LSO and LYSO crystals excited by UV light, X-ray and γ-ray, *IEEE Trans. Nucl. Sc.*, 55, 1759-1766.

Melcher,C.L. & Schweitzer,J.S. (1992) Cerium-doped Oxyorthosilicate: A Fast, Efficient New Scintillator, *IEEE Trans. Nucl. Sc.*, 39, 1759-1766.

National Electrical Manufacturers Association (2001) Performance Measurements of Positron Emission Tomographs, NEMA Standard Publications NU-2-2001.

Yokoyama,M. (2009) Application of Hamamatsu MPPCs to T2K neutrino detectors, *Nucl.Instr.Meth.A*, 610, 128.

Saveliev V., Golovin V. (2000), Silicon Avalanche Photodiodes on the basis of Metal-Resistor-Semicinductor (MRS) Structures, *Nucl.Instr.Meth.A*, 442, 223-229.

Saveliev V. (2010), Silicon Photomultiplier - New Era of Photon Detection, Advances in Optical and Photonic Devices, Ki Young Kim (Ed.), ISBN: 978-953-7619-76-3, *InTech*.

SensL (2010), http://www.sensl.com.

Stewart,A.G.; Saveliev,V.; Bellis,S.J.; Herbert,D.J.; Hughes,P.J. & Jackson,J.C. (2008) Performance of 1 mm^2 Silicon Photomultiplier, *IEEE J. Quantum Electron*, 44(2), 157.

Strul,D.; Santin,G.; Breton,V. & Morel,C. (2003) *Nucl.Phys.B*,125,75-79.

Thompson,M. (2006) Particle Flow Calorimetry at the International Linear Collider, *Pramana journal of physics*, 69,6, 1101-1107.

Toshikaza Hakamata et al. (2006), Photomultipliers Tubes, Basics and Applications, Hammamatsu Photonics K.K., Electron Tube Division, Japan.

Tsang W.T. (Ed.) (1985) Semiconductors and Semimetals: Lightwave Communication Technology, Part D, Photodetectors, 1-309, Academic Press Inc., ISBN :978-0-12-752153-4.

Yamamura Kazuhisa et al. (2009) Production and Development status of MPPC, Proceedings of Science, PoS (PD09) 017.